INTERNATIONAL SERIES OF MONOGRAPHS IN

NATURAL PHILOSOPHY

GENERAL EDITOR: D. TER HAAR

VOLUME 60

DISLOCATIONS AND PLASTIC DEFORMATION

DISLOCATIONS AND PLASTIC DEFORMATION

BY

I. KOVÁCS C. Sc. (Phys.)

Head of the Institute of General Physics,
Eötvös Loránd University, Budapest

AND

L. ZSOLDOS C. Sc. (Phys.)

Associate Professor, Institute of Solid State Physics,
Eötvös Loránd University, Budapest

PERGAMON PRESS

OXFORD NEW YORK TORONTO

SYDNEY BRAUNSCHWEIG

Pergamon Press Ltd., Headington Hill Hall, Oxford
Pergamon Press, Inc., Maxwell House, Fairview Park, Elmsford, New York 10523
Pergamon of Canada Ltd., 207 Queen's Quay West, Toronto 1
Pergamon Press (Aust.) Pty. Ltd., 19a Boundary Street, Rushcutters Bay, N.S.W. 2011, Australia
Vieweg & Sohn GmbH, Burgplatz 1, Braunschweig

Co-edition of Pergamon Press, Oxford and
Akadémiai Kiadó, Budapest

First edition 1973

Library of Congress Cataloging in Publication Data

Kovács, István, 1913-
Dislocations and plastic deformation.

(International series of monographs in natural philosophy, v. 60)
Published in 1965 under title: Diszlokációk és képlékeny alakváltozás.
Bibliography: p.
1. Dislocations in crystals. 2. Plasticity.
I. Zsoldos, Lehel, joint author. II. Title.
QD921.K67813 1973 548'.842 73-6995
ISBN 0-08-017062-5

This book is the revised and enlarged version of the Hungarian original

Diszlokációk és képlékeny alakváltozás

Published by Műszaki Könyvkiadó, Budapest

Translated by

Dr. Z. Morlin

Translation revised by

Dr. A. D. Durham

Printed in Hungary

Contents

Preface

The physical and technological properties of solid matter are basically influenced by the existence of lattice defects. The mechanical properties of materials, for example the process of plastic deformation, can be understood only on the basis of dislocation theory. A large amount of experimental and theoretical knowledge has been acquired in this field, particularly during the last two decades. As a result of research efforts, a lot of problems have now been resolved, but there are other problems like the work-hardening and plastic properties of alloys which are not very well understood at present.

The aim of this book is first of all to provide a short and, where possible, precise explanation of dislocation theory. This purpose is served by the first six chapters, in which the properties of dislocations and of the point defects both in crystals and in an elastic continuum are discussed. In Chapters 7 and 8 some applications of dislocation theory are given which show, for instance, the difficulties involved in understanding the hardening of alloys and the work-hardening of pure metals. Chapter 9 gives a short survey of the effect of heat treatment on the defect structure in metals.

In the appendices some basic elements of crystallography and elasticity theory are summarized, in order to provide a basis for understanding dislocation theory.

The book is based on lectutes delivered by the authors at the Eötvös Loránd University, Budapest. The English edition is a revised and enlarged version of the first Hungarian edition published in 1965.

It is a pleasure to acknowledge the assistance that we have received from many people in writing the manuscript. Particular thanks are due to Prof. E. Nagy for initiating and constantly stimulating our work; to Dr. Z. Morlin for the translation, and to Dr. A. D. Durham for the revision of the text, and finally to the following people for providing the photos: Prof. S. Amelinckx, Prof. E. N. da C. Andrade, Prof. W. Boas, Dr. D. G. Brandon, Dr. W. C. Dash, Dr. U. Essmann,

Prof. P. Feltham, É. Harta, Prof. P. B. Hirsch, Dr. J. A. Hugo, Prof. D. Kuhlmann-Wilsdorf, Prof. A. R. Lang, Dr. S. Mader, Prof. T. E. Mitchell, I. Neubauer, Dr. V. A. Philips, Dr. J. Ogilvie, Dr. P. B. Price, Dr. H. Saka, Prof. A. Seeger, Prof. B. Šestak, Prof. E. Suito, Dr. N. Uyeda, Dr. M. Wald, Dr. D. H. Warrington and Prof. H. G, F. Wilsdorf.

I. Kovács and L. Zsoldos

Chapter 1

The concept of dislocation

1.1. Structural defects

The elastic properties of solid bodies can be completely interpreted both by the mechanics of continua (bodies of continuous material distribution) and by the elastic theory of crystals. However, if the force applied is large the body may eventually become plastically distorted. The theoretical investigation of plastic deformation may be carried out in two ways: in one it is restricted to a descriptive (phenomenological) treatment which establishes experimentally the properties characteristic of the material under investigation, and applies the equations thus obtained to the solution of actual problems, taking into account the corresponding boundary conditions. However, this theory, though describing the mechanical changes occurring under given circumstances, does not reveal the cause and nature of the phenomenon since it does not include in its considerations the atomic arrangement of matter and the crystalline structure of solid bodies. It is, for instance, a well-known experimental fact that copper wires harden considerably on bending (work-hardening) and soften

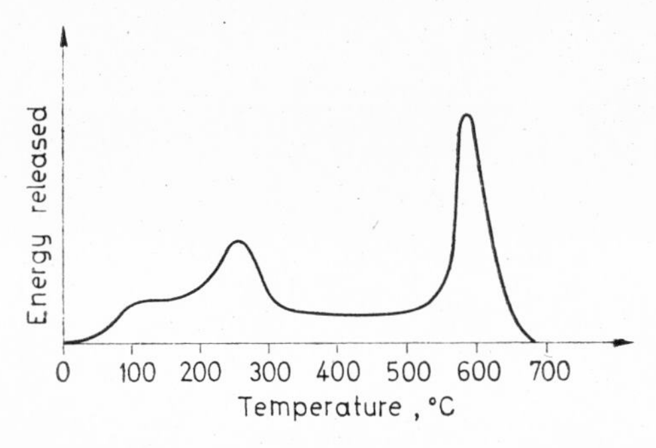

Fig. 1.1. The temperature-dependence of the release of energy stored during the deformation in Ni

on being heated up to a suitable temperature (heat treatment). It is quite clear that this and other similar phenomena are connected with the structural properties of matter. The following observation is even more indicative of this: part of the mechanical work supplied during plastic deformation of a body is transformed into heat and another part is stored in the body. This latter energy is released during heating up to a high enough temperature. The temperature dependence found by Clarebrough *et al.* [1] for the stored energy release from nickel is shown in Fig. 1.1. In order to understand this phenomenon a set of questions must be considered. What is the mechanism which transforms part of the mechanical work into heat energy during the deformation? How does energy become stored in a body, and why is it released in stages during the heat treatment? What changes occur at the temperatures which correspond to the maxima? Satisfactory answers to all these questions may be obtained only if the structural properties are known exactly.

1.2. Deformation of single crystals

The mechanism of deformation can be most conveniently studied on single crystals. If a tensile load is applied to a single-crystal rod the deformation obtained is neither homogeneous nor isotropic. Some parts of the crystal become displaced with respect to each other by gliding along their crystallographic planes (the glide planes), whereas regions between consecutive glide planes remain unchanged (Fig. 1.2). The extent of the glide amounts to 0.1–1 μm and the distance between two consecutive glide planes is several microns. The time of a single glide in a crystallographic plane is approximately 0.01 sec or less, after which no further glide takes place in this plane.

Investigations carried out with a great variety of materials have shown that those crystallographic planes, in which the atoms are arranged in a close-packed layer, always act as glide planes and that within these planes, only well-defined glide directions exist. Thus, for example, in a hexagonal single crystal the glide usually takes place along the (001) basal plane in the direction of the three hexagonal diameters (Fig. 1.3).

The active glide plane of face-centred cubic (fcc) crystals is usually one of the $\{111\}$ planes (four possibilities), and the glide takes place in the $\langle 110 \rangle$ direction (three possibilities). Hence one has twelve different but equivalent glide possibilities.

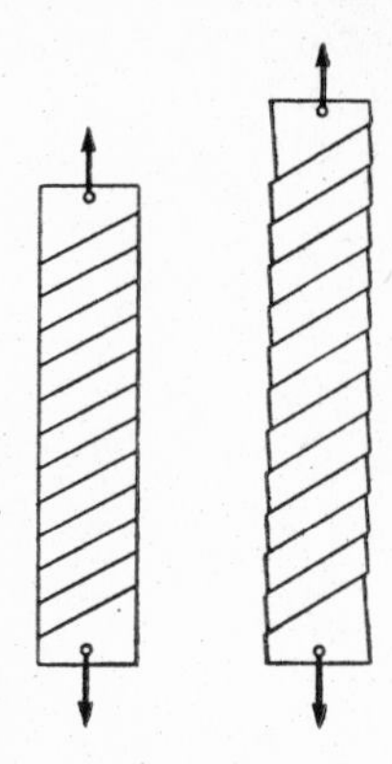

FIG. 1.2. Formation of glide lamellae

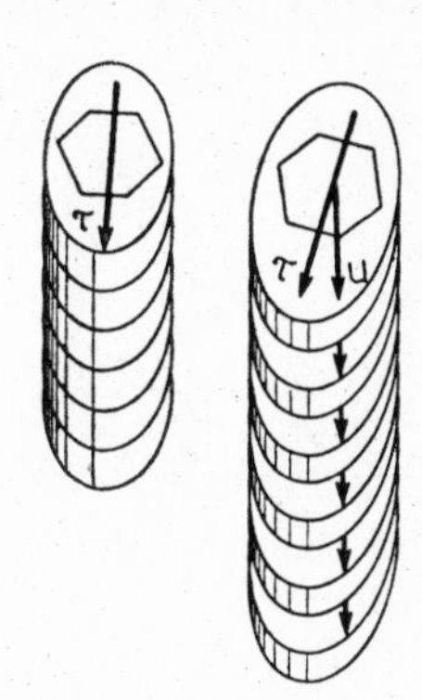

FIG. 1.3. Glide planes and directions (**u**) in a hexagonal system

Experience shows that glide is initiated only if the stress acting in any of the glide directions attains a critical value characteristic of the material under investigation. This stress, called *critical resolved shear stress*, is generally 10^{-5}–10^{-3} μ, where μ is the shear or torsional modulus.

In the case of a fcc lattice the shear stress may attain its critical value in two or three directions simultaneously. In these cases the glide takes place in several directions at the same time.

1.3. The critical shear stress

The phenomenon described above can only be explained in terms of the atomic structure of the crystalline material, even so, the formation of the glide bands is not quite clear. The difficulties increase if one attempts to calculate theoretically the value of the critical resolved shear stress.

Consider the model depicted in Fig. 1.4. Let us assume that the body becomes homogeneously deformed by a shear stress τ so that two atomic layers slip with regard to each other. The correspondence between the displacement x and the stress τ may be obtained by the following reasoning: if the displacement is equal to one atomic distance b (i.e. the atoms again occupy their normal sites after the slip), no stress is needed to maintain this state, and thus the stress is zero.

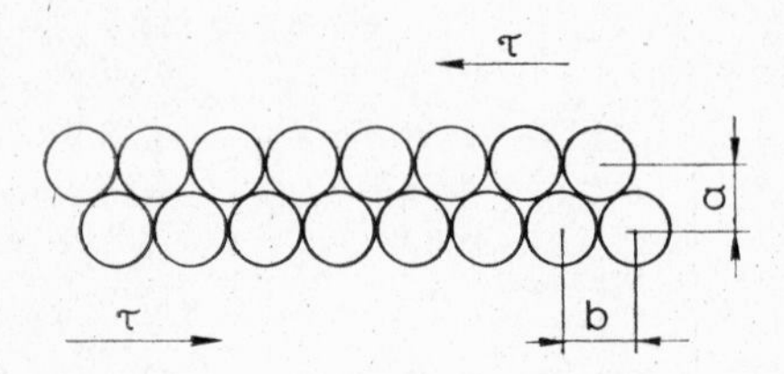

FIG. 1.4. For the derivation of equation (1.1)

For reasons of symmetry the stress is also zero if the atomic layers are displaced by only half an atomic distance. Furthermore, on both sides of the zero sites the forces differ only in sign. Consequently, the stress necessary to effect a displacement x is a periodic, odd function of x with period b. For simplicity, one may assume that $\tau(x)$ is a pure sine function

$$\tau(x) = K \sin \frac{2\pi x}{b}. \tag{1.1}$$

The value of the constant K is obtained from the condition that (1.1) has to transform into Hooke's law for small x values, i.e. $\tau(x) \cong K\, 2\pi x/b \cong \mu x/a$ where a is the layer distance. Thus

$$\tau(x) = \frac{\mu b}{2\pi a} \sin \frac{2\pi x}{b}. \tag{1.2}$$

The critical shear stress at which the lattice becomes unstable is the maximum value of (1.2).

$$\tau_{cr} = \tau\left(\frac{b}{4}\right) = \frac{b}{a}\frac{\mu}{2\pi}. \tag{1.3}$$

Since $a \approx b$, one obtains $\tau_{cr} \approx \mu/2\pi$. If a law other than a simple sinusoidal force law is assumed, according to the calculation of Mackenzie [2], one obtains for a glide in the basal plane of a close-packed hexagonal lattice the value $\tau_{cr} \geqq \mu/30$.

Both values, however, are several orders of magnitudes larger than those obtained by experiment. This deviation cannot be explained by thermal motion, because measurements carried out at very low temperatures also yield surprisingly low critical stresses. The values compiled in Table 1.1 were obtained [3] from data measured on Zn and Cd single crystals extrapolated from 1.2 °K down to 0 °K.

TABLE 1.1. *Calculated and measured critical shear stress for Zn and Cd*

Material	σ_{cr} at 0 °K	$\frac{\mu}{30}$ at room temperature
Zn	80 p/mm^2	63.3×10^3 p/mm^2
Cd	150 p/mm^2	130×10^3 p/mm^2

Thus, the above considerations are not verified by experiment. However, since the crystalline structure of solid bodies is unquestionable, the error can only lie in the assumption of perfect crystals and the supposition that the total glide takes place at the same time along the whole glide surface.

1.4. Lattice defects

In addition to plastic deformation, many other properties of the solid state cannot be explained by assuming a perfect crystal lattice. A great variety of phenomena, e.g. the diffusion processes within solids, or the considerable deviations from the theoretical values of the measured intensities of X-rays diffracted from crystals, seem to prove that real crystals have a great number of defects. The most simple lattice disorders are vacant lattice sites and interstitial atoms [4]. Both are created by thermal motion or other effects, with the result that atoms vacate their original sites leaving behind *empty lattice sites* (vacancies) and going to the surface or becoming wedged amongst other atoms, thus occupying interstitial positions in the lattice (Fig. 1.5). Considerable knowledge has accumulated about these types of defect. It can be shown that they are thermodynamically stable formations whose concentration increases exponentially with temperature, perhaps attaining at the melting point a value of 10^{-4}.

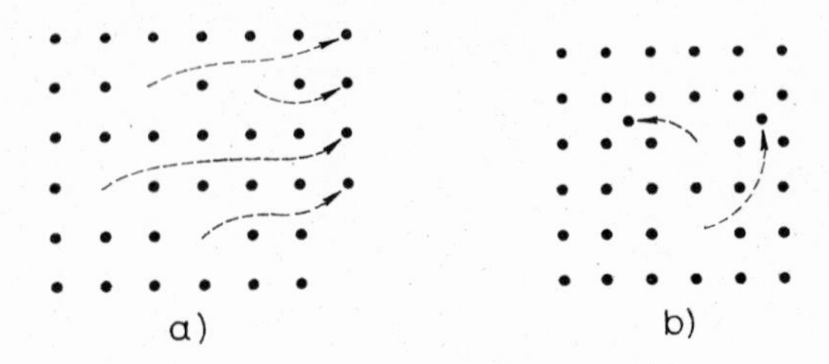

FIG. 1.5. (a) Schottky defect. Some of the atoms have diffused to the surface leaving behind a vacancy. (b) Frenkel defect. The atoms leaving their lattice positions move to some other place within the lattice

Diffusion processes can be accounted for by these defects though they give no answer to problems connected with plastic deformation [5].

The vacancies and interstitial atoms are *point defects*, since they extend over a relatively small area and are surrounded by "good material" even if a great number of vacancies cluster together. (The concept of "good material" is defined in Section 1.6.)

The process of plastic deformation is presumed to consist of slips which do not take place simultaneously in the whole glide plane; hence, this contains slipped and still unslipped areas. The boundary between these two areas clearly creates another crystal defect. The boundary is either a closed loop or a half loop extending from surface to surface which sweeps along the glide plane during the deformation. This new type of defect is a line defect or a *dislocation*.

1.5. Dislocations in the lattice

The concept of dislocation was first applied by Taylor and independently of him by Orowan and Polányi in 1934 to the successful description of the phenomena of plastic deformation [6] though similar types of defect had been assumed long before this.

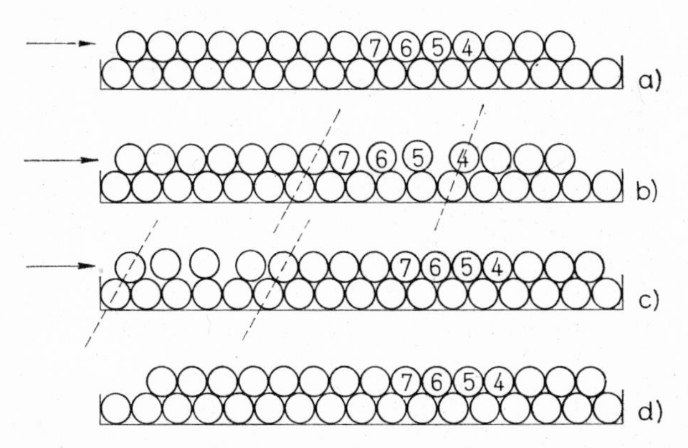

FIG. 1.6. Andrade's model

The structure of dislocations can be most simply demonstrated by the model of Andrade [7] shown in Fig. 1.6. We replace the individual atomic rows with rigid cylinders. On both sides of the upper row a rubber tape is fixed to the centres of the bases of the cylinders. Thus the members of the upper row are held by an elastic force.

Now we try to make the two rows of cylinders slip on each other by pushing the left side of the upper row in the direction of the arrow. Simultaneous displacement can be achieved only by applying a relatively very large force. Let us rearrange the right end of the upper row of cylinders by placing the first four cylinders one place to the right. As a result the adjacent few cylinders become more or less displaced from their regular sites creating a misfit part. To the left of this, however, the cylinders remain in their original position. At the misfit part a stress is created in the rubber tape. In the rearranged section four cylinders occupy as much room as five in the lower row. The range of misfit is a *dislocation* (Fig. 1.6b).

If even a small force is applied, the stretch of the rubber between the 6th and 7th cylinders decreases. The 4th cylinder pulls the 5th towards itself. At the same time the stress increases between the 5th and 6th cylinders, with the result that the 6th cylinder jumps into the hole on its right dragging the 7th out of its place, and so on.

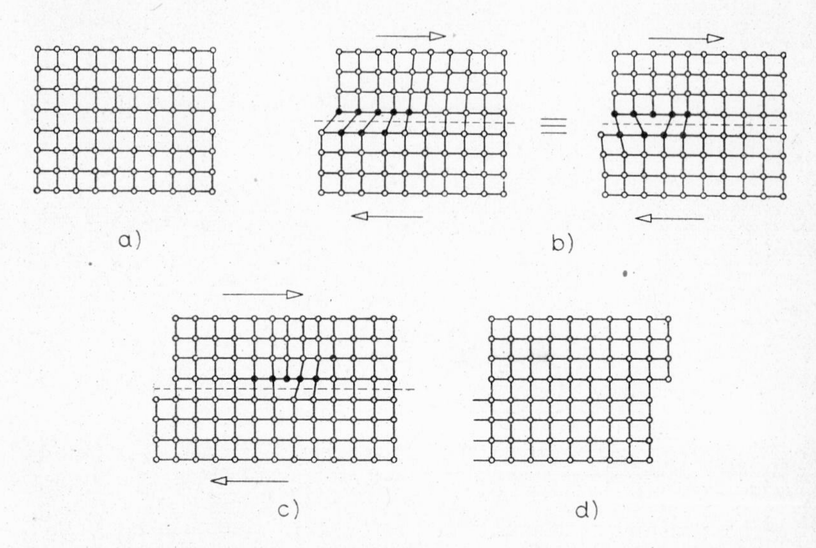

FIG. 1.7. (a) Lattice domain without defects. (b) Dislocation at the edge of the lattice immediately after its formation. (c) Dislocation in the centre of lattice. (d) The state of the lattice after the dislocation has moved out

As a result, the *misfit part runs opposite to the shear stress* along the row of cylinders. At the same time the upper row glides a further distance of one cylinder-diameter. Thus, without applying the large force necessary for the simultaneous glide of the row of cylinders, the upper row slips with respect to the lower one. A shear stress which moves only a few cylinders (the dislocation) is sufficient to effect the glide, though, of course, more time is necessary for the whole process than if the glide were carried out simultaneously. In reality the situation is somewhat more complicated, since the atoms of the lattice are held together by three-dimensional elastic forces, i.e. the lower lattice plane is not fixed, and consequently the atoms

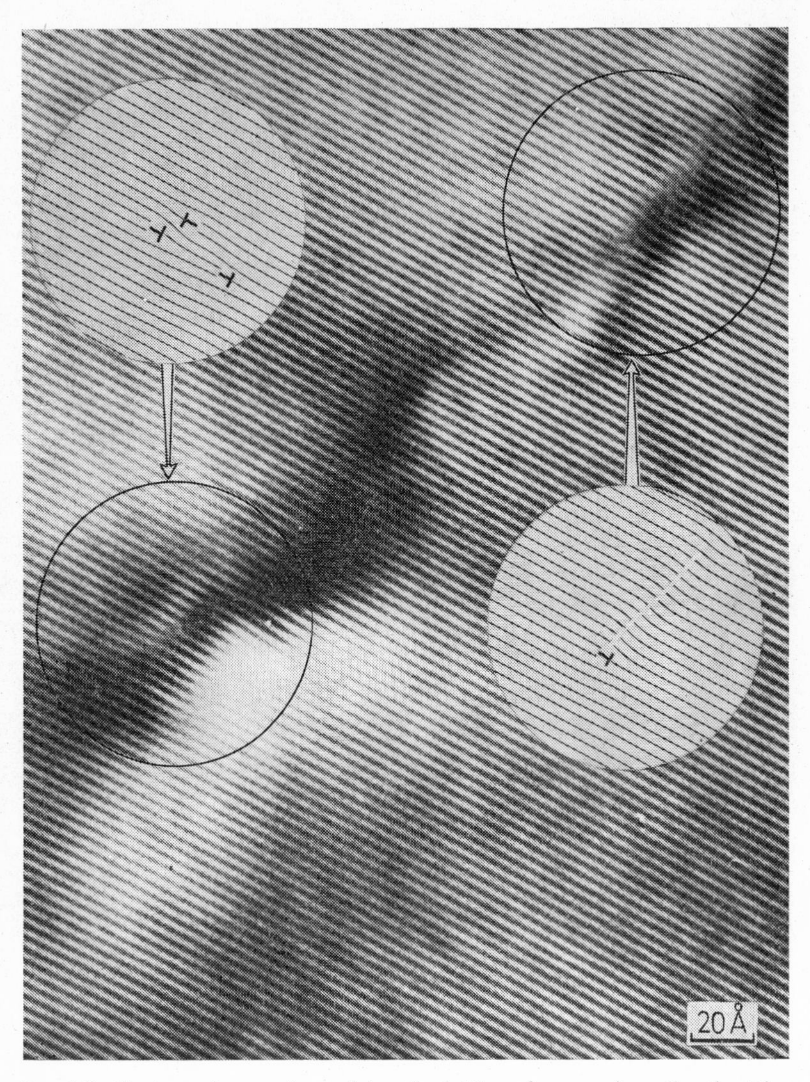

FIG. 1.8. Electronmicrograph resolving the lattice of germanium directly. The distance of the resolved (111) planes is 3.27 Å. At the left lower part of the picture three dislocations can be seen relatively close to each other. The encircled sketch shows their exact positions [V. A. Philips and J. A. Hugo, *Acta Met.* **18** (1970) 123]

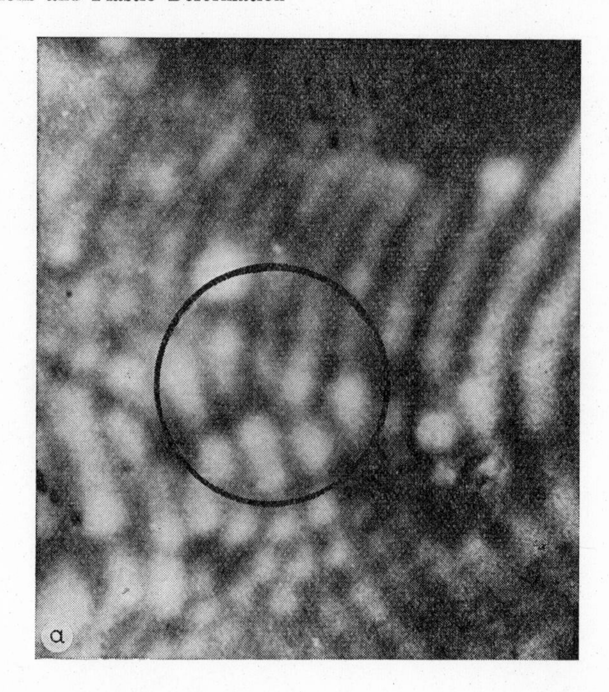

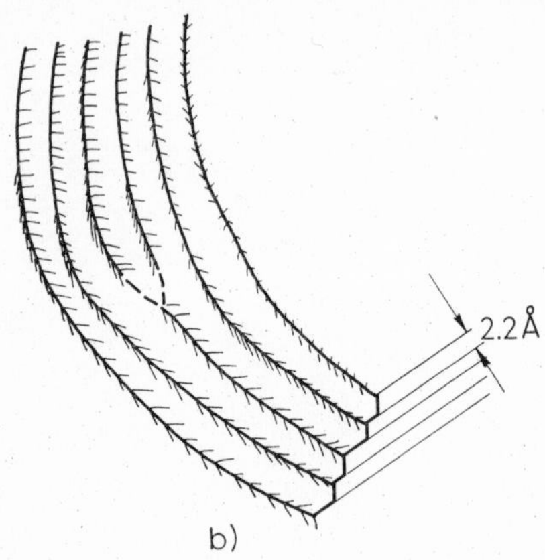

Fig. 1.9. Picture taken with a field ion-microscope showing a Mo tip containing one dislocation. The height of the steps formed by the atomic rows is **2.2. Å** [D. G. Brandon and M. Wald, *Phil. Mag.* **6** (1961) 1035]

of this plane are somewhat pressed together because of the dilatation of the upper row.

The defect discussed is an *edge dislocation.* It must not be forgotten that the processes take place in three dimensions, and consequently, e.g. in Fig. 1.6, the dislocation continues in both directions regardless of whether cylinders or atomic rows are placed normal to the plane of the drawing, thereby creating a *dislocation line.*

Before discussing the properties of dislocations in some detail, let us consider another model. Figure 1.7a depicts a perfect lattice (a section of a three-dimensional crystal). Let us suppose that for some reason the upper left side of the lattice slips one atomic distance, thus creating an edge dislocation. In reality dislocations created by spontaneous glide due to an external stress do not occur frequently; on the other hand, depending upon the conditions of crystallization, even the most perfect crystals contain a certain number of dislocations, which (as will be seen in Chapter 5) may become sources of new dislocations. It can be seen that the lattice is compressed above the glide plane and expanded below it. An alternative way of forming an edge dislocation is to make a cut normal to the glide plane into the lattice as far as the glide plane, and insert an extra atomic layer into the cut up to the dislocation line. The presence of such an extra half plane has actually been observed electron-microscopically or with the aid of field-emission microscopy in some instances (Figs 1.8 and 1.9). During the deformation the edge dislocation sweeps through the crystal creating a slip of one atomic distance (Figs 1.7b–d).

In drawings the edge dislocations are denoted by the symbol ⊥ (if they cut the plane of diagram). This lies on the glide plane while its shank points towards the "extra atomic layer" (Fig. 1.10).

The behaviour and motion of dislocations are well demonstrated by the bubble model [8]. This is a monolayer raft of soap bubbles of small, uniform size (with a diameter of 1–2 mm), floating on a liquid. The atoms are represented by the individual bubbles in this model.

1.6. Burgers vector

If a dislocation traverses a crystal a plastic deformation, i.e. glide, is created. Consequently, a dislocation may be characterized by the magnitude and direction of a slip and the concept of a dislocation can be defined quite exactly [9].

The vector defining the position of the repeating unit (unit cell) of a perfect single crystal (Fig. 1.10a) is given by the equation

$$\mathbf{r} = n_1\mathbf{a} + n_2\mathbf{b} + n_3\mathbf{c}\,, \tag{1.4}$$

where n_1, n_2 and n_3 are integers, and the vectors **a**, **b** and **c** represent the primitive translations of the lattice. With simple lattice structures – e.g. with most pure metals – the repeating units are single atoms (ions). In this case the vector **r** is simply the position vector of the atoms.

For real crystals, equation (1.4) applies only approximately because of the lattice distortions. If, however, the atoms are only slightly displaced from their perfect lattice sites, every atom of the real lattice belongs unambiguously to some ideal lattice site defined by equation (1.4). Thus, an ideal crystal may be correlated to the real crystal. The region where this is possible is termed good material, and where not possible, bad material.

Let us consider a closed curve in a perfect crystal such that every part of it has its equivalent in the good material surrounding the dislocation. For the sake of simplicity let the curve consist of a series of steps from atomic site to atomic site so that equivalent steps may easily be found in the good material of the real crystal. (For example, in Fig. 1.10 five steps are made to the right, four downwards, five to the left and again four upwards.) Such curves always exist and by constructing their image in the real lattice a new curve (the Burgers circuit) may be unequivocally fitted to the closed curve of the ideal crystal, which runs throughout in good material. This new curve, however, is not necessarily closed (Fig. 1.10b). The vector pointing from the end to the beginning of this curve is termed the *Burgers vector* and is denoted by **b**. If the Burgers circuit is not closed it encircles one or more dislocations.

The Burgers vector defines the slip caused by one moving dislocation, or the resulting slip if more dislocations are involved. The Burgers circuit encircling a point defect or a series of point defects is always closed.

The result of a Burgers circuit is independent of the path of the circuit, only the sign of the vector changing according to the clockwise or anticlockwise direction of the circuit. This uncertainty concerning the sign can easily be resolved by some *arbitrary agreement* on the sense of the circuit. An exact description of a dislocation, of course,

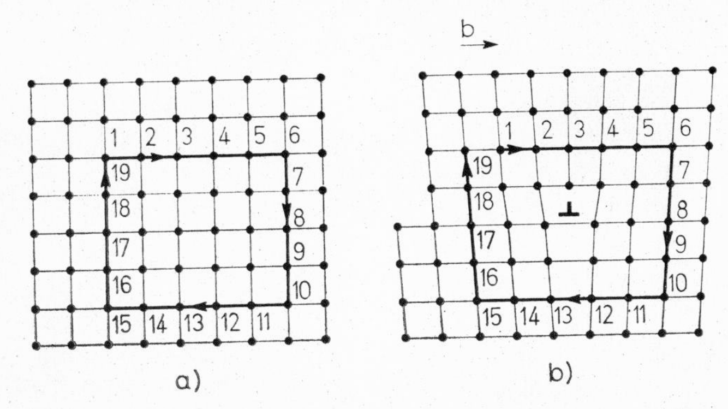

FIG. 1.10. Burgers circuit in (a) a perfect and (b) a dislocated lattice

also involves the definition of the exact direction of the dislocation line. As a first step *the positive direction of advance along a dislocation line is arbitrarily chosen.* The direction of any section of a dislocation is then given by a vector $\Delta\mathbf{l}$ which points in the positive direction and $|\Delta\mathbf{l}| = \Delta l$. Hereby, the sense of the Burgers circuit is fixed by this vector $\Delta\mathbf{l}$ in the following way: *looking towards positive* $\Delta\mathbf{l}$ *the sense of the circuit is anticlockwise.* The Burgers vector $\mathbf{b}$ so obtained and the line element $\Delta\mathbf{l}$ characterize the dislocation segment considered. It will be seen later that with every phenomenon which depends upon the direction of the Burgers vector, $\mathbf{b}$ and $\Delta\mathbf{l}$ appear together in the mathematical description. Consequently, the above definition is independent of the choice of $\Delta\mathbf{l}$ and yields unambiguous results. It is also clear that there is no sense in speaking of the direction of the Burgers vector alone without giving too the sense of $\Delta\mathbf{l}$, since by changing the sign of $\Delta\mathbf{l}$ the sign of $\mathbf{b}$ also changes. Nevertheless, dislocations of parallel and opposite Burgers vectors are sometimes discussed. This, however, implicitly contains the condition that their *line elements are unidirectional.*

In the case depicted in Fig. 1.10b, **b** is a lattice vector and is perpendicular to the dislocation line, which must be visualized as normal to the plane of the diagram. We shall see later (Chapter 6) that there exist dislocations whose Burgers vector is not a lattice vector. Dislocations defined by Burgers vectors which are also lattice vectors are called *perfect dislocations,* whereas dislocations whose Burgers vectors are not lattice vectors are *partial or imperfect dislocations.*

A dislocation is also conceivable whose Burgers vector is not perpendicular to the dislocation line, but lies parallel to it. This dislocation is called a *screw dislocation* (Figs 1.11a and b).

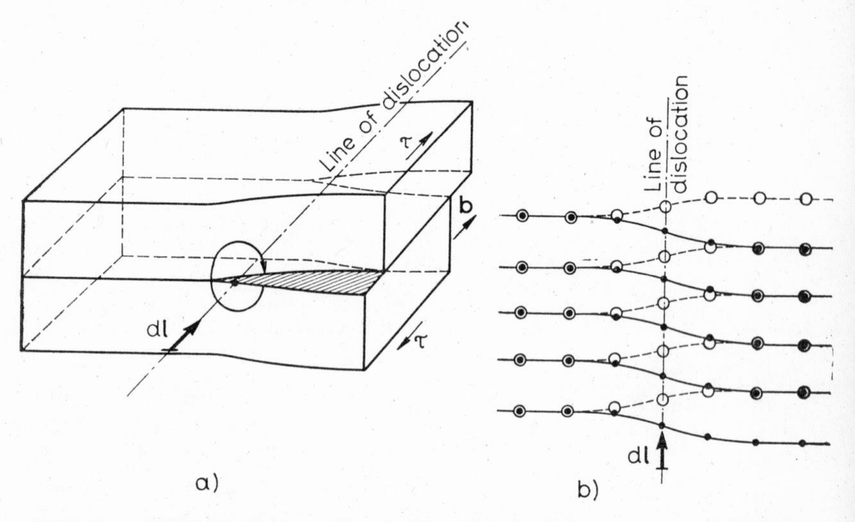

FIG. 1.11. (a) Perspective picture showing a glide creating a screw dislocation. (b) The row of atoms in a screw dislocation. ● atoms above the plane, and ○ atoms below the plane of the drawing

With the above convention the Burgers vector of a right-handed screw dislocation lies in the direction of the vector of the line element, whereas with a left-handed screw dislocation the directions of the two vectors are opposite. Finally, Fig. 1.12 depicts a dislocation beginning at one side of the crystal as an edge dislocation, this, however, gradually becomes transformed into a screw dislocation by the changing direction of the dislocation line. It is easy to see that the same Burgers vector is obtained whichever part of the dislocation is encircled.

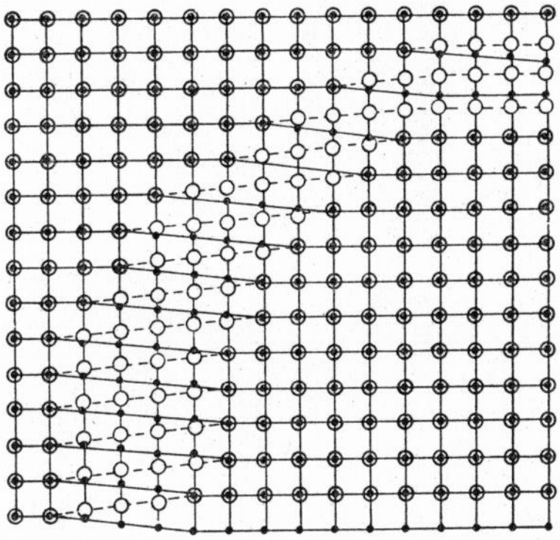

FIG. 1.12. Conversion of a pure screw dislocation to a pure edge dislocation during the change of direction of the dislocation line. o atoms above the plane, and ● atoms below the plane of the drawing

With the aid of the Burgers circuit and Burgers vector a simple model can be given to construct general dislocations. A cut is made along a plane of a perfect crystal and a closed circuit intersecting the plane of cut in one point is considered. The two sides of the cut are displaced relative to each other along a vector **b** and reconnected (perhaps by removing or adding material at the sides of the cut). In this way a dislocation is produced, since a Burgers circuit with a closure failure **b** has been created. The Burgers vector **b** is equal to the displacement vector. As regards plastic deformation, only those displacements are considered which give rise to finite differentiable stresses. These conditions are fulfilled if the relative displacement of the cut surfaces corresponds to a rigid body displacement [10]. In this case the relative displacement of the adjacent points on the two sides of the cut is given by the equation

$$\delta u_i = b_i + \sum_{j=1}^{3} d_{ij} x_j, \qquad (i = 1,\ 2,\ 3) \tag{1.5}$$

where x_j denotes the coordinates of the point, b_i is the relative translation of this point and d_{ij} stands for the relative rotation of the cut surface. Since d_{ij} is antisymmetrical by definition, a general dislocation may be described by six constants, or in other words it consists of six elementary dislocations. Each of these can be characterized by one constant, the other five constants then being zero. Such an elementary dislocation may be described in the following way (Fig. 1.13).

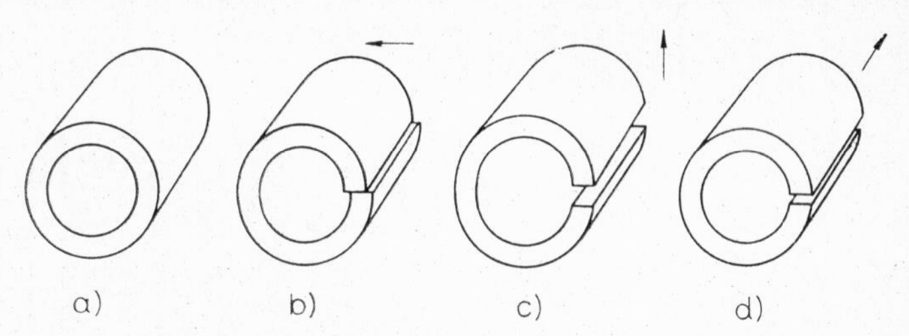

FIG. 1.13. Deformations creating dislocations

Consider a cylindrical ring which is cut along one of its generatrices. By displacing one side of the cut, a dislocation line is produced. The six dislocations may be classed into three types:

1. Translation perpendicular to the axis of the cylinder (Figs 1.13b, c): edge dislocation.
2. Translation parallel to the axis of the cylinder (Fig. 1.13d): screw dislocation.
3. Relative rotation of the cut surface. This type of dislocation cannot be produced in plastic bodies since at infinity their stress does not vanish.

1.7. Conservation of the Burgers vector

From the definition of dislocations it follows that the *Burgers vector is constant* along a dislocation line; in other words, *if two Burgers circuits can be transformed into each other by moving them continuously in good material they result in identical Burgers vectors.* This becomes quite clear if one considers the fact that dislocations are produced if two parts of the material slip with respect to each other along a part of a glide plane. The Burgers circuit around a dislocation forming the boundary of the slipped part can only be "pulled" along the dislocation by moving it in good material if the relative displacement of the slipped surfaces constitutes a lattice vector which is the same everywhere.

From this it follows that a dislocation cannot terminate within a crystal. If this were the case, **b** would abruptly vanish by moving

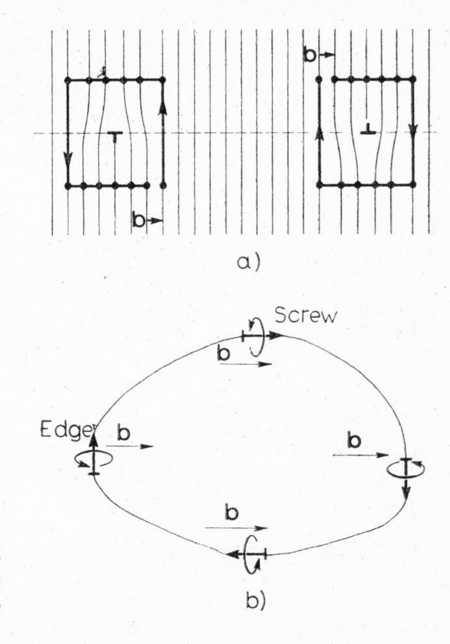

FIG. 1.14. (a) Section of a closed dislocation loop. (b) Burgers circuit along a closed dislocation loop

the Burgers circuit beyond the end of the dislocation line. Consequently, a dislocation line may terminate in only three ways:

(a) It extends (not necessarily in a straight line) from crystal surface to crystal surface. **b** does not change with change of the direction of the dislocation line, and consequently it generally makes an angle ϑ with the tangent of the dislocation line which varies between 0° and 90° ($0° \leqq \vartheta \leqq 90°$). Pure edge or screw dislocations occur fairly infrequently (see Peierls model).

(b) The dislocation line forms a closed loop. Along the loops the Burgers vector is constant and therefore the dislocation has in each case a pure edge and a pure screw character in at least two places (Fig. 1.14b). Although the Burgers vector is identical everywhere along the loops, in the two opposite sectors of the loop the line elements are antiparallel and consequently the two segments are physically different. If, for example, the loop is examined in a section normal to the pure edge segments (Fig. 1.14a), the extra half-planes belonging to the two segments are on the two different sides of the glide plane. These two dislocation segments thus have signs opposite to each other. The situation is quite the same with screw-type segments.

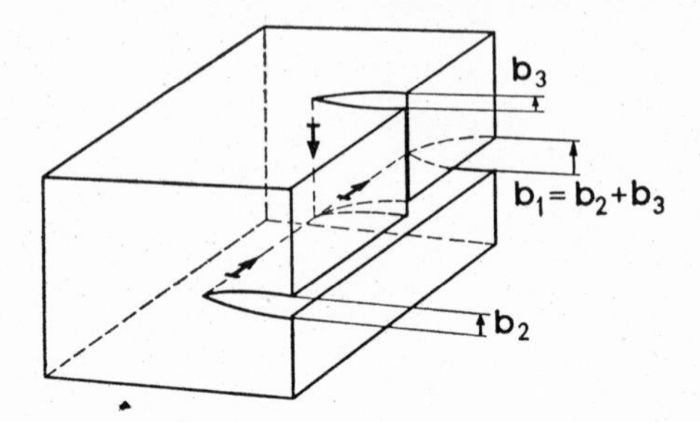

FIG. 1.15. Model of the production of a node

(c) The dislocation line ends in a node. (If three or more dislocations intersect each other, the point of intersection is a node.) A node can be produced by cutting an elastic body and displacing one part of the cut surfaces by a vector $\mathbf{b}_1$ and the other part by $\mathbf{b}_2$ ($\mathbf{b}_1 \neq \mathbf{b}_2$), relative to each other. Now, however, a second cut must of necessity

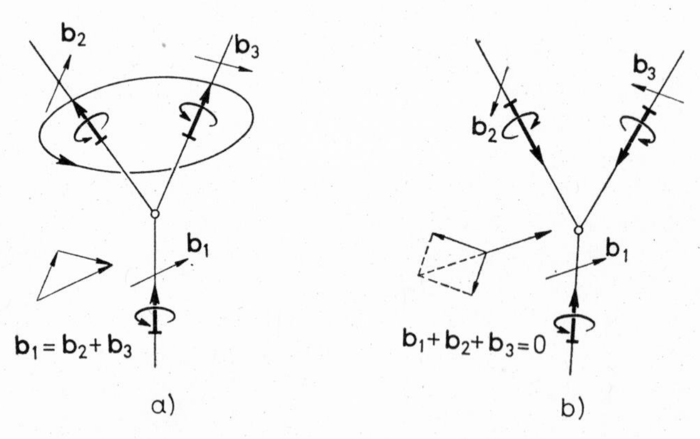

FIG. 1.16. The conservation of the Burgers vector on branching

be made with displacement $\mathbf{b}_3$ (Fig. 1.15). In this way three dislocations are produced which intersect at one point, the node. The fixing of the direction of the line element of one of the dislocations creating the node does not determine the direction of the other line elements. If these are so chosen that the positive directions after the node are the continuation of that before the node, the sum of the Burgers vectors after the branching is equal to the sum of the Burgers vectors before the node (Fig. 1.16a). This is the natural consequence of the conservation of the Burgers vector, since after the node the dislocations can be enclosed by one single extended Burgers circuit. If, on the other hand, all the line elements are directed towards the node (Fig. 1.16b), the sum of the Burgers vectors of the dislocations ending into the node is zero. (This resembles Kirchhoff's first law according to which the sum of the currents flowing into a node is zero.)

1.8. The origin of dislocations and their distribution in the crystal

Though the presence of dislocations does not mean a thermodynamically stable condition (Section 2.7) as has already been pointed out, *natural and artificial crystals always contain a certain number of dislocations. Their origin may always be traced back to the conditions of crystal growth.* In general one of the following mechanisms of production of dislocations must be considered:

(a) Growth of dislocations already present in the nucleus and terminating on the solid–liquid boundary.

(b) Multiplication of dislocations of the seed crystal (see Chapter 5) due to thermal stresses induced by a temperature gradient; the loops continue to grow when they reach the liquid–solid boundary (this occurs in crystals which grow from the melt).

(c) Bad-quality epitaxy due to the contamination of the nucleus (oxide layer, polycrystalline layers). Small solid grains falling into the melt or solution have the same effect. If a small dislocation density is wanted, particular care must be taken as to the quality and surface of the seed crystal (especially if the crystal grows from the melt).

(d) Multiplication induced by thermal stresses originating from surface defects (e.g. scratches) on the crystal.

(e) The impurity atoms always present are not usually incorporated into the solid phase. Consequently the contamination of the melt (solution) increases, with the result that in the last stage of crystallization the coefficient of thermal expansion changes. This leads to considerable internal stresses and to a multiplication of dislocations.

(f) As a consequence of the condensation of excess vacancies, closed dislocation loops are formed (see Section 6.9). This effect is found to be particularly strong in quenched material.

(g) Multiplication of dislocations of other origin by thermal stresses.

In spite of the great variety of sources, *from certain substances* (Si, Ge) *single-crystals can be grown whose insides are practically free of dislocations.* With a properly selected growth direction (the angle between the glide plane of the growing dislocation and the direction of crystal growth should be large) the dislocations originating in the seed grow out to the surface where they are no longer troublesome. The rest of the dislocations move towards the surface as a result of their absorption by the excess vacancies. The only exceptions are screw dislocations which grow parallel to the growth direction. These dislocations take the shape of a helical line (see Sections 4.6 and 4.8).

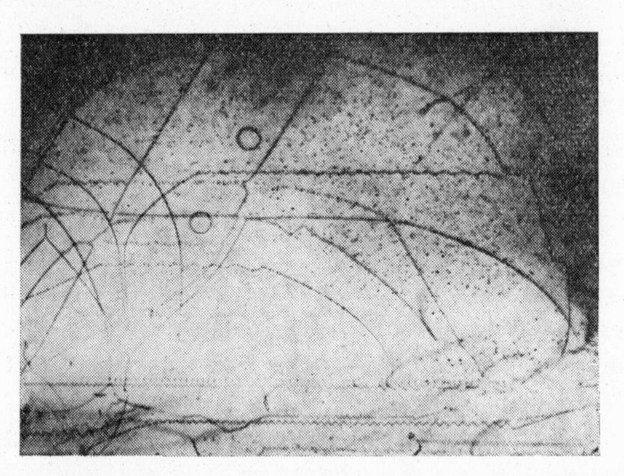

FIG. 1.17. Dislocations in Si decorated with copper and gold. The horizontal helices of the ⟨110⟩ direction can be well observed. The diameter of the large loops lying in the (111) plane is approximately 50 μm [W. C. Dash, *J. Appl. Phys.* **30** (1959) 459]

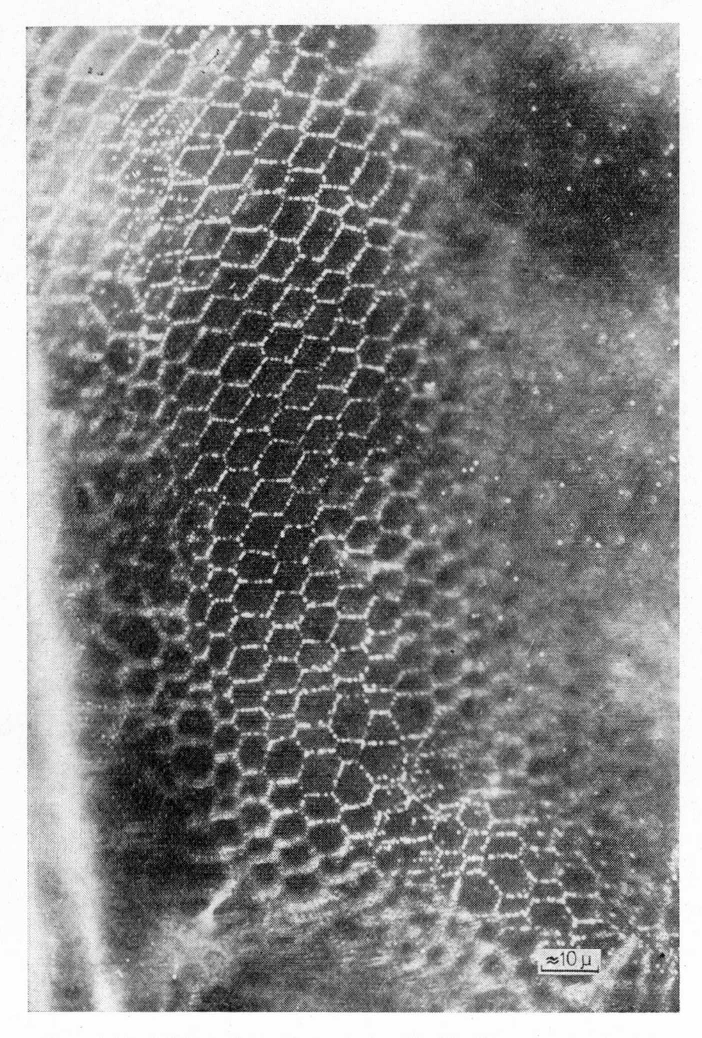

FIG. 1.18. Dislocation network in NaCl. Decorated picture taken with an optical microscope [S. Amelinckx, *Phil. Mag.* **1** (1966) 269]

According to distribution, the dislocations may be classed in three groups:

1. *Spatially distributed dislocations.* These are disordered and frequently interconnected forming a network (Figs 1.17 and 1.18).

2. *Low-angle grain boundaries.* It is shown in Section 2.5.4 that parallel edge dislocations forming rows, as depicted in Fig. 1.19, are

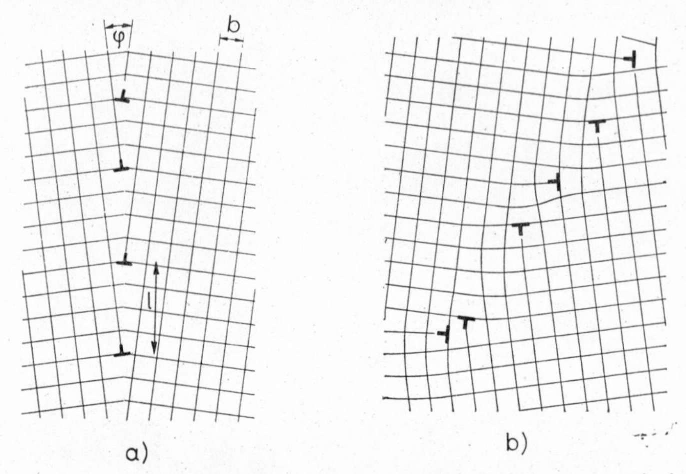

FIG. 1.19. Dislocation structures in symmetrical and asymmetrical tilted small-angle grain boundaries

extremely stable. However, because of the nearly periodical arrangement of the extra half-planes the volume elements separated by these two-dimensional "dislocation walls" are somewhat rotated relative to each other (tilt boundary). It can be seen directly that in a symmetrical case (Fig. 1.19a) the angle of tilt is

$$\phi = \frac{b}{l}. \tag{1.6}$$

This was first proved by Vogel and coworkers on Ge crystals [12]. The average distance of the dislocations was determined by an etching technique, and the misorientation of the two crystal halves of an order of minutes was measured by X-ray diffraction. Since this pioneer work, the correctness of equation (1.6) has also been proved several times electron-microscopically (Fig. 1.20a).

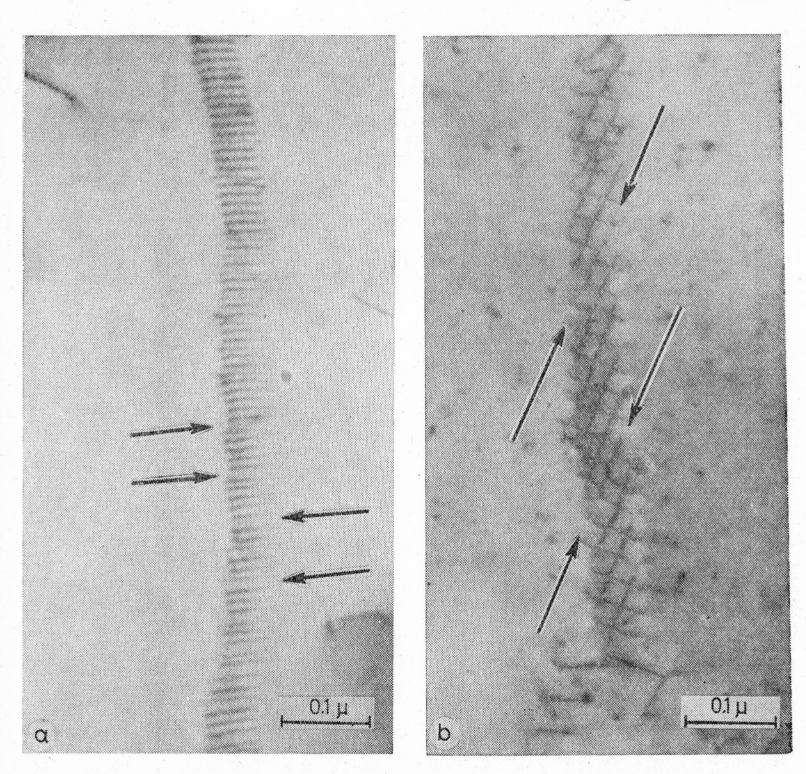

FIG. 1.20. (a) Tilt and (b) twist boundaries in Pt. The arrows indicate the positions of dislocations of other Burgers vectors which in this setting do not give any contrast [S. Amelinckx, *J. Nucl. Mat.* **6** (1962) 46]

In the previous case the axis of rotation was parallel to the dislocations, i.e. to the grain boundary. A small twist around an axis perpendicular to the grain boundary is realized if parallel screw dislocations are present. However, since the relative displacement of the two sides of the grain boundary in this case can be described only as a resultant of two shears of different directions (Fig. 1.21), these *twist boundaries* always contain two sets of screw dislocations of *different Burgers vectors*. These two sets usually form a two-dimensional network (Fig. 1.20b).

3. *Dislocations produced by plastic deformation and arranged in a glide plane.* The dislocation content of a crystalline material is given by the *dislocation density* which is the total length of dislocations in 1 cm^3 of the material. In most practical cases, however, only those dislocations can be counted which emerge on the surface, and hence it is more suitable to consider the surface density. In the case of parallel dislocations this is defined by the number of dislocations intersecting a unit surface normal to the dislocations. In the case of a random distribution of directions, only a part of the density as related to a given surface (the number of dislocations intersecting the surface in any direction) is considered. For dislocations oriented perfectly randomly, the volume density is just twice the value of the apparent surface density [13].

In the case of a surface distribution, "line density" is used analogously to the volume distribution and surface density.

With metal crystals the density after heat-treatment and without any plastic deformation is 10^7–10^8 dislocations per cm^2, which may increase after a strong deformation up to 10^{10}–10^{11} dislocations per cm^2. For ionic crystals the density is somewhat less. With simple methods the relatively low value of 10^4–10^5 dislocations per cm^2 may be attained. A very low value may be obtained with carefully grown Si and Ge single crystals, whose dislocation density can be reduced to 1–100 dislocations per cm^2 in single crystals with diameters of a few centimetres.

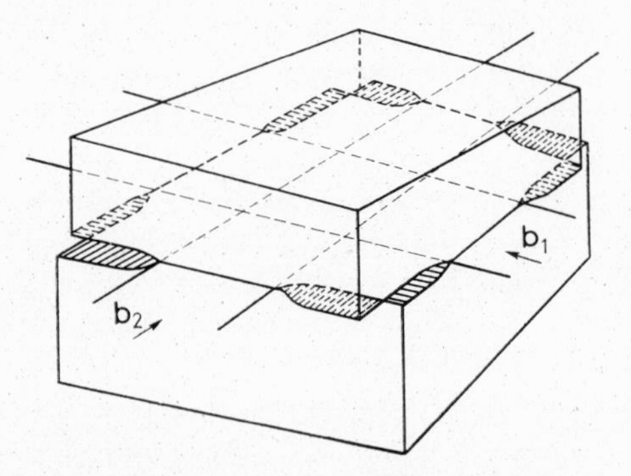

FIG. 1.21. The relative rotation of two half-crystals due to the presence of two perpendicular sets of screw dislocations

Chapter 2

Dislocations in an elastic continuum

2.1. Fundamentals of the continuum theory

Dislocations were discovered by the study of the atomic structure of solid material. In the first chapter, this type of crystal defect was briefly reviewed, and the characteristics of the dislocations were defined. If the quantitative properties are to be investigated by a crystallographic approach, great difficulties are encountered. Thus, if the distortions caused by the dislocations and the stress field around them are to be described, not only the forces acting between the atoms must be known, but the interaction of each atom with the others must also be considered. Though reasonable approximations and simplifications may facilitate the solution, one is faced with considerable mathematical difficulties. It can readily be understood that though dislocations are typical crystal defects, the study of their quantitative properties was first restricted to models of elastic continua which can be treated more easily mathematically. These models, of course, do not possess the exact properties of the dislocations, but nevertheless they may be fairly good approximations. The following reasoning is based on the linear theory of elasticity which assumes the validity of Hooke's law. Accordingly, stresses generated by small deformations are described as linear functions of these deformations. From this it follows that the volume around the core of a dislocation must be omitted from the calculations since here the relative displacement and the deformation connected with it are so large that Hooke's law is no longer valid. This, however, does not constitute a serious problem, because in most practical cases the effects originating in the core of the dislocation may be neglected.

On the other hand, it may be asked if there is any sense in investigating the crystal and its defects by an elastic model, since it is known that crystals do not behave elastically even at stresses which are several orders of magnitude smaller than the theoretical elastic limit ($\tau_{cr} \approx \mu/30$), but become plastically deformed or even fractured. This contradiction, however, is only apparent. Any deviation from elastic behaviour is caused precisely by lattice defects. The fact that the crystal as a whole is no longer elastic below a certain stress limit, which is smaller than the theoretical value, is due to these defects. However, any lattice defect is necessarily embedded into a part of the crystal which is free of defects. The actual defect generates elastic deformations in an area which – being free of defects – is elastic up to the theoretical limit.

Thus, the elastic stresses inside the crystal must always be regarded as results of elastic deformations generated by lattice defects in *the perfect matrix* around these defects. For this material free of defects the elastic continuum model may clearly be applied with good results.

Our calculations are limited to isotropic bodies. Though most crystals are anisotropic, the difficulties would be considerable if one tried to allow for anisotropy in the mathematical treatment. Many of the results of the calculations depend upon the anisotropic effect only to a small degree, and thus a satisfactory approximation can be obtained to describe the quantitative properties of the dislocations. The fundamental concept of elasticity, which is necessary to understand this chapter, is summarized in Appendix B.

2.2. The definition of dislocations in an elastic continuum

It was shown in Section 1.6 that a dislocation is generated if the crystal is cut along one of its planes, and the two sides obtained by the cut are displaced by **b** relative to each other. This definition can easily be generalized to apply to a continuum. Consider an arbitrary closed curve C in the continuum. Make a cut into the material along the surface F ending at the curve considered. Displace the material at the two sides of the cut surface relative to each other, e.g. the lower part relative to the upper one, by **b**. After this reconnect the two sides (perhaps by adding or removing material). In this way a dislocation is created along the curve C, because this curve constitutes a boundary between the slipped and unslipped regions of the body (Section 1.4). This definition is of general character, and may be applied to any dislocation of any shape. It can be proved that the properties of a dislocation are independent of the selection of the cut surfaces, and depend only on their relative displacement and the curve C [1].

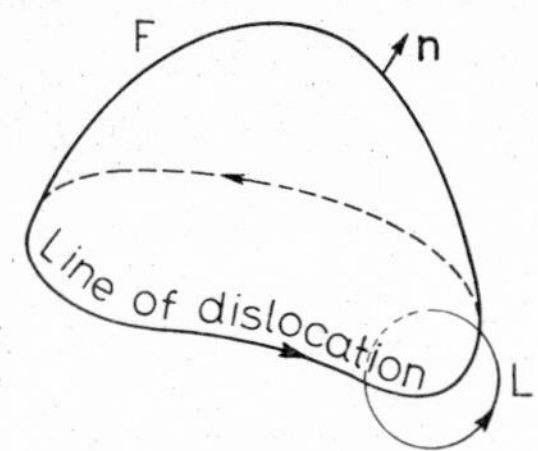

FIG. 2.1. The definition of the positive direction along the Burgers circuit with respect to the normal vector of the cut surface

One can easily see that in a body containing a dislocation the displacement vector $\mathbf{u}$ is not a single-valued function. Let us consider before the creation of a dislocation two non-coincident points A_1 and A_2 of the cut surfaces, which during the deformation transform into the same point A (such a point can always be found if the cut surfaces are displaced parallel to each other). A_1 and A_2 become displaced in different ways during the deformation, and consequently the displacement of the point A will be different if it is regarded as a displacement of the point A_1 or A_2. This value multiplicity may be used to introduce the analogy of the Burgers circuit as defined for crystals. Let the change of the displacement vector between the points defined by the vectors $\mathbf{r}$ and $\mathbf{r} + d\mathbf{r}$ be $d\mathbf{u}$, and take the integral

$$\oint_L d\mathbf{u} = \oint_L \frac{d\mathbf{u}}{d\mathbf{r}} d\mathbf{r} ,$$

where L is a closed curve within the body. If L trespasses on a surface along which the displacement is not unequivocal, the value of the above integral is not zero, and the curve L then clearly encircles a dislocation. The value of the line integral is equal to the relative displacement of the cut surfaces, and thus the Burgers vector of the dislocation encircled by the curve L (equivalent to the closure failure in the crystal) is

$$\mathbf{b} = \oint_L \frac{d\mathbf{u}}{d\mathbf{r}} d\mathbf{r} . \tag{2.1}$$

In order to define the Burgers vector unequivocally the direction of the circuit is fixed as depicted in Fig. 2.1. It is quite clear that the definition given by (2.1) is equivalent to the previous definition.

2.3. The energy of dislocations

When a body is elastically deformed, work must be performed against the stresses created by this elastic deformation. In this way energy becomes stored in the body. It can be proved that the value of this energy per unit volume is [2]

$$u = \frac{1}{2} \sum_{i,\,k=1}^{3} \sigma_{ik}\varepsilon_{ik}\,, \tag{2.2}$$

where ε_{ik} and σ_{ik} denote the kth element of the ith row of the strain and stress tensor respectively (see Appendix B). Let us suppose that two dislocations A and B are present in the material. Let the respective strain and stress tensors be ε_{ik}^{A}, ε_{ik}^{B} and σ_{ik}^{A}, σ_{ik}^{B}, respectively. It follows from the linearity of the equations of elastic equilibrium that the resultant strain and stress tensors are

$$\begin{aligned} \varepsilon_{ik} &= \varepsilon_{ik}^{A} + \varepsilon_{ik}^{B}, \\ \sigma_{ik} &= \sigma_{ik}^{A} + \sigma_{ik}^{B}. \end{aligned} \tag{2.3}$$

Let us calculate the elastic energy of a body containing two dislocations by equations (2.2) and (2.3)

$$\begin{aligned} U = \frac{1}{2}\sum_{i,k=1}^{3} \int \sigma_{ik}\varepsilon_{ik}\,dV &= \frac{1}{2}\sum_{i,k=1}^{3} \int \sigma_{ik}^{A}\varepsilon_{ik}^{A}\,dV + \\ + \frac{1}{2}\sum_{i,k=1}^{3} \int \sigma_{ik}^{B}\varepsilon_{ik}^{B}\,dV &+ \frac{1}{2}\sum_{i,k=1}^{3} \int (\sigma_{ik}^{A}\varepsilon_{ik}^{B} + \sigma_{ik}^{B}\varepsilon_{ik}^{A})\,dV\,, \end{aligned} \tag{2.4}$$

where the integrals should be taken over the whole volume of the body. The first term on the right side of equation (2.4) depends only on the A, and the second term on the B dislocation. The total elastic energy is given by one of them if the other is absent. Thus the expression

$$U_s = \frac{1}{2} \sum_{i,k=1}^{3} \int \sigma_{ik} \varepsilon_{ik} dV \tag{2.5}$$

is called the *self-energy of the dislocation*. The third term of (2.4) acts only if both dislocations are present. If the expressions (B.18) and (B.19) are summarized in the equation

$$\sigma_{ik} = 2\mu\varepsilon_{ik} + \lambda\Theta\delta_{ik} \qquad (i, k = 1, 2, 3) \tag{2.6}$$

(δ_{ik} is the Kronecker delta, which has the value 1 if $i = k$, and is zero for $i \neq k$), it can easily be seen that

$$\sum_{i,k=1}^{3} \sigma_{ik}^{A} \varepsilon_{ik}^{B} = \sum_{i,k=1}^{3} \sigma_{ik}^{B} \varepsilon_{ik}^{A} . \tag{2.7}$$

With this, the third term of (2.4) may be written in the following form:

$$U_i = \sum_{i,k=1}^{3} \int \sigma_{ik}^{A} \varepsilon_{ik}^{B} dV . \tag{2.8}$$

This energy clearly comes from the interaction of two dislocations, and is an *energy of interaction*.

According to (2.5) and (2.8), the self-energy and the energy of interaction can be calculated by first determining the respective strain and stress tensors, and then integrating the expressions of the type (2.7) obtained over the whole volume of the body. In many instances, however, the calculations can be carried out much more simply if the cut surfaces are known. Thus it is recommended to express the energies of (2.5) and (2.8) by integrals taken over the cut surfaces. This can be done in the following way. Let us consider, for example, the expression (2.5). According to definition (B.4):

$$\varepsilon_{ik} = \frac{1}{2} \left(\frac{\partial u_i}{\partial x_k} + \frac{\partial u_k}{\partial x_i} \right) . \tag{2.9}$$

Substituting in (2.5) one obtains

$$U_s = \frac{1}{4} \sum_{i,k=1}^{3} \int \sigma_{ik} \frac{\partial u_i}{\partial x_k} dV + \frac{1}{4} \sum_{i,k=1}^{3} \int \sigma_{ik} \frac{\partial u_k}{\partial x_i} dV =$$

$$= \frac{1}{2} \sum_{i,k=1}^{3} \int \sigma_{ik} \frac{\partial u_i}{\partial x_k} dV.$$

The indices i and k in the second term are interchangeable, since in so doing only the sequence of the terms within the sum are changed. By applying the rule for the differentiation of the product one obtains

$$\sum_{i,k=1}^{3} \sigma_{ik} \frac{\partial u_i}{\partial x_k} = \sum_{i,k=1}^{3} \frac{\partial}{\partial x_k} (\sigma_{ik} u_i) - \sum_{i,k=1}^{3} u_i \frac{\partial \sigma_{ik}}{\partial x_k} =$$

$$= \sum_{k=1}^{3} \frac{\partial}{\partial x_k} \sum_{i=1}^{3} \sigma_{ik} u_i - \sum_{i=1}^{3} u_i \sum_{k=1}^{3} \frac{\partial \sigma_{ik}}{\partial x_k} = \operatorname{div} (\boldsymbol{\sigma}\, \mathbf{u}) - \mathbf{u} \operatorname{Div} \boldsymbol{\sigma}.$$

According to (B.12), however, if body forces are not operative, $\operatorname{Div} \boldsymbol{\sigma} = 0$, since

$$U_s = \frac{1}{2} \int \operatorname{div} (\boldsymbol{\sigma}\, \mathbf{u})\, dV. \tag{2.10}$$

From this it is already apparent that by applying Gauss's theorem, U_s can be expressed by a surface integral. However, it must be considered that this theorem may only be applied if $\boldsymbol{\sigma}\, \mathbf{u}$ is continuous

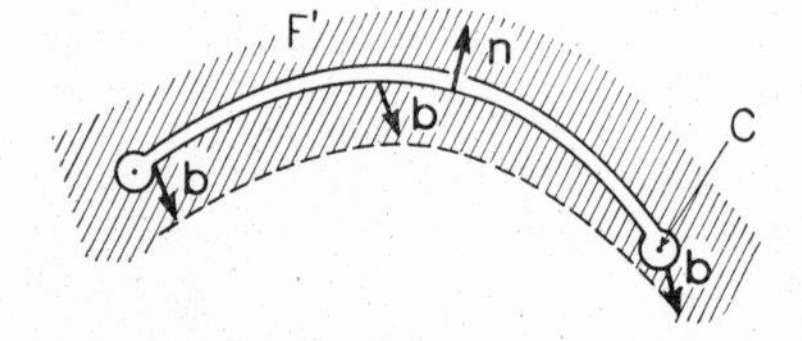

FIG. 2.2. Section of a closed surface embracing a dislocation lying along a closed curve and the cut surface

and single-valued everywhere in the volume investigated. This condition is not satisfied along the cut surface and hence cut surfaces must be excluded from the investigated volume element by enclosing them with a closed surface F' (Fig. 2.2). Let us extend the volume integral over the whole space. This can be done because the elastic energy is zero outside the body. In this case one must calculate the integral (2.10) for a volume V' one of whose limiting surfaces is at infinity, and the other is the surface F' which was chosen. (This integral naturally yields a value $U_s' \neq U_s$, because one part of the body has been excluded with the surface F'.) Since the stresses must be zero at infinity, it follows that

$$U_s' = \frac{1}{2}\int_{V'} \operatorname{div}(\boldsymbol{\sigma}\,\mathbf{u})\,dV = \frac{1}{2}\int_{F'} (\boldsymbol{\sigma}\,\mathbf{u})\,d\mathbf{F}'.$$

Let the surface F' converge to F, then one obtains the limiting value

$$U_s = \frac{1}{2}\int_{\pm F} (\boldsymbol{\sigma}\,\mathbf{u})\,d\mathbf{F}.$$

Here the integration must be carried out for both sides of the cut surface. However, the difference of the displacement vector **u** on both sides of the cut surface is exactly the Burgers vector **b** of the dislocation. Thus one obtains

$$U_s = \frac{1}{2}\,\mathbf{b}\int_F \boldsymbol{\sigma}\,d\mathbf{F}, \tag{2.11}$$

or with the previous notation

$$U_s = \frac{1}{2}\sum_{i,k=1}^{3} b_i \int \sigma_{ik}\,dF_k, \tag{2.12}$$

where b_i represents the ith component of the Burgers vector, and dF_k is the kth component of the surface-element vector. By analogous reasoning one obtains for the energy of interaction:

$$U_i = \mathbf{b}^A \int_{F_A} \boldsymbol{\sigma}^B\,d\mathbf{F}, \tag{2.13}$$

or

$$U_i = \sum_{i,k=1}^{3} b_i^A \int_{F_A} \sigma_{ik}^B\,dF_k, \tag{2.14}$$

where F_A denotes the cut surface belonging to dislocation A.

2.4. Forces acting on dislocations

It was shown in the previous section that if two dislocations are present in one body, besides their self-energy, they necessarily give rise to an energy originating from their interaction. This means that whenever one wants to produce a dislocation in a body already containing dislocations, work must be performed by deformation corresponding to the self-energy of the dislocation, and also against the stress-field of the already existing dislocations. From all this it follows that a similar situation arises if one wants to produce a dislocation in a body in which a stress-field exists which was generated not by dislocations but by external forces. The stress tensor $\boldsymbol{\sigma}^B$ of formula (2.13) may originate not only from a dislocation, but also from external forces. This means, on the other hand, that any motion of a dislocation in a body under the effect of an external force – i.e. the increase of the slipped region – requires an energy input excess and the change of the self-energy as well. Extraneous stresses which are present in the material impede the motion of dislocations and exert a force on them. (This force shows a close analogy with the forces by which a magnetic field influences a current flowing in a conductor.)

The magnitude of the force on a line element $d\mathbf{l}$ of a dislocation can easily be determined. Let the stress present in the material and not originating from the dislocation under investigation be $\boldsymbol{\sigma}$, the Burgers vector of the dislocation be $\mathbf{b}$, and let the line element $d\mathbf{l}$ of the dislocation move by a distance $d\mathbf{s}$. Since, as was stated in Section 2.2, the properties of the dislocations are independent of

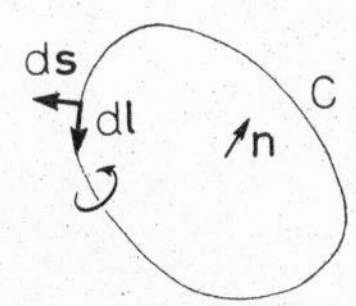

FIG. 2.3. For the derivation of equation (2.15)

the choice of the cut surface, the surface $d\mathbf{s} \times d\mathbf{l}$ swept by the dislocation may be regarded as the increment of the cut surface (Fig. 2.3). Thus, according to equation (2.13), the change of the interaction energy is

$$dU_i = \mathbf{b}\,\boldsymbol{\sigma}\,(d\mathbf{s} \times d\mathbf{l}) = -\,(d\mathbf{l} \times d\mathbf{s})(\boldsymbol{\sigma}\,\mathbf{b}). \tag{2.15}$$

Using the identity of the mixed products of the vectors $(\mathbf{a}\times\mathbf{b})\mathbf{c} =$ $= (\mathbf{c}\times\mathbf{a})\mathbf{b}$, one obtains from (2.15):

$$dU_i = -[(\boldsymbol{\sigma}\,\mathbf{b})\times d\mathbf{l}]\,d\mathbf{s}. \tag{2.16}$$

It is well known from mechanics that in a conservative (or potential) field the force is expressed by the equation

$$\mathbf{f} = -\operatorname{grad} U = -\frac{\partial U}{\partial \mathbf{s}},$$

where U denotes the potential of the field. (A field is termed conservative if the law of the conservation of mechanical energy is valid in that field. The elastic stress-field is such a conservative field.)

From (2.16) the force acting on a line element $d\mathbf{s}$ of the dislocation is

$$d\mathbf{f} = (\boldsymbol{\sigma}\,\mathbf{b})\times d\mathbf{l}. \tag{2.17}$$

If this expression is compared with the Biot–Savart law: $d\mathbf{f} = 1/c(\mathbf{B}\times d\mathbf{l})$, the analogy mentioned above becomes quite clear. The idea of a force acting on dislocations was introduced by Koehler in 1941 [3]. Equation (2.17) was derived by Peach and Koehler [4] in 1950 and by Nabarro [5] in 1951.

2.5. Properties of individual straight dislocations

Some characteristic data of pure edge and pure screw dislocations can be determined by applying the general equations obtained in the previous section. Let us consider an elastic cylinder of infinite length and diameter R. Let the z-axis of the rectangular coordinate system coincide with the axis of the cylinder. Make a cut into the cylinder along the plane $y = 0$ as far as the z-axis. If the cut surfaces so obtained are displaced relative to each other by a distance b in the directions x, y or z and then reconnected, an edge or screw dislocation of infinite length is produced (see Appendix, Figs C.1, C.2). In the case of a perfect cylinder, infinite stresses clearly result along the z-axis if the cut surfaces are displaced by a constant distance b. In order to avoid this, a cylinder of radius r_0 is removed from the centre of the body. This procedure is permissible, since dislocations actually have such cavities (see Section 4.8).

In order to determine the stress-field around the dislocations, the equilibrium conditions (B.26) must be established. It is quite clear that in the case of equilibrium when no external forces act on the body, forces perpendicular to the boundary surfaces may not act. Consequently the solution of (B.26) must satisfy the following conditions:

(a) the displacement vector must be discontinuous along the cut surfaces (when intersecting the cut surface it must change by **b**);
(b) the perpendicular stress components must vanish along the boundary.

The solutions satisfying these requirements are given in Appendix C.

2.5.1. *The stress-field and energy of an edge dislocation*

If an edge dislocation has been produced by displacing the cut surfaces in the x direction the non-vanishing stress components can be expressed in a rectangular coordinate system by the equations (Appendix C.1):

$$\sigma_{xx} = -\frac{\mu b}{2\pi(1-\nu)}\left\{\frac{y(3x^2+y^2)}{(x^2+y^2)^2} - \frac{3y}{r_0^2+R^2} + \frac{r_0^2R^2}{r_0^2+R^2}\,\frac{y(y^2-3x^2)}{(x^2+y^2)^3}\right\}, \tag{2.18a}$$

$$\sigma_{yy} = \frac{\mu b}{2\pi(1-\nu)}\left\{\frac{y(x^2-y^2)}{(x^2+y^2)^2} + \frac{y}{r_0^2+R^2} + \frac{r_0^2R^2}{r_0^2+R^2}\,\frac{y(y^2-3x^2)}{(x^2+y^2)^3}\right\}, \tag{2.18b}$$

$$\sigma_{xy} = \frac{\mu b}{2\pi(1-\nu)}\left\{\frac{x(x^2-y^2)}{(x^2+y^2)^2} - \frac{x}{r_0^2+R^2} - \frac{r_0^2R^2}{r_0^2+R^2}\,\frac{x(x^2-3y^2)}{(x^2+y^2)^3}\right\}, \tag{2.18c}$$

$$\sigma_{zz} = -\frac{\nu\mu b}{\pi(1-\nu)}\left\{\frac{y}{x^2+y^2} - \frac{2y}{r_0^2+R^2}\right\}. \tag{2.18d}$$

Those terms of these expressions which contain the radius r_0 of the cavity and the radius R of the cylinder are stresses necessary to satisfy the boundary conditions. These contribute only close to the surfaces. If the distance from the surface is large enough, the stress-field of an edge dislocation is approximately described by the first terms. In the literature most often only these are given, since the studied bodies are usually not cylindrical and the dislocation is not in the centre. The exact calculation satisfying the boundary conditions is difficult in general, and solutions are known for only a few special cases [3, 6, 7]. In the case of large dimensions, however, with dislocations distant from the surface and with a small core radius $r_0 \approx b$ and therefore, $r_0/R \approx 0$ (these conditions are satisfied for the majority of the dislocations in a macroscopic body), a good approximation can be obtained for a body of any shape if only the first terms of (2.18) are considered. Introducing the polar coordinates r and ϑ with $x = r \cos \vartheta$ and $y = r \sin \vartheta$, the non-zero components of the stress-field of the edge dislocations to be studied are to a good approximation

$$\sigma_{xx} = -\frac{\mu b}{2\pi(1-\nu)} \frac{\sin \vartheta(2 + \cos 2\vartheta)}{r}, \tag{2.19a}$$

$$\sigma_{yy} = \frac{\mu b}{2\pi(1-\nu)} \frac{\sin \vartheta \cos 2\vartheta}{r}, \tag{2.19b}$$

$$\sigma_{xy} = \frac{\mu b}{2\pi(1-\nu)} \frac{\cos \vartheta \cos 2\vartheta}{r}, \tag{2.19c}$$

$$\sigma_{zz} = -\frac{\nu \mu b}{\pi(1-\nu)} \frac{\sin \vartheta}{r}. \tag{2.19d}$$

It can be seen that every stress component can be written in the form $f(\vartheta)/r$ and consequently the stress distribution around a dislocation can be well demonstrated with polar diagrams. In Figs 2.4a, b, c, and d the absolute values of the $r\sigma_{ik}$ quantities are given in $\mu b/2\pi(1-\nu)$ units for various ϑ directions ($i, k = x, y, z$). The signs of the quantities are designated by (+) or (−) in the relevant region. One can see that the shear stress σ_{xy} has its maximum value in the glide plane ($y = 0$).

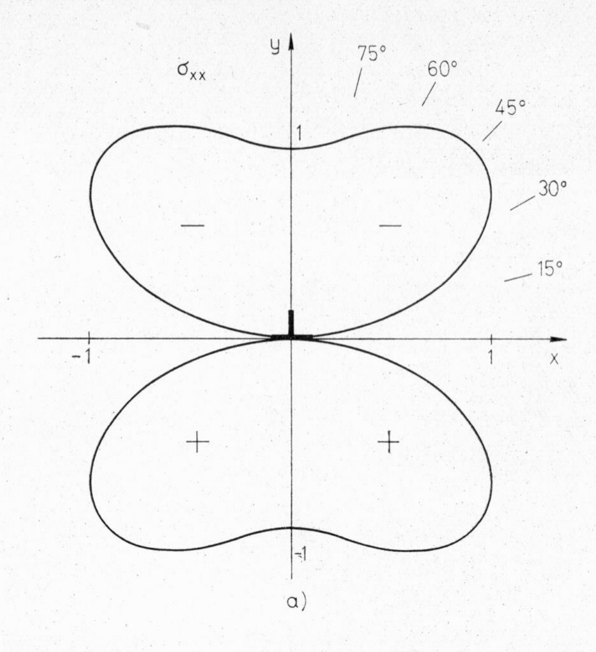

a)

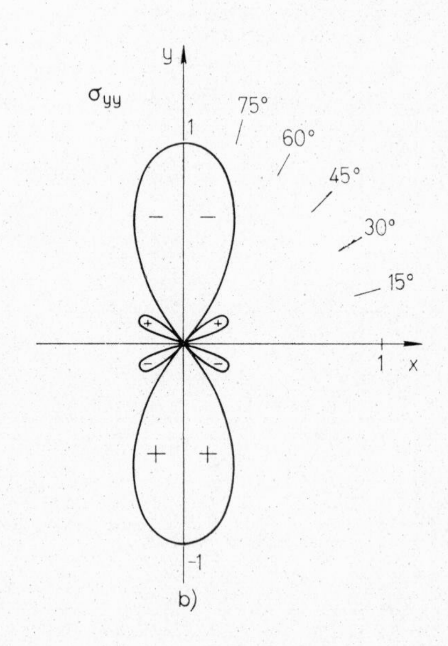

b)

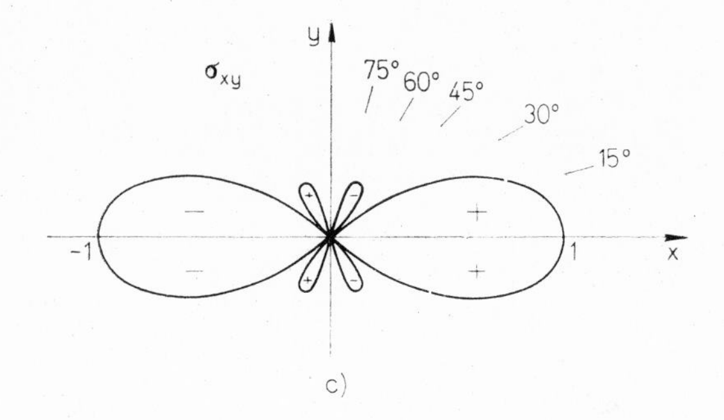

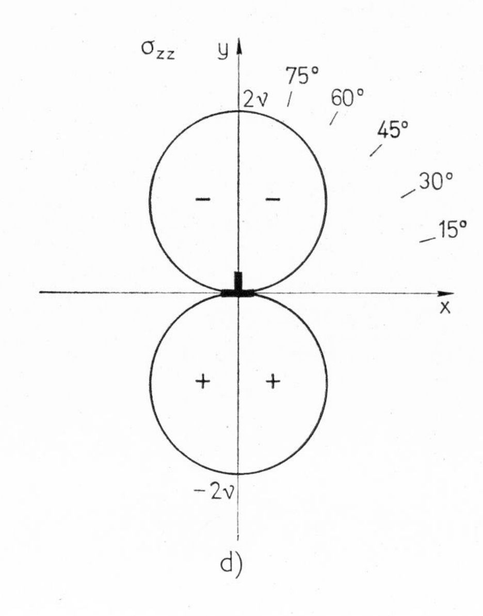

FIG. 2.4. (a) The σ_{xx} component of the stress-field of an edge dislocation. (b) The σ_{yy} component of the stress-field of an edge dislocation. (c) The σ_{xy} component of the stress-field of an edge dislocation. (d) The σ_{zz} component of the stress-field of an edge dislocation

The stress-field of an edge dislocation was first studied by Volterra [9], later by Burgers [10] and Koehler [3]. Mann [11] generalized their results for the Burgers vectors depending on position. Such cases may be reduced to Volterra dislocations distributed in the glide plane. Eshelby extended these investigations to anisotropic media [12].

Let us investigate the elastic energy of an edge dislocation. The Burgers vector is now $\mathbf{b} = (b, 0, 0)$ and the cut surface is the plane $y = 0$, and thus the vector of a surface element in (2.11) is $d\mathbf{F} = (0, dx\,dz, 0)$. With these values

$$\mathbf{b}(\boldsymbol{\sigma} d\mathbf{F}) = \mathbf{b}\left[\begin{pmatrix} \sigma_{xx} & \sigma_{xy} & 0 \\ \sigma_{yx} & \sigma_{yy} & 0 \\ 0 & 0 & \sigma_{zz} \end{pmatrix}\begin{pmatrix} 0 \\ dx\,dz \\ 0 \end{pmatrix}\right] = b\sigma_{xy}\ dx\,dz\,.$$

By combining this result with (2.11), the energy per unit length of a dislocation is

$$U = \frac{b}{2}\int_{r_0}^{R} \sigma_{xy}|_{y=0}\,dx =$$

$$= \frac{\mu b^2}{4\pi(1-\nu)}\int_{r_0}^{R}\left[\frac{1}{x} - \frac{x}{r_0^2 + R^2} - \frac{r_0^2 R^2}{(r_0^2 + R^2)x^3}\right]dx\,,$$

that is

$$U = \frac{\mu b^2}{4\pi(1-\nu)}\left(\ln\frac{R}{r_0} - \frac{R^2 - r_0^2}{R^2 + r_0^2}\right). \tag{2.20}$$

Our result shows that the energy of the dislocation diverges logarithmically if $R \to \infty$, or $r_0 \to 0$. This latter case is easily understandable, since if the radius of the cavity is zero, infinite stresses must be applied to move the cut surface to a distance $\mathbf{b}$ and this means that the work performed must be infinitely large. However, it is more difficult to understand that the energy per unit length of the dislocation increases beyond all limits if the dimensions of the body perpendicular to the dislocation line tend to infinity ($r_0 \neq 0$). The reason for this is the infinite length of the dislocation. Since a dislocation cannot terminate inside the body, a dislocation of finite length can be produced only along a closed curve or else it may start and terminate at the surface. If the distance between these parts of the surface is finite, the specific energy of the dislocation is finite even if the body has an infinite extension in some other direction [13]. The energy of a dislocation lying along a circle in a medium which is infinitely extended in every direction also has a finite value (Section 2.6).

Let us estimate numerically the magnitude of the energy in (2.20). If it is assumed that $R = 1$ cm, $r_0 = 10^{-7}$ cm and $b = 2.5 \times 10^{-8}$ cm, and if $\mu = 4 \times 10^{11}$ dyne/cm and $\nu = 0.34$, one obtains that the energy of an edge dislocation is approximately 5×10^{-4} erg/cm, or 8 eV per atomic plane intersecting the dislocation. It has been pointed out in the introduction to this chapter that the cavity of the dislocation affects the results to only a small degree. This is borne out by consideration that with the above values more than half of the energy of (2.20) is accumulated in a space outside the volume with radius 10^{-4} cm.

2.5.2. *Energy and stress-field of a screw dislocation*

According to calculations carried out in Appendix C.2, those stress components of a screw dislocation which are not zero are

$$\sigma_{xz} = -\frac{\mu b}{2\pi}\left(\frac{y}{x^2+y^2} - \frac{2y}{r_0^2+R^2}\right), \tag{2.21}$$

$$\sigma_{yz} = \frac{\mu b}{2\pi}\left(\frac{x}{x^2+y^2} - \frac{2x}{r_0^2+R^2}\right). \tag{2.22}$$

The first terms of the above expressions refer to the stresses arising from a displacement in the z direction. They satisfy the prescribed boundary conditions independently of the dimensions of the cylinder. Terms containing the dimensions of the cylinder are only included in the above expressions because a screw dislocation produced by a displacement in the z direction alone is still not stable. The stresses which originate in this way exert a torque in the plane normal to the dislocation line. This can be counter-balanced by considering too the displacements u_x and u_y which give rise to a torque of the same magnitude but of opposite sign. Since these corrective stresses can be neglected for dislocations which are far from the surface, th stress-field of a screw dislocation is to a good approximation

$$\sigma_{xz} = -\frac{\mu b}{2\pi}\frac{y}{r^2}, \quad \sigma_{yz} = \frac{\mu b}{2\pi}\frac{x}{r^2}, \tag{2.23}$$

where $r^2 = x^2 + y^2$.

With (2.11) and (2.22) the energy per unit length of a screw dislocation is

$$U = \frac{b}{2}\int_{r_0}^{R} \sigma_{yz}|_{y=0}\, dx = \frac{\mu b^2}{4\pi}\int_{r_0}^{R}\left(\frac{1}{x} - \frac{2x}{r_0^2 + R^2}\right) dx =$$

$$= \frac{\mu b^2}{4\pi}\left(\ln\frac{R}{r_0} - \frac{R^2 - r_0^2}{R^2 + r_0^2}\right). \tag{2.24}$$

With the values of the previous section the energy per unit length of a screw dislocation is approximately 3.5×10^{-4} erg/cm or 5 eV per atomic plane.

The calculated values are, of course, not exact; they merely yield an estimation of the order of magnitude. Consequently no serious error is made if the second term in (2.20) and (2.24) is neglected. Thus, the energy per unit length of a dislocation may be given by the following expression:

$$U = \frac{\mu b^2}{4\pi a}\ln\frac{R}{r_0}, \tag{2.25}$$

where $a = (1 - \nu)$ for an edge dislocation and $a = 1$ for a screw dislocation.

2.5.3. Forces acting on straight dislocations

If the body containing a dislocation also contains another stress-field

$$\boldsymbol{\sigma} = \begin{pmatrix} \sigma_{xx} & \sigma_{xy} & \sigma_{xz} \\ \sigma_{yx} & \sigma_{yy} & \sigma_{yz} \\ \sigma_{zx} & \sigma_{zy} & \sigma_{zz} \end{pmatrix} \tag{2.26}$$

produced by either some inner or an outer source, according to (2.17) a force

$$\mathbf{F} = (\boldsymbol{\sigma}\,\mathbf{b}) \times \mathbf{t} \tag{2.27}$$

acts on unit length of the dislocation, where **t** is a tangential unit vector pointing in the positive direction of the dislocation line. Applying (2.27) to a straight-edge dislocation with **b** = $(b, 0, 0)$ and **t** = $(0, 0, 1)$, one obtains for the components of the force the expressions

$$F_x = \sigma_{xy} b\,, \quad F_y = -\sigma_{xx} b\,, \quad F_z = 0\,. \tag{2.28}$$

Similarly for a screw dislocation with **b** = $(0, 0, b)$ and **t** = $(0, 0, 1)$

$$F_x = \sigma_{yz} b\,, \quad F_y = -\sigma_{xx} b\,, \quad F_z = 0\,. \tag{2.29}$$

These expressions clearly demonstrate that the value of the force-component F_x operative in the glide plane of the edge dislocation which causes it to move is determined only by the shear stress acting along the direction of the Burgers vector. With a screw dislocation the situation is different, since this dislocation has no preferred glide plane. The result of this is that the glide always begins along that plane in which the shear stress attains the value necessary to move the dislocation. However, as the screw dislocation is usually a part of a curved dislocation, and the glide plane is fixed by the edge components attached to it, in this case too the shear stress acting in the glide plane as determined by the total dislocation line must be taken into account. Therefore, if the moving force of a dislocation is to be determined it is usually enough to consider only the stress component produced by the external forces, which acts in the glide plane and glide direction. Let this component be denoted by τ, then the force per unit length moving the dislocation (which is independent of the character of the dislocation) may be written as

$$F = \tau\, b. \tag{2.30}$$

This expression was first derived by Mott and Nabarro [14].

2.5.4. The interaction of straight and parallel dislocations

The forces acting between parallel, straight dislocations are simply determined with the aid of equation (2.27). Let us suppose that the dislocations are far from the surface, so that the surface effects may be neglected. First the forces acting between edge dislocations are investigated. The position of the dislocations is depicted in Fig. 2.5. In this case the force per unit length acting between the two dislocations is

$$\mathbf{F} = (\boldsymbol{\sigma}\, \mathbf{b}_1)\times \mathbf{t}_1,$$

where $\boldsymbol{\sigma}$ represents the stress-field of the dislocation lying in the z-axis with Burgers vector pointing in the positive x direction, $\mathbf{b}_1$ and $\mathbf{t}_1$ are the Burgers vector and the tangential unit vector, respectively, of the dislocation passing through the point x, y, i.e.

$$\boldsymbol{\sigma} = \begin{pmatrix} \sigma_{xx} & \sigma_{xy} & 0 \\ \sigma_{xy} & \sigma_{yy} & 0 \\ 0 & 0 & \sigma_{zz} \end{pmatrix},$$

and

$$\mathbf{b}_1 = b\,(\cos\alpha,\ \sin\alpha,\ 0), \qquad \mathbf{t}_1 = (0, 0, 1).$$

With these terms the components of the force are

$$F_x = b\,(\sigma_{xy}\cos\alpha + \sigma_{yy}\sin\alpha), \qquad F_y = -b\,(\sigma_{xx}\cos\alpha + \sigma_{xy}\sin\alpha),$$

$$F_z = 0.$$

The force in the direction of the Burgers vector $\mathbf{b}_1$ is given by the equation

$$F_b = \mathbf{F}\frac{\mathbf{b}_1}{b} = b\left[\sigma_{xy}(\cos^2\alpha - \sin^2\alpha) + (\sigma_{yy} - \sigma_{xx})\sin\alpha\ \cos\alpha\right].$$

Substituting the first terms of equations (2.18) into this equation for the force exerted in the glide plane, the following expression is obtained:

$$F_b = \frac{\mu b^2}{2\pi(1-\nu)}\frac{x}{r^4}\left[(x^2 - y^2)\cos 2\alpha - 2xy\ \sin 2\alpha\right]. \quad (2.31)$$

In the following part a few special cases are investigated in more detail.

(a) The Burgers vectors are perpendicular to each other, i.e. $\alpha = \pm\pi/2$. In this case the force is

$$F_{b\perp} = -\frac{\mu b^2}{2\pi(1-\nu)}\frac{x(x^2 - y^2)}{r^4}. \quad (2.32)$$

$F_{b\perp}$ acts in this case in the glide plane parallel to the y-axis. Its direction corresponds to the direction of the Burgers vector if $F_{b\perp} > 0$, and is opposite to it when $F_{b\perp} < 0$. Fixing the value of x, (2.32) may be expressed by the equation

$$F_{b\perp} = -\frac{\mu b^2}{2\pi(1-\nu)x} \frac{1-y^2/x^2}{(1+y^2/x^2)^2}.$$

By plotting the value $F_{b\perp}$ in $\mu b^2/2\pi(1-\nu)x$ units as a function of y/x, Fig. 2.6a is obtained. It can be seen from the figure that there is a stable configuration for the two cases, $\alpha = -\pi/2$ and $\alpha = \pi/2$,

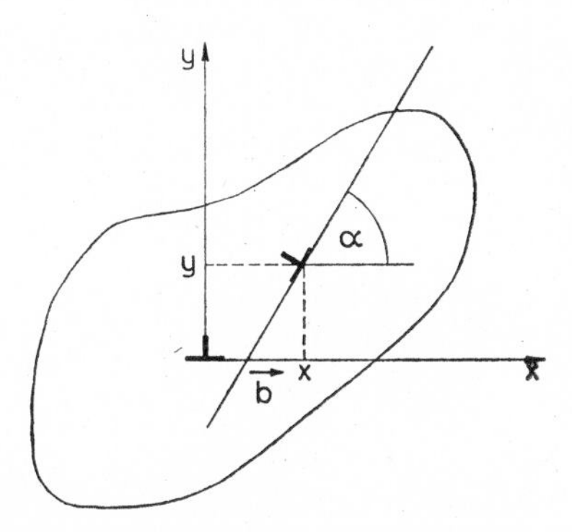

FIG. 2.5. Parallel edge dislocations in general position

when the force is zero. In this case $|x| = |y| = \xi$ (Fig. 2.6b and c). The two configurations are equivalent to each other. For $\alpha = -\pi/2$, the two dislocations repel one another, if $0 < y < x$, or $-x > y > > -\infty$.

(b) The Burgers vectors are parallel, i.e. $\alpha = 0$ or π. The glide plane in this case is parallel to the x-axis, and thus the value of y may be fixed. Taking into consideration (2.31) one obtains

$$F_{b\|} = -\frac{\mu b^2}{2\pi(1-\nu)y} \frac{x/y(1-x^2/y^2)}{(1+x^2/y^2)^2}.$$

Figure 2.7a depicts the position dependence of $F_{b\|}$. It can be seen from the Figure that dislocations of the same sign repel each other if $|x| > |y|$, and attract one another if $|x| < |y|$. No force is exerted if $x = 0$, or $x = y$. The first condition represents a stable equilibrium whereas the second one is unstable. Edge dislocations of the same sign have a stable configuration if the dislocation lines are in the plane perpendicular to the glide plane (Fig. 2.7b).

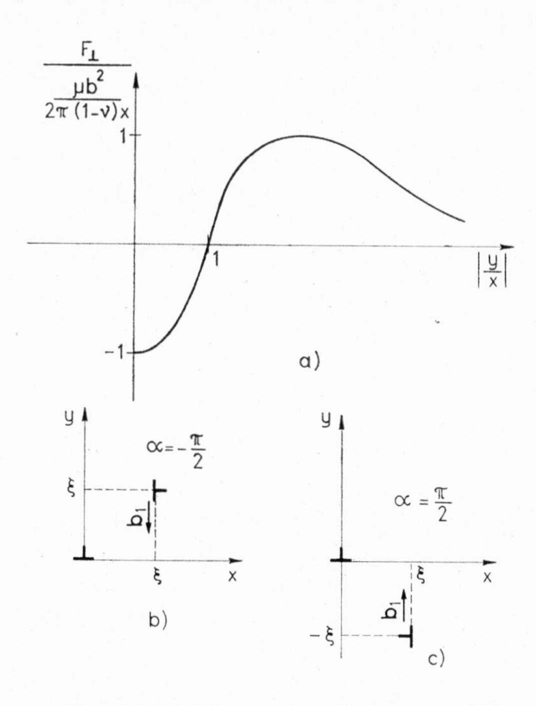

FIG. 2.6. (a) Force acting between parallel edge dislocations of perpendicular Burgers vectors. (b) The stable configuration of parallel edge dislocations in the case $\alpha = -\pi/2$. (c) The stable configuration of parallel edge dislocations in the case $\alpha = \pi/2$

For Burgers vectors of opposite sign the parallel edge dislocations attract each other if $|x| > |y|$, and repel one another when $|x| < |y|$. The equilibrium is stable if $x = y = \xi$ (Fig. 2.7c), and unstable if $x = 0$.

It follows from equations (2.28) and (2.29) that no interaction exists between parallel edge and screw dislocations. For two parallel screw dislocations from (2.23) and (2.29) one obtains the expressions

$$F_x = \frac{\mu b^2}{2\pi} \frac{x}{x^2 + y^2}, \quad F_y = \frac{\mu b^2}{2\pi} \frac{y}{x^2 + y^2}, \tag{2.33}$$

or in polar coordinates

$$F_r = \frac{\mu b^2}{2\pi r}, \quad F_\vartheta = 0. \tag{2.34}$$

Only radial forces are exerted between screw dislocations. The force is attractive with opposite Burgers vectors and repulsive if the signs of the Burgers vectors are the same. The radial force cannot be zero and hence no stable configurations are formed by a single set of parallel screw dislocations.

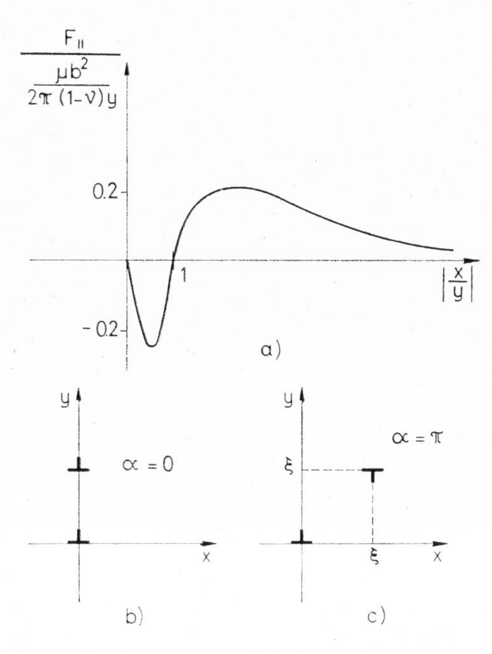

FIG. 2.7. (a) Force acting between parallel edge dislocations of parallel Burgers vectors. (b) Stable configuration of parallel edge dislocations with parallel Burgers vectors. (c) Stable configuration of parallel edge dislocations with antiparallel Burgers vectors

2.5.5. *Dislocation dipoles*

Stable configurations produced by parallel edge dislocations with antiparallel Burgers vectors are termed dislocation dipoles. A considerable number of dislocation dipoles are produced by the movement of dislocations and the operation of dislocation sources (Section 5.4). Thus, it is worth while to investigate the properties of these stable configurations in detail.

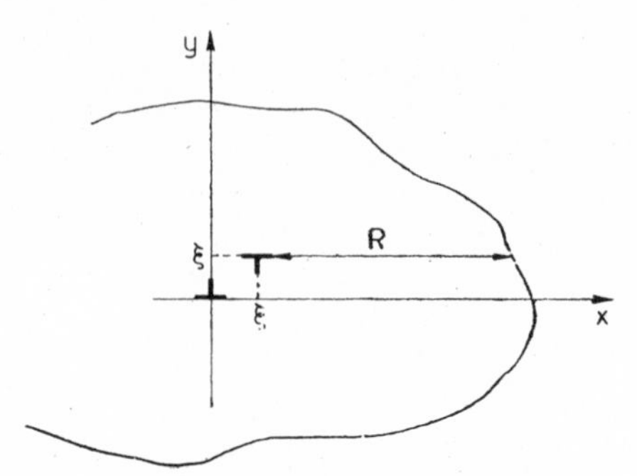

FIG. 2.8. Dislocation dipole at a distance R from the surface

If the dipole is at a large distance from the surface (Fig. 2.8), one may assume that this distance for both dislocations is R and the surface effects can be neglected. Then from (2.13) the interaction energy per unit length may be expressed by the integral

$$U_i = -b \int_{\xi}^{R} \sigma_{xy}|_{y=\xi}\, dx,$$

where σ_{xy} represents the stress originating from the dislocation lying in the z-axis and the right-hand term must be integrated over the cut surface of the other dislocation, i.e. over that part of the plane defined by $y = \xi$ for which $\xi \leqq x \leqq R$. By applying the first term of (2.18) and with the notation $w = x^2/\xi^2$ one obtains

$$U_i = \frac{\mu b^2}{4\pi(1 - \nu)} \int_{1}^{R^2/\xi^2} \frac{1 - w}{(1 + w)^2}\, dw\,.$$

is assumed that $R \gg \xi$, thus after integration

$$U_i \cong -\frac{\mu b^2}{2\pi(1-\nu)}\left\{\ln\frac{R}{\sqrt{2}\,\xi}-\frac{1}{2}\right\}. \tag{2.35}$$

This is the binding energy of the dipole. The self-energy of the dipole is

$$U^{\text{dipole}} = 2U + U_i\,.$$

Substituting (2.25) and (2.35) one obtains

$$U^{\text{dipole}} = \frac{\mu b^2}{2\pi(1-\nu)}\left\{\ln\frac{\sqrt{2}\,\xi}{r_0}+\frac{1}{2}\right\}. \tag{2.36}$$

Thus the self-energy of the dipole is independent of the dimensions of the body. This means that the stress-field of the dipole is strongly localized. The stress-field of an isolated dislocation at infinity tends to zero with $1/r$, whereas that of a dipole tends to zero with $1/r^2$.

The stress-field of a dipole is given by

$$\sigma_{ij}^{\text{dipole}} = \overset{1}{\sigma}_{ij} + \overset{2}{\sigma}_{ij} = \sigma_{ij}(\mathbf{r}) - \sigma_{ij}(\mathbf{r}-\boldsymbol{\xi}),$$

where $\sigma_{ij}(\mathbf{r})$ represents the stress-field of a dislocation lying in the z-axis, and $\boldsymbol{\xi} = (\xi, \xi, 0)$. At a great distance from the dipole $|\mathbf{r}| \gg$ $\gg |\boldsymbol{\xi}|$, and thus

$$\sigma_{ij}^{\text{dipole}}(\mathbf{r}) = \xi\left(\frac{\partial\sigma_{ij}}{\partial x}+\frac{\partial\sigma_{ij}}{\partial y}\right) \to \frac{1}{r^2}, \quad \text{if} \quad r \to \infty.$$

Similarly to the dislocation dipoles, more complicated configurations (tripoles, quadrupoles, etc.) may also be studied [15, 16].

2.5.6. *The energy of mixed dislocations*

Straight dislocations with not purely edge or screw character are called mixed dislocations. Their energy may be calculated by the following reasoning: a mixed dislocation of the Burgers vector $(b_x, 0, b_z)$ is produced, for example, with the aid of the model described in Sections 2.5.1 and 2.5.2 by giving the two sides of the cut surface an offset b_x in the x, and b_z in the z direction. It was shown in Section 2.5.4 that no interaction exists between parallel edge and screw dislocations, and consequently the energy of a mixed disloca-

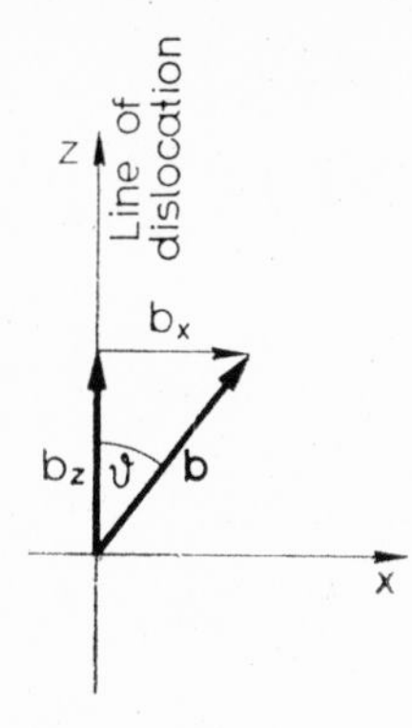

FIG. 2.9. The Burgers vector of a mixed dislocation

tion is the sum of the self-energies of the two dislocations. Let the angle between the Burgers vector and the dislocation line be ϑ; then according to Fig. 2.9

$$b_x = b \sin \vartheta,$$

$$b_z = b \cos \vartheta, \qquad b = |\mathbf{b}|.$$

Applying (2.25), the desired energy is

$$U(\vartheta) = \frac{\mu}{4\pi} \ln \frac{R}{r_0} \left\{ \frac{b^2 \sin^2 \vartheta}{1 - \nu} + b^2 \cos^2 \vartheta \right\} =$$

$$= \frac{\mu b^2}{4\pi(1 - \nu)} \ln \frac{R}{r_0} (1 - \nu \cos^2 \vartheta) =$$

$$= \frac{\mu b^2}{8\pi(1 - \nu)} \ln \frac{R}{r_0} [(1 - \nu) - \nu \cos 2\vartheta]. \qquad (2.37)$$

2.5.7. *The interaction of straight, non-parallel dislocations*

Only the interaction between straight and either pure edge or pure screw dislocations is discussed. The resulting (integrated) interaction between two dislocations lying along two deviating straight

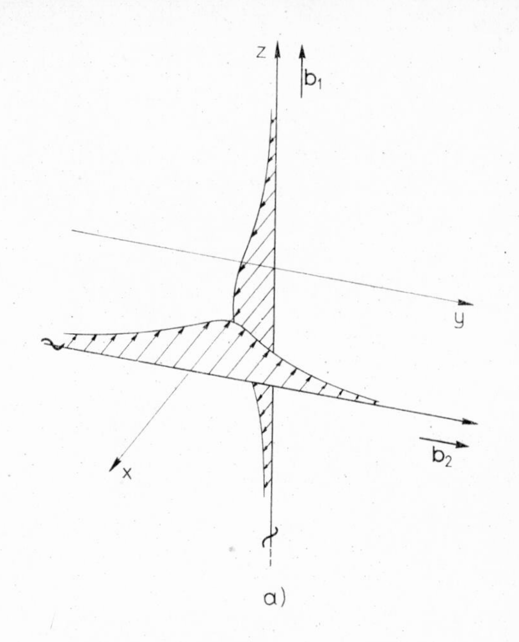

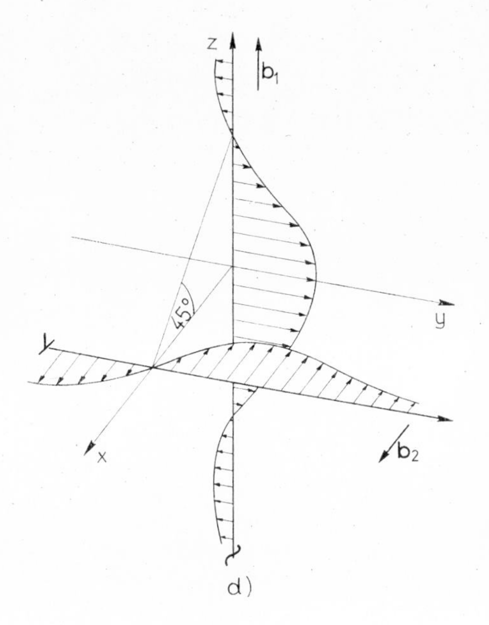

FIG. 2.10. Interaction of perpendicular dislocations. The positive direction of the line element is indicated by an arrow drawn on the dislocation lines. In the figures showing the interaction of the perpendicular edge dislocations, the positions of the extra half-planes are also indicated

(a) Perpendicular screw dislocations. (b) Perpendicular edge dislocations with parallel glide planes. (c) Perpendicular edge dislocations with parallel Burgers vectors. (d) Perpendicular edge and screw dislocations with perpendicular Burgers vectors. (e) Perpendicular edge dislocations with perpendicular glide planes. (f) Perpendicular edge and screw dislocations with parallel Burgers vectors

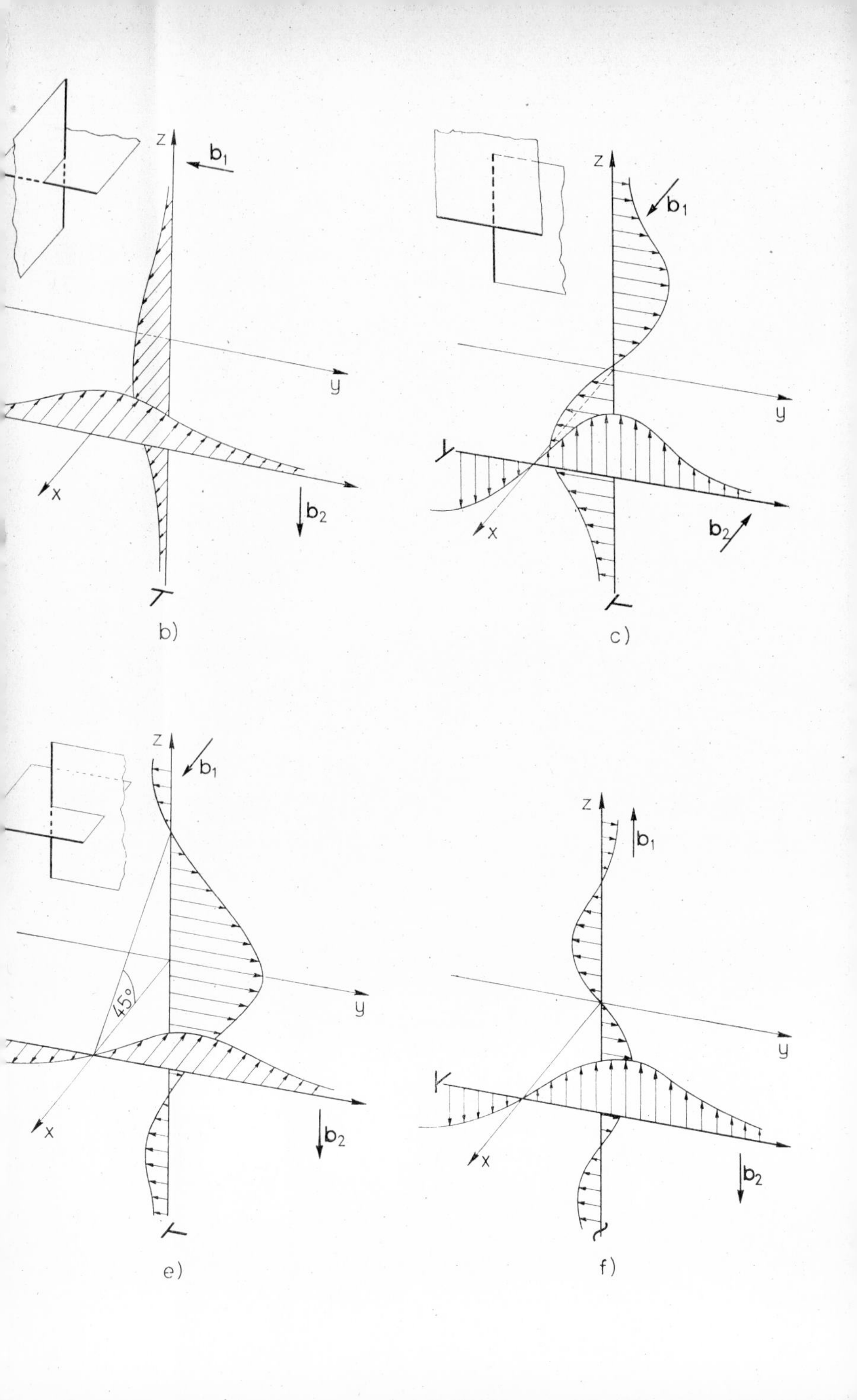

z
b_1
y
x
b_2
b)
z
b_1
y
x
b_2
c)
z
b_1
45°
y
x
b_2
e)
z
b_1
y
x
b_2
f)

lines is usually not large, because only short dislocation segments may come relatively close to each other. On the other hand, the local interacting forces near to the shortest connecting lines may approach the values of those forces which act between parallel dislocations. Consequently their effect cannot be neglected.

It will be seen that in this case the interaction of two dislocations does not follow Newton's third law. This indicates that the concept of the force acting on a dislocation is not identical with the mechanical concept of force.

With straight dislocations perpendicular to each other, the calculations can easily be carried out using the relations (2.17). The following results are summarized after Hartley and Hirth [17]:

(a) Screw dislocations of the same type (character) which are perpendicular to one another attract each other along the shortest line of connection (Fig. 2.10a). Dislocations of this type repel one another when they are parallel, and the interaction vanishes if the angle between the two dislocation lines is 45°. If the sign of one of the dislocations changes, the sign of the force of interaction also changes. As a result of the attraction between screw dislocations of the same character they can form a stable two-dimensional network (twisted grain boundary, Section 1.1.8).

(b) In the case of parallel glide planes (Fig. 2.10b), perpendicular edge dislocations produce attractive or repulsive interactions normal to the glide plane. The edge dislocations attract each other if their extra half-planes are inserted into the material from opposite directions. This motion of the dislocations, however, is possible only by the emission or absorption of point defects (non-conservative motion, Section 4.6.).

(c) In the case of perpendicular edge dislocations of parallel Burgers vectors, the force of interaction is also normal to the glide plane, and perpendicular to their shortest connecting line (Fig. 2.10c). In addition, a resultant torque is generated which strives to turn the dislocations into an antiparallel position. The result of the change of the sign of the force of interaction is that during the (non-conservative) motion the vacancies created at one side are absorbed by the other. Since the energy needed for the formation and motion of the vacancies close to the core of the dislocation is presumably much less than in the perfect matrix, this motion may already occur at a relatively low temperature.

(d) In the case of perpendicular Burgers vectors, edge and screw dislocations which are perpendicular to each other produce an interaction (Fig. 2.10d) which strives to change the edge dislocation into a screw dislocation [case (a)] while the original screw dislocation becomes distorted.

(e) For the interaction of perpendicular edge dislocations with glide planes normal to each other the situation is quite similar (Fig. 2.10e). In this case a formation similar to (b) is aimed at, but here the forces are normal to the glide plane.

(f) For perpendicular edge and screw dislocations of parallel Burgers vectors (Fig. 2.10f), the shear stress effective on the edge dislocation strives to turn it into a screw dislocation antiparallel to the interacting screw dislocations.

Of course, in most metals the dislocations are usually not perpendicular to one another; one can see, however, from the above considerations that the interaction of non-parallel dislocations lying in different planes results in rather complicated motions during which the originally straight dislocation sections may become bent. This presumably plays an important role in the mechanism of work-hardening (Chapter 9). The general formula for forces of interaction between dislocations intersecting at any angle was derived by Li [18], who also carried out calculations to determine the interaction between two-dimensional dislocation networks (small-angle grain boundaries) and individual dislocations [19].

2.6. Dislocation loops

Straight dislocations are infrequent in real crystals. The mechanism of the formation of a dislocation usually results in dislocation loops, that is in dislocations lying along a curve which is closed within the crystal. The loops are usually formed along well-defined crystallographic planes. According to the relative position of the Burgers vector and the plane of the loop two basic types of dislocation loop can be distinguished:

1. slip loops, if the Burgers vector lies in the plane of the loop, and
2. prismatic loops, if the Burgers vector is normal to the plane of the loop.†

Slip loops are usually generated when dislocation sources become active (Chapter 5), whereas prismatic loops are due to the condensation of either vacancies or interstitial atoms.

The discussion of the quantitative properties of loops is much more complicated than that of straight dislocations [1, 20–22]. A detailed mathematical treatment is not within the scope of this book, and only a few results are summarized.

The displacement formed around a dislocation lying along an arbitrary curve C was first determined by Burgers [23]. According to his result the displacements are given by the line integrals along the curve C. The value of the displacement in a point of an infinite body defined by the vector $\mathbf{r}$ is given by the following equation:

$$\mathbf{u} = -\frac{b\Omega(\mathbf{r})}{4\pi} - \frac{1}{4\pi}\oint_C \frac{\mathbf{b}\times d\mathbf{l}'}{|\mathbf{r}-\mathbf{r}'|} - \frac{\lambda+\mu}{4\pi(\lambda+2\mu)} \operatorname{grad} \oint_C \frac{\mathbf{b}\times(\mathbf{r}-\mathbf{r}')}{|\mathbf{r}-\mathbf{r}'|} d\mathbf{l}', \qquad (2.38)$$

where $d\mathbf{l}'$ denotes the vector of the line element of the curve C in the point $\mathbf{r}'$ and Ω is the solid angle filled up by the surface F laid on the curve C from the point $\mathbf{r}$. This means that (2.38) depends only upon the curve C and the vector $\mathbf{b}$. Since every property of the dislocations can be derived from $\mathbf{u}$ [1], the properties of a dislocation lying along an arbitrary curve C do not depend upon the cut surface.

If the displacement field is known the energy of a dislocation loop can also be expressed by a line integral. Let the glide plane be the $z = 0$ plane and let the Burgers vector point in the direction of the x-axis. The energy of the *slip loop* is then [24]

$$U_g = -\frac{\mu b^2}{8\pi}\oint_C\oint_C \left\{\frac{dx\,dx'}{|\mathbf{r}-\mathbf{r}'|} + \frac{1}{1-\nu}\frac{dy\,dy'}{|\mathbf{r}-\mathbf{r}'|}\right\}. \qquad (2.39).$$

† The mixed loops are sometimes also referred to as prismatic in the literatur

If the variables $dx = \cos \vartheta \, dl$ and $dy = \sin \vartheta \, dl$ are introduced, one obtains

$$U_g = -\frac{\mu b^2}{8\pi} \oint_C \left\{ \cos \vartheta \oint_C \frac{dx'}{|\mathbf{r} - \mathbf{r}'|} + \frac{\sin \vartheta}{1 - \nu} \oint_C \frac{dy'}{|\mathbf{r} - \mathbf{r}'|} \right\} dl. \quad (2.40)$$

With the aid of this expression the energy per unit length of a dislocation is defined by the equation

$$U'_g = \frac{dU_g}{dl} = -\frac{\mu b^2}{8\pi} \left\{ \cos \vartheta \oint_C \frac{dx'}{|\mathbf{r} - \mathbf{r}'|} + \right.$$

$$\left. + \frac{\sin \vartheta}{1 - \nu} \oint_C \frac{dy'}{|\mathbf{r} - \mathbf{r}'|} \right\}. \quad (2.41)$$

The energy per unit length of a line element, however, cannot be derived in a single-valued manner from (2.40) because the same total energy is obtained if the scalar product of the gradient of an arbitrarily differentiable $\psi(\mathbf{r})$ function and a tangential unit vector is added to U'_g. The physical reason for this is that the various parts of the dislocation loops interact with one another, and consequently the effect of any individual part of the loop cannot be studied independently.

Thus, the energy per unit length is not an unambiguous concept, though for reasons of symmetry with straight dislocations its uniqueness is apparently valid. However, it does seem to be practicable to apply (2.41) in other cases too, but some care must be taken when calculating with it.

For circular slip loops of radius $\mathbf{r}$, the following equation can be derived from (2.41) [24]:

$$U'_{gc} = -\frac{\mu b^2}{4\pi} \left(\cos^2 \vartheta + \frac{\sin^2 \vartheta}{1 - \nu} \right) \left(\ln \frac{8r}{r_0} - 2 \right). \quad (2.42)$$

Comparing this with (2.37) one can see that the same result would have been obtained if every line element dl had been regarded as a screw dislocation segment of the length $dl \cos \vartheta$ and an edge dislocation segment of length $dl \sin \vartheta$.

For a *prismatic loop* in the $z = 0$ plane with perpendicular Burgers vector the energy per unit length is

$$U'_p = -\frac{\mu b^2}{8\pi(1-\nu)}\left\{\cos\vartheta \oint_C \frac{dx'}{|\mathbf{r}-\mathbf{r}'|} + \sin\vartheta \oint_C \frac{dy'}{|\mathbf{r}-\mathbf{r}'|}\right\}, \quad (2.43)$$

and for a circular, prismatic loop

$$U'_{pc} = -\frac{\mu b^2}{8\pi(1-\nu)}\left(\ln\frac{8r}{r_0} - 2\right). \quad (2.44)$$

It can be shown that parallel slip and prismatic loops do not interact with one another. Consequently, the total energy of an arbitrary circular dislocation loop defined by the Burgers vector $\mathbf{b} = (b_x, b_y, b_z)$ is given by the equation

$$U_c = \int_0^{2\pi} (U'_{gc} + U'_{pc})\, r d\vartheta .$$

From this one obtains

$$U_c = -\frac{\mu r}{2(1-\nu)}\left[\left(1 - \frac{\nu}{2}\right)(b_x^2 + b_y^2) + b_z^2\right]\left(\ln\frac{8r}{r_0} - 2\right). \quad (2.45)$$

2.7. The effect of dislocations on the thermal properties of crystals

In the previous sections the properties of the elastic deformations around the dislocations have been investigated. These deformations always develop under given thermodynamic conditions (e.g. at constant temperature and pressure) when the defect is created, and consequently they result in a change of many of the thermodynamic parameters of the crystal. These changes may be examined by considering the thermodynamic relations.

When a crystal is elastically deformed the work done is

$$\Delta L = -p\Delta V + U_e$$

where ΔV is the change of volume caused by the deformation, p the external pressure, and U_e the elastic energy stored in the body.

By applying the first law of thermodynamics, the thermodynamic parameters of the crystal and U_e may be related by the equation

$$T\Delta S = \Delta U - \Delta L = \Delta U + p\Delta V - U_e .$$

If the crystal is deformed at constant pressure and temperature (which is actually the case with every static lattice defect)

$$U_e = \Delta(U - TS + pV) = \Delta G. \tag{2.46}$$

Hence the elastic energy stored in the body alters its Gibbs potential. This was first realized by Holder and Granato [25] and is of great importance. As has been seen, the properties of lattice defects are well described by the theory of elastic deformation. If the elastic energy is known, the thermodynamic parameters may also be derived with the aid of (2.46).

Let the contribution of an arbitrary defect to G be g. In the case of n independent defects

$$G = G_t + ng - TS_k ,$$

where G_t is the Gibbs energy of the perfect crystal and S_k the configurational entropy. If on the change of n the quantity G takes up an extreme value, the defect in question may also exist in the crystal in thermal equilibrium. Vacancies and interstitial atoms are examples of defects of this type.

Cottrell has shown [26] that for dislocations the contribution of the configurational entropy is so small that G increases monotonously with n, and consequently the dislocations are thermally unstable defects. This important result may be understood in the following way. If some defect can exist in thermal equilibrium this means, of necessity, that the defect can be generated spontaneously by thermal fluctuations. If an energy of formation E_F is necessary to create the defect, according to statistical physics the probability of such an event is proportional to $\exp(-E_F/kT)$. At room temperature kT is about 1/40 eV, and consequently the thermal generation of a defec- with a formation energy of 1–2 eV may be assumed with some probt ability.

In the case of dislocations—because of the localized nature of the energy fluctuations—only the possibility of a thermal generation of dislocation loops has to be considered. However, the energy of a dislocation loop depends upon the size of the loop. In order to calculate with realistic dimensions one should examine loops which can expand upon the external stress τ. This means that at a given stress τ the slipped area encircled by the loop increases.

For simplicity, consider an otherwise perfect crystal containing a circular loop. Let us determine the energy necessary to develop a loop which may be expanded by an external stress comparable with the flow stresses observed.

Let a shear stress τ be effective in the glide plane in the direction of the Burgers vector. It follows from equation (2.30) that if the dislocation line sweeps an area dA the work done by the external forces is

$$dL = \tau b \, dA.$$

Denote the energy of the loop by U_k. The external forces increase the slip if the work necessary to increase the area dA encircled by the loop is greater than the increase of the energy of the loop, i.e. if $\tau b dA > dU_k$. If this inequality is not fulfilled the stress acting on the loop cannot increase its radius. In the limit, the equation

$$d(U_k - \tau b A) = 0 \tag{2.47}$$

must hold, which means that in this case the function

$$U = U_k - \tau b A$$

takes up an extreme value. Using the value $A = r^2\pi$ and (2.45), the function to be investigated has the following form:

$$U = K\mu b^2 r \left(\ln \frac{8r}{r_0} - 2\right) - r^2 \pi \tau b\,, \tag{2.48}$$

where $K = (2 - \nu)/4(1 - \nu)$. This function has a maximum if

$$r = r_{cr} = \frac{K\mu b}{2\pi\tau} \left(\ln \frac{8r_{cr}}{r_0} - 1\right). \tag{2.49}$$

This critical radius is the smallest one which can still be enlarged by the applied stress. Using (2.48) and (2.49), the activation energy to generate a loop of this radius in a perfect crystal is

$$U_a = \frac{1}{2} K\mu b^2 r_{cr} \left(\ln \frac{8r_{cr}}{r_0} - 2 \right). \tag{2.50}$$

The results may be summarized as follows:

(a) A given external stress can initiate the glide only if a loop of a radius larger than that given by (2.49) is present in the material.

(b) If the crystal is free of dislocations, besides the work of the external force, an additional energy of magnitude given by (2.50) is necessary to nucleate slip; this is the formation energy of a dislocation loop in the perfect crystal. The values of this energy together with the critical radii for various τ values are compiled in Table 2.1.

TABLE 2.1. *Activation energy and critical radius of dislocation loops as a function of the applied stress*

Applied stress in units of $\mu/2\pi$	Energy of activation in eV	Critical radius in cm
1.72×10^{-4}	1.12×10^{5}	10^{-3}
1.36×10^{-3}	8.3×10^{3}	10^{-4}
1.00×10^{-2}	5.4×10^{2}	10^{-5}
6.40×10^{-2}	22.5×10	10^{-6}

(The data were calculated with $r_0 = 2b$, $b = 2.5\times 10^{-8}$ cm and $\mu b^3 = 1$ eV.) One can see that the formation energy of a dislocation loop with realistic dimensions is several orders of magnitude larger than that of a defect formed with finite probability. For this reason dislocations cannot be created by thermal fluctuations.

2.7.1. *The Gibbs potential of an isotropic body*

The change of the Gibbs energy of a body of volume v containing a homogeneous, elastic deformation can be expressed by applying equation (2.2) as

$$\Delta G = \frac{v}{2} \sum_{i,k=1}^{3} \sigma_{ik}\varepsilon_{ik}. \tag{2.51}$$

The change of the various thermodynamic parameters of the body can be expressed as the derivatives of ΔG. With fixed deformations ε_{ik}, ΔG depends on the thermodynamic variable p and T on the volume v and the elastic constants. Thus it is possible to transform (2.51) so that the deformations ε_{ik} enter the equation as the coefficients of the elastic constants. We introduce the deviatoric strain tensor characteristic of pure shear by the definition

$$'\varepsilon_{ik} = \varepsilon_{ik} - \frac{1}{3}\Theta\delta_{ik}.$$

Using this definition and equation (2.6), (2.46) may be written in the following form:

$$\Delta G = v\left\{\mu(p, T)\sum_{i,k=1}^{3} {'\varepsilon_{ik}}\,{'\varepsilon_{ik}} + \frac{1}{2}K(p, T)\,\Theta^2\right\}, \qquad (2.52)$$

where $K = \lambda + 2\mu/3$ is the bulk modulus. The first term (U_s) in (2.52) gives the elastic energy density stored in pure shear strains, whereas the second term (U_d) refers to the energy density due to pure dilatations.

From (2.52) the thermodynamic parameters of an elastically deformed, isotropic body may be expressed with the elastic constants and their derivatives. As an example, let us investigate the change of volume of an isotropic medium. According to definition, using (2.52), one can obtain for the relative change of volume

$$\frac{\Delta v}{v} = \left(\frac{\mu'}{\mu} - \frac{1}{K}\right)U_s + \left(\frac{K'}{K} - \frac{1}{K}\right)U_d, \qquad (2.53)$$

where the primes in μ' and K' denote differentiations according to p. This expression, which gives the macroscopic change of volume of a body containing the elastic energy $U_s + U_d$, was first derived by Zener [27]. It can be seen that the macroscopic change of volume is not zero even in the case of pure shear. This follows from the thermodynamical treatment. If the problem is considered purely from the mechanical point of view, (2.53) can be obtained only by applying the equations of non-linear elasticity theory.

Our result permits the estimation of the change of volume caused by dislocations. Applying (2.53), the change of volume caused by a straight dislocation of unit length is

$$\Delta V = \left(\frac{\mu'}{\mu} - \frac{1}{K}\right)\int U_s \, dV + \left(\frac{K'}{K} - \frac{1}{K}\right)\int U_d \, dV, \qquad (2.53a)$$

where the integrals are the energy contributions per unit length of the dislocation. For screw dislocations the deformations are pure shears, and thus in this case the change of volume is given by the first term of (2.53a). For edge dislocations the situation is somewhat more complicated. It is easy to see, however, that the first term of (2.53a) gives a correct estimation of the order of magnitude in this case too. Let us investigate therefore the energy which is stored in dilatations. Its value is

$$U_d^{\text{edge}} = \frac{K}{2}\int \Theta^2 dV = \frac{1}{18K}\int (\sigma_{xx} + \sigma_{yy} + \sigma_{zz})^2 \, dV .$$

Using (2.19) and substituting $dV = r \, dr \, d\vartheta$, one obtains

$$U_d^{\text{edge}} = \frac{\mu b^2}{12\pi} \frac{(1+\nu)(1-2\nu)}{(1-\nu)^2} \ln \frac{R}{r_0} . \qquad (2.54)$$

With $\nu = \frac{1}{3}$, this gives $U_d^{\text{edge}} = 2/9 U_{\text{total}}^{\text{edge}}$, which means that nearly 80% of the energy of an edge dislocation is stored in shear strains.

Thus it is enough to investigate the first term of (2.53a) to estimate the order of magnitude of the change of volume introduced by the dislocations. For fcc metals one obtains nearly one atomic volume per atomic length of a dislocation [25]. If the dislocation density is N cm/cm^3, the relative macroscopic change of volume is

$$\frac{\Delta V}{V} \cong \frac{N}{b} \Theta \cong N b^2 .$$

In the case of large deformations $N \approx 10^{11}$–10^{12} cm^{-2}, which means a relative volume change of 10^{-4}–10^{-3}. This value is in good agreement with the experimental data [28].

2.8. The line tension of dislocations

It is a natural consequence of the non-vanishing self-energy of the dislocations that the dislocation line tends to take a form which minimizes its energy under the given conditions. To characterize this property of the dislocations, Mott and Nabarro [14] introduced the concept of line tension. This is perfectly analogous to the surface tension of liquids which follows from the non-vanishing self-energy of the free surface.

The line tension is the force acting tangentially on the two end-points of a line element dl at the position of investigation (Fig. 2.11). This never vanishes, and thus the curved dislocation section can be kept in equilibrium only by some suitable external stress. Its magnitude can easily be determined. Let the radius of curvature of the investigated line element dl be r_g, its central angle $d\varphi$, and the line tension T. The shear stress τ in the glide plane acting in the direction of the Burgers vector results in a force $\tau b\, dl$ on the line element. At the same time, from Fig. 2.11 the force originating from the line tension is $2T \sin d\varphi/2 \approx T\, d\varphi = Tdl/r_g$.

Thus, the stress necessary to maintain the equilibrium is

$$\tau = \frac{T}{br_g}. \tag{2.55}$$

If the energy of the dislocation were independent of its direction, the line tension T would be identical with the energy per unit length in the same way as the surface tension and specific surface energy of

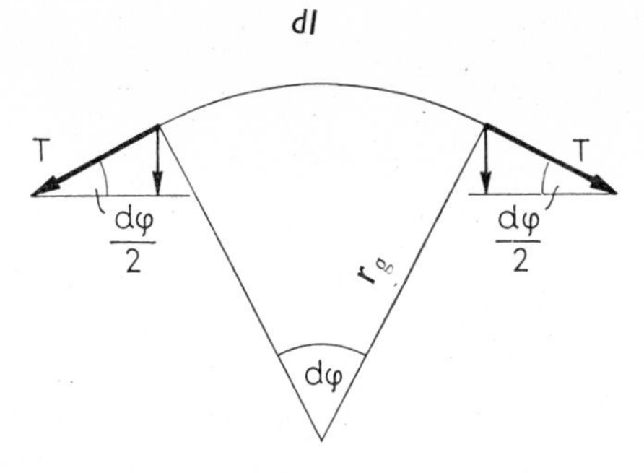

FIG. 2.11. The determination of the line tension

liquids. However, the energies of screw and edge dislocations differ essentially from each other. The character of a dislocation changes continually along a curved dislocation; consequently, the energy per unit length changes from point to point along the curve, resulting in a further tangential force.

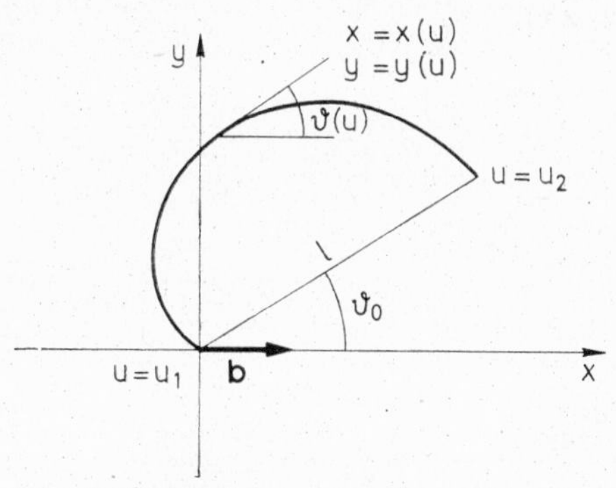

FIG. 2.12. Bowing out of a dislocation segment with fixed end-points as a result of an external stress

The exact determination of line tension is only possible by investigating the equilibrium shape of the dislocation line due to a given external stress. If the equilibrium shape and the external stress are known, the line tension may be calculated from (2.55).This method, however, is rather tedious and leads to results whose mathematical treatment is difficult [24].

Instead it is more advisable to make suitable simplifying assumptions. The simplest general and analytical result may be obtained by the procedure of de Wit and Koehler [29]. It is assumed that a single-valued energy per unit length belongs to every element of the dislocation line which depends only upon the angle between the line element $d\mathbf{l}$ and the Burgers vector $\mathbf{b}$. This assumption is not evident and, as has been shown in Section 2.6, is only approximately valid. However, attempts to determine the line tension in an exact way have shown that the various approximations do not give essentially different results [30].

In order to determine the relation between the line tension and the energy of a dislocation, let us investigate by de Wit and Koehler's method the properties of a straight dislocation of length l_0 between two pinning points. Such dislocation segments play a basic role in the multiplication processes of dislocations (see Chapter 5). The theory developed is based on the following ideas. If no external force acts, the dislocation segment is at rest and its shape does not change. If, however, a suitable shear stress is applied to the crystal, force is exerted on the dislocation, and the segment will tend to bow out between the pinning points forming finally a curve for which equation (2.55) is satisfied for every point. If the function describing this equilibrium curve is determined, the value of the radius of curvature can be calculated at every point of the curve. A comparison of the equation so obtained with (2.55) gives the value of the line tension.

Let the glide plane be the x, y plane and let the Burgers vector point in the x-direction. With no external stresses the dislocation segment lies along the line $y = x \tan \vartheta_0$. Let the parametric system of equations describing the equilibrium curve in the case of external stress be

$$x = x(u), \qquad y = y(u). \tag{2.56}$$

Since this curve must go through the fixed end points (Fig. 2.12)

$$\begin{aligned} x(u_1) &= 0, & y(u_1) &= 0, \\ x(u_2) &= l \cos \vartheta_0, & y(u_2) &= l \sin \vartheta_0. \end{aligned} \tag{2.57}$$

The differential equation of the equilibrium shape of the curve can be determined by variation calculus. Let us assume that the equilibrium shape of the curve changes so that the values $x = x(u)$ and $y = y(u)$ take the values

$$x = x(u) + \delta x(u), \qquad y = y(u) + \delta y(u). \tag{2.58}$$

Here δx and δy are infinitesimal variations which are arbitrary everywhere except at the end-points, where

$$\delta x(u_1) = \delta x(u_2) = \delta y(u_1) = \delta y(u_2) = 0.$$

If an external shear stress τ is active in the direction of the Burgers vector, from (2.27) the components of the force acting on the line element dl of the dislocation are

$$dF_x = -\tau b \sin \vartheta \, dl(u),$$

$$dF_y = \tau b \cos \vartheta \, dl(u).$$

With the notations $x' = dx/du$, $y' = dy/du$ from Fig. 2.12 one obtains the following expressions:

$$dl(u) = \sqrt{x'^2 + y'^2} \, du,$$

$$\sin \vartheta = \frac{y'}{\sqrt{x'^2 + y'^2}}, \qquad \cos \vartheta = \frac{x'}{\sqrt{x'^2 + y'^2}}. \tag{2.59}$$

Using (2.59) the work of the external forces when the curve changes its shape according to (2.58) can be written as

$$\delta U_F = \int_{u_1}^{u_2} (dF_x \, \delta x + dF_y \, \delta y) = \tau b \int_{u_1}^{u_2} (-y' \delta x + x' \delta y) du. \tag{2.60}$$

Since the curve is displaced from its equilibrium position the work done by the external forces (δU_F) during the displacement must be equal to the change of the self-energy (δU_S) of the dislocation. If for instance $\delta U_S > \delta U_F$ is supposed, the dislocation may move at the expense of its self-energy against the external forces beyond its equilibrium position, which would mean that the dislocation was not at equilibrium. The inequality $\delta U_F > \delta U_S$ would similarly be contradictory. Consequently, the condition of equilibrium is

$$\delta(U_F - U_S) = 0. \tag{2.61}$$

According to our assumption, the self-energy of the dislocation is

$$U_S = \int_{u_1}^{u_2} U dl(u), \tag{2.62}$$

where

$$U = U(\vartheta) = U(x', y').$$

Making use of this equation and applying variation calculus, from (2.62) one obtains

$$\delta U_S = \delta \int_{u_1}^{u_2} U(x', y') \sqrt{x'^2 + y'^2}\, du =$$

$$= - \int_{u_1}^{u_2} \left(\frac{dF_{x'}}{du} \delta x + \frac{dF_{y'}}{du} dy \right) du, \tag{2.63}$$

where

$$F = U(x', y') \sqrt{x'^2 + y'^2}, \quad F_{x'} = \frac{\partial F}{\partial x'}, \quad F_{y'} = \frac{\partial F}{\partial y'}. \tag{2.64}$$

The condition of equilibrium may be written with (2.60) and (2.63) in the following form:

$$\delta(U_F - U_S) = \int_{u_1}^{u_2} \left\{ \left(\frac{dF_{x'}}{du} - \tau b y' \right) \delta x + \right.$$

$$\left. + \left(\frac{dF_{y'}}{du} + \tau b x' \right) \delta y \right\} du = 0 .$$

This latter equation holds for arbitrary δx and δy only if

$$\frac{dF_{x'}}{du} - \tau b y' = 0, \qquad \frac{dF_{y'}}{du} + \tau b x' = 0 . \tag{2.65}$$

These are the differential equations of the equilibrium shape of the dislocation line. Their general solutions are [29]

$$y = C_1 + \frac{1}{\tau b} \left[\cos \vartheta U(\vartheta) - \sin \vartheta \frac{dU}{d\vartheta} \right],$$

$$x = C_2 - \frac{1}{\tau b} \left[\sin \vartheta U(\vartheta) - \cos \vartheta \frac{dU}{d\vartheta} \right], \tag{2.66}$$

where C_1 and C_2 are constants of integration. The results are also valid for the anisotropic cases, since U is only assumed to be a differentiable function of ϑ.

The radius of curvature of the curve defined by the equations $x = x(\vartheta)$, $y = y(\vartheta)$ is determined by the following expression:

$$r_g = \frac{(x'^2 + y'^2)^3}{(x'y'' - x''y')^2},$$

where

$$x' = \frac{dx}{d\vartheta}, \qquad y' = \frac{dy}{d\vartheta}.$$

Carrying out the prescribed mathematical operations on (2.66), one obtains for the radius of curvature the expression

$$r_g = \frac{U + \dfrac{d^2U}{d\vartheta^2}}{\tau b}. \tag{2.67}$$

Comparing this equation with (2.55), one obtains the relations between the line tension and the energy of dislocation

$$T(\vartheta) = U(\vartheta) + \frac{d^2U}{d\vartheta^2}. \tag{2.68}$$

For straight dislocations U depends upon ϑ according to (2.37). In this case

$$T(\vartheta) = \frac{\mu b^2}{8\pi(1-\nu)} \ln\frac{R}{r_0}\left[(2-\nu) + 3\nu\cos 2\vartheta\right]. \tag{2.69}$$

If the Burgers vectors of the edge and screw components of a mixed dislocation are equal, then $\vartheta = 45°$, and consequently the last term in the bracket of (2.69) is zero. Introducing the notation $b/\sqrt{2} = b_1$, the line tensions of a pure edge and screw dislocations are

$$T_{\text{edge}} = \frac{\mu b_1^2}{4\pi(1-\nu)} \ln\frac{R}{r_0},$$

$$T_{\text{screw}} = \frac{\mu b_1^2}{4\pi} \ln\frac{R}{r_0}. \tag{2.70}$$

Thus in this case the line tension is equal to the energy per unit length.

The equilibrium shape of a dislocation segment whose end-points are fixed clearly cannot develop at an arbitrarily large external stress. There exists a critical stress above which the dislocation cannot be in equilibrium. In this case the bowing out of the segment increases continuously, until the dislocation line finally comes off from the pinning points and a dislocation loop is formed. Consequently, if the stress is large enough a dislocation segment between pinning points may become the source of new dislocations (Frank–Read source). The role of this source and the knowledge of the critical stress are of basic importance in the glide mechanism.

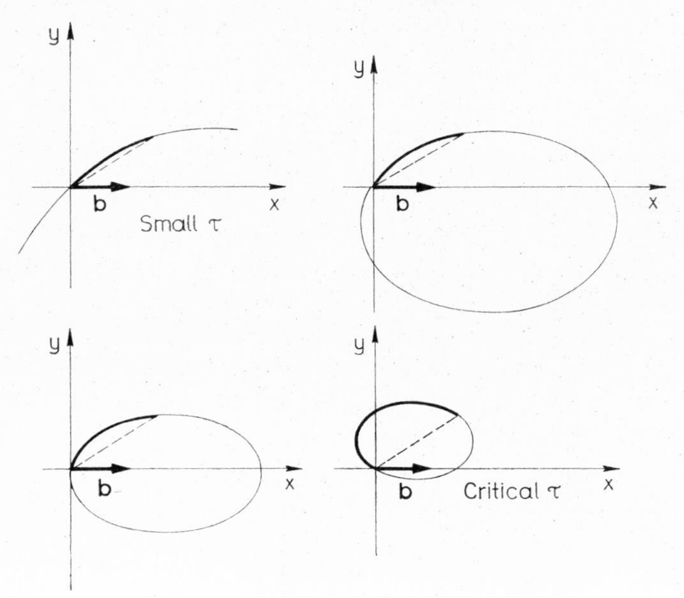

FIG. 2.13. The effect of an increasing stress on the equilibrium shape of a bowed-out dislocation segment

In order to determine the critical stress, let us go back to the discussion in (2.66). We restrict ourselves only to the case where $U + d^2U/d\vartheta^2 > 0$ for each ϑ. This condition is always fulfilled if the material is isotropic. De Wit and Koehler have shown that in this case the equilibrium dislocation line is a part of a closed curve which has no cusp and does not cross itself. If the stress is small, the diameter of the curve is large, the diameter decreasing with increasing stress.

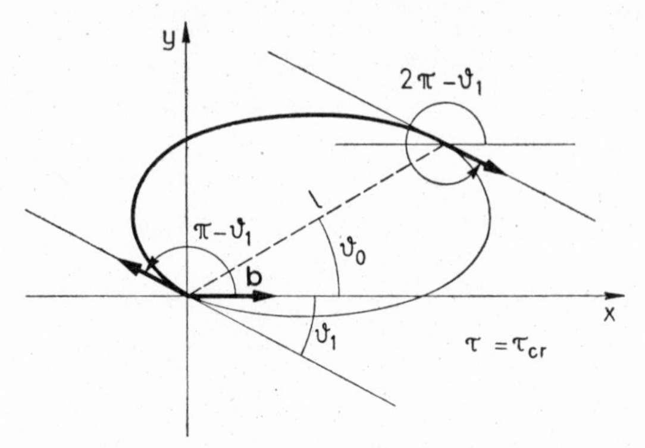

FIG. 2.14. The equilibrium shape of a bowed-out dislocation in the case of the critical stress

As a consequence the curve can no longer contain the pinning points above a certain critical stress, i.e. if this critical stress is reached no equilibrium can be maintained. Figure 2.13 depicts the variations of the curves if the stress increases. In order to determine the critical stress let us assume that the energy of the dislocation line produced by the stress may be written as [31]:

$$U(\vartheta) = \sum_n \alpha_{2n} \cos 2n\,\vartheta. \tag{2.71}$$

This equation is satisfied for an isotropic medium and for example for straight dislocations in fcc crystals. It follows from (2.71) that

$$U(n\pi - \vartheta) = U(\vartheta). \tag{2.72}$$

It is easy to see that with these conditions the curve is symmetrical to the $X = x - C_2$ and $Y = y - C_1$ coordinate axes. At the critical stress the initial dislocation segment is a diameter of the curve. The tangents at the pinning points of this curve are parallel because of the symmetry mentioned previously. Applying (2.66) to the pinning points and using the expressions (2.71) and (2.72) with the notations of

Fig. 2.14, one obtains the equation

$$\frac{1}{2}\tau_{cr}\, bl\cos\vartheta_0 = \sin\vartheta_1 U(\vartheta_1) + \cos\vartheta_1\left(\frac{dU}{d\vartheta}\right)_{\vartheta_1},$$

$$\frac{1}{2}\tau_{cr}\, bl\sin\vartheta_0 = \cos\vartheta_1 U(\vartheta_1) - \sin\vartheta_1\left(\frac{dU}{d\vartheta}\right)_{\vartheta_1}. \tag{2.73}$$

From these the critical stress is

$$\tau_{cr} = \frac{2U(\vartheta_1)}{bl\sin(\vartheta_0 + \vartheta_1)}. \tag{2.74}$$

2.9. Boundary condition problems

Disregarding a few special cases, the effect of the surfaces of the body containing the dislocations has so far not been considered. This is usually a permissible approximation. In certain cases, however, the surface effect may be considerable, since the elastic energy of the dislocations close to the surface depends upon the mutual positions of the dislocations and the surface. If the dislocation moves relative to the surface the energy of the crystal changes, and thus the surface exerts a "force" on the dislocation. In order to calculate this force, the stress-field of a given dislocation must be determined so that the stresses normal to the surface vanish at the boundary. The solution of this problem is rather difficult and laborious. For the sake of simplicity we restrict ourselves in the following treatment to screw dislocations only. Moreover, the core surface of the dislocation is not included in the boundary conditions (the stress-field is determined only for the case $r_0 = 0$). With these conditions the method of the "image dislocation" yields simply the results in some special cases. First let us investigate the simplest case. Consider an elastic half-space bordered by the (y, z) plane of the coordinate system, and a screw dislocation parallel to the free surface (along the z-axis) at a distance ξ from the surface (Fig. 2.15a). With these conditions the stress-field must be such that the stresses parallel to the x-direction be zero along the plane $x = 0$.

In order to understand the image dislocation method let us consider the following problem. Let us determine the stress-field of two screw dislocations with parallel and opposite Burgers vectors, which

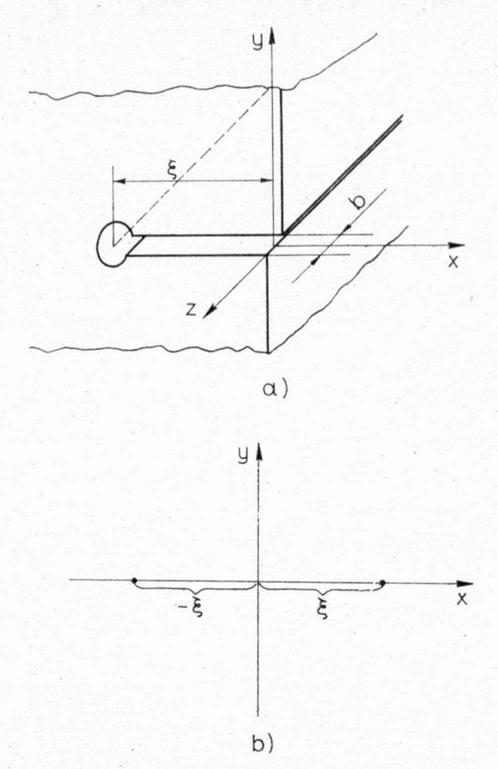

FIG. 2.15 (a) Screw dislocation in an elastic half-space. (b) Parallel screw dislocations of opposite Burgers vectors in an infinite medium

are at a distance of 2ξ from one another in an infinite medium (Fig. 2.15b). Since the equations of the elastic continuum are linear the resultant displacement according to (C.14) is

$$u_z = \frac{b}{2\pi} \operatorname{arc\,tan} \frac{y}{x+\xi} - \frac{b}{2\pi} \operatorname{arc\,tan} \frac{y}{x-\xi}, \qquad (2.75)$$

where ξ indicates that the investigated dislocations are not in the centre of the coordinate system. This means that if for a single dislocation the coordinate system has been displaced by the vector $(\xi, 0, 0)$, $(x - \xi)$ must be inserted instead of x into the equation. From (2.75) the stresses are expressed by the following equations:

$$\sigma_{xz} = \mu \frac{\partial u_z}{\partial x} = \frac{2\mu b \xi x y}{\pi[(x+\xi)^2 + y^2][(x-\xi)^2 + y^2]},$$

$$\sigma_{yz} = \mu \frac{\partial u_z}{\partial y} = \frac{\mu b}{2\pi}\left[\frac{x+\xi}{(x+\xi)^2 + y^2} - \frac{x-\xi}{(x-\xi)^2 + y^2}\right]. \qquad (2.76)$$

However, since σ_{xz} is zero along the $x = 0$ plane, (2.75) also solves the original problem, because on cutting the body into two halves along the $x = 0$ plane, which is free of normal stresses, the stress distribution does not change and the half-space on the right side can be removed. The second term in (2.75) may then be regarded as a displacement induced by an "image dislocation". (Similarly, for example, the potential-field of a point charge and a conducting plane may be determined with the aid of an "image charge".)

The method of image dislocation may be outlined in general according to Eshelby [32]. Consider a body containing a given dislocation and bordered by a finite surface S. The stresses which develop in the body and also satisfy the boundary conditions may be determined in the following way. Let us produce an array of dislocations in an infinite continuum so that the equivalent of the body under discussion be surrounded by the surface S. Denote the displacements and stresses induced by the dislocations in this infinite body by $\mathbf{u}^\infty(\mathbf{r})$ and $\boldsymbol{\sigma}^\infty(\mathbf{r})$, respectively. If the vector $\boldsymbol{\xi}$ defines the position of the dislocation, according to the problem discussed previously one has to introduce the quantities $\mathbf{u}^\infty(\mathbf{r} - \boldsymbol{\xi})$, $\boldsymbol{\sigma}^\infty(\mathbf{r} - \boldsymbol{\xi})$. Remove the material outside the surface S so that the stress distribution $\boldsymbol{\sigma}^\infty$ remains unaltered in the remaining body. The values of the forces normal to the surface S are then generally not zero. In order to attain this, produce suitable displacements in the body to which no dislocations can be assigned (the displacements have no singularities in the body), though together with them the normal components of the stresses become zero along the surface. The resultant displacements and stresses are

$$\begin{aligned} \mathbf{u} &= \mathbf{u}^{\infty}(\mathbf{r}-\boldsymbol{\xi}) + \mathbf{u}^{k}(\mathbf{r},\boldsymbol{\xi}), \\ \boldsymbol{\sigma} &= \boldsymbol{\sigma}^{\infty}(\mathbf{r}-\boldsymbol{\xi}) + \boldsymbol{\sigma}^{k}(\mathbf{r},\boldsymbol{\xi}), \end{aligned} \tag{2.77}$$

where $\mathbf{u}^k(\mathbf{r},\boldsymbol{\xi})$ and $\boldsymbol{\sigma}^k(\mathbf{r},\boldsymbol{\xi})$ denote the image displacement and the stresses, respectively. The stress must be so chosen that

$$(\boldsymbol{\sigma}\,\mathbf{n})_S = \left(\sum_{j=1}^{3} \sigma_{ij} n_j\right)_S = 0$$

along the surface S if $\mathbf{n}$ is the unit vector normal to the surface. With this equation one obtains from (2.77)

$$\left(\sum_{j=1}^{3} \sigma_{ij}^{k} n_j\right)_S = -\left(\sum_{j=1}^{3} \sigma_{ij}^{\infty} n_j\right)_S . \tag{2.78}$$

Since

$$\frac{\partial \sigma_{ij}^{\infty}}{\partial x_l} = -\frac{\partial \sigma_{ij}^{\infty}}{\partial \xi_l}$$

(where x_l denotes one of the x, y, z coordinates and ξ one of the ξ_x, ξ_y, ξ_z coordinates), one obtains from (2.78)

$$\left(\sum_{j=1}^{3} \frac{\partial \sigma_{ij}^{k}}{\partial \xi_l} n_j\right)_S = \left(\sum_{j=1}^{3} \frac{\partial \sigma_{ij}^{\infty}}{\partial x_l} n_j\right)_S . \tag{2.79}$$

If the stress distribution σ_{ij}^{∞} is known the stresses arising from the image displacements must fulfil the condition (2.79). Thus the image displacements should be determined accordingly.

It follows from the above treatment that the elastic energy of the crystal is apparently a function of the vector $\boldsymbol{\xi}$ characterizing the position of the dislocation, $U = U(\boldsymbol{\xi})$. The dependence of the energy on the position of the dislocation results in a force whose magnitude is given by the equation

$$\mathbf{f} = -\frac{\partial U}{\partial \boldsymbol{\xi}} = -\operatorname{grad}_{\xi} U . \tag{2.80}$$

This force, which characterizes the interaction between the dislocation and the free surface, is called the *image force.*

In the previous problem (Fig. 2.15a) the energy of the dislocation can be determined in a simple way. By applying (2.76) and (2.12), provided that $\xi \gg r_0$, one comes to the following result:

$$U(\xi) = \frac{\mu b^2}{4\pi} \ln \frac{2\xi}{r_0}. \tag{2.81}$$

From this the force acting on the dislocation is

$$f = \frac{\mu b^2}{4\pi\xi}. \tag{2.82}$$

This force tends to move the dislocation towards the surface. It should be noted that the image force does not attract the dislocations towards the surface in every case.

Eshelby has shown [6] that the energy of a screw dislocation in a cylinder has a local minimum in the centre of the cylinder. This means that here the image force repels the dislocation from the surface.

It is mathematically more complicated to fulfil the boundary conditions in the case of edge dislocations. Several works deal with this problem, only a few being listed as examples [3, 13, 33–35].

2.10. The motion of dislocations in an elastic continuum

The motion of a dislocation in an otherwise perfect body of finite dimensions is impeded by the image force due to the change of the elastic energy, as defined in the previous section. If, however, the energy of the dislocation is not a function of its position (e.g. in the case of an infinite medium), the image force is zero and the dislocation moves at a constant velocity. This consequence of the continuum model is clearly not realized in crystalline material, since in the crystals the dislocations are distributed over a periodic potential field (Chapter 4), and work must be done to displace any dislocation to one atomic distance even if its elastic energy is the same in the new position as in the previous one. Consequently, a dislocation can be moved in the crystal only by some suitable force. In spite of this fact the investigation of the motion of a dislocation in a continuum yields valuable information. It will be seen that the motion of a dis location is analogous to the motion of a particle in the special theory of relativity. This fact was first pointed out by Frenkel and Kontorova [36] in 1938 in a study of a one-dimensional dislocation model. These investigations were extended to the three-dimensional case by Frank [37] and Eshelby [38] and to anisotropic materials by Bullough and Bilby [39] and Teutonico [40]. More recently an increasing number of papers deal with the motion of curved dislocations [41, 42].

Let us investigate the motion of a straight screw dislocation in an infinite medium. If the dislocation starts moving, according to the above-mentioned facts it moves at a constant velocity after the driving force has come to an end. The stress-field of this dislocation can be determined by solving the equation of motion (B.25).

Neglecting the body forces the equations of motion of the dislocations is

$$\frac{\partial^2 u_z}{\partial x^2} + \frac{\partial^2 u_z}{\partial y^2} = \frac{1}{c_t^2}\frac{\partial^2 u_z}{\partial t^2}, \tag{2.83}$$

where $c_t^2 = \mu/\rho$ is the transverse velocity of sound in the medium. The solution must become static for $v = 0$. Assuming that the dislocation moves in the direction of the positive x-axis, transform (2.83) in the following way

$$x_1 = \frac{x - vt}{(1 - v^2/c_t^2)^{1/2}} = \frac{1}{\beta}(x - vt). \tag{2.84}$$

Because of the constant velocity $\partial^2/\partial t^2 = v^2 \partial^2/\partial x^2$, and so (2.84) transforms (2.83) into

$$\frac{\partial^2 u_z}{\partial x_1^2} + \frac{\partial^2 u_z}{\partial y^2} = 0. \tag{2.85}$$

(2.85) is formally identical with the static equation, and thus the solution is

$$u_z = \frac{b}{2\pi} \arctan \frac{y}{x_1}, \tag{2.86}$$

which is clearly the static solution for $v = 0$. The stress terms are now of the same form as with dislocations in rest, but instead of x one has to write $(x - vt)/(1 - v^2/c_t^2)^{1/2}$. The quantity $(x - vt)$ indicates that the dislocation moved in the co-ordinate system set to the material. On the other hand, the factor $(1 - v^2/c_t^2)$ shows that the stress-field was decreased in the direction of motion (contraction). This effect is analogous to the relativistic contraction, which is not surprising considering that the wave equation (2.84) describing the motion of the dislocation was solved with the "Lorentz transformation". The analogy extends also over the energy of dislocation. By (2.2) the elastic, i.e. potential energy is

$$U_{\text{pot}} = \frac{\mu}{2} \iint_{-\infty}^{\infty} \left[\left(\frac{\partial u_z}{\partial x} \right)^2 + \left(\frac{\partial u_z}{\partial y} \right)^2 \right] dxdy =$$

$$= \frac{\mu}{2\beta} \left\{ \iint_{-\infty}^{\infty} \left(\frac{\partial u_z}{\partial x_1} \right)^2 dx_1 dy + \beta^2 \iint_{-\infty}^{\infty} \left(\frac{\partial u_z}{\partial y} \right)^2 dx_1 dy \right\}, \tag{2.87}$$

because

$$\partial / \partial x = \frac{1}{\beta} \partial / \partial x_1 .$$

The moving dislocation, however, has also a kinetic energy, whose value per unit volume is $(\rho/2)\,(\partial u_z/\partial t)^2$.

The total kinetic energy is

$$U_{\text{kin}} = \frac{\rho}{2} \iint_{-\infty}^{\infty} \left(\frac{\partial u_z}{\partial t} \right)^2 dxdy = \frac{\mu v^2}{2c_t^2} \iint_{-\infty}^{\infty} \left(\frac{\partial u_z}{\partial x_1} \right)^2 dx_1 dy . \tag{2.88}$$

Thus, the total energy of the dislocation is

$$U = U_{\text{pot}} + U_{\text{kin}} =$$

$$= \frac{1}{\beta} \left\{ \frac{\mu}{2} \iint_{-\infty}^{\infty} \left[\left(\frac{\partial u_z}{\partial x_1} \right)^2 + \left(\frac{\partial u_z}{\partial y} \right)^2 \right] dx_1 dy + \frac{v^2}{c_t^2} \iint_{-\infty}^{\infty} \left[\left(\frac{\partial u_z}{\partial x_1} \right)^2 - \right. \right.$$

$$\left. \left. - \left(\frac{\partial u_z}{\partial y} \right)^2 \right] dx_1 dy \right\} .$$

The integrals of this expression refer to the coordinate system fixed to the dislocation, and thus the first integrals give the "rest energy" of the dislocation. Because of the infinite extension of the medium for reasons of symmetry, the value of the second integral is zero. Consequently, for the total energy one obtains the "relativistic" expression

$$U = \frac{U_0}{\sqrt{1 - v^2/c_t^2}}, \tag{2.89}$$

which is analogous to the Einstein equation. It is evident from (2.89) that a screw dislocation cannot move with a larger velocity than the transverse velocity of sound in the medium.

The study of the motion of an edge dislocation requires more complicated calculations. The reason for this is mainly that the dilatation for an edge dislocation is not zero, and consequently longitudinal displacements also occur, whose velocity $c_l^2 = (\lambda + \mu)/\rho$ differs from the transverse velocity. The exact solution of the problem was given by Eshelby in 1949 [38]. Calculating with the continuum model, he came to the result that the upper limit of the velocity of an edge dislocation too is the transverse velocity of sound. Eshelby also extended his investigations to the Peierls model (see Chapter 4), which takes into account the atomic structure of the material. As will be seen later, a very characteristic parameter, the width of a dislocation, can be defined with this model. This width to some extent characterizes the shear stress of the dislocation in the glide plane. In connection with the Peierls model Eshelby found that the width of an edge dislocation becomes zero if its velocity attains the approximate value of $0.9c_t = c_r$, i.e. the velocity of the Rayleigh surface waves. From this he concluded that the upper limit for an edge dislocation is not c_t, as may be calculated from the continuum model, but instead the smaller Rayleigh velocity. Weertman, however, pointed out that there is not sufficient physical reason to regard c_r as the maximum velocity of the edge dislocation if the dislocation width becomes zero [43]. At this velocity the energy is still finite (even in the case of the Peierls model it becomes infinite only at c_t); at the same time the width and also the shear stress which is active in the glide plane become zero. If the width is investigated at $c_r < v < c_t$, a negative value is obtained, and the shear stress in the glide plane is also negative. (The negative width expresses this fact.) The change of sign of the shear stress in question, however, has the interesting effect that high velocity edge dislocations with parallel Burgers vectors (in contrast to the static case) attract one another. On the other hand, this may play an important role in the mechanism of the fracture of crystals when very large velocities occur. Weertman has also shown that at the velocity limit of this anomalous behaviour (at the Rayleigh velocity) the potential and kinetic energies of the edge dislocation are equal, and on further increase of the velocity the kinetic energy becomes larger than the potential energy. By comparing (2.87) and (2.88) one can see that this cannot occur with screw dislocations, since this type of dislocation does not show any anomalous behaviour. This is supported by the investigation of a screw dislocation by the Peierls model, which shows that in this case the width is zero only at $v = c_t$ when the energy becomes infinite. According to the investigations of Weertman [44] and Teutonico [40], the anisotropic effects generally decrease the velocity interval relating to the anomalous behaviour of the edge dislocation (the Rayleigh and the transverse sound velocities are closer to each other).

Finally, the effects impeding the motion of high-speed dislocations are discussed briefly. Taking into consideration the crystalline structure too, the dislocations can move only by a continuously active external stress. At a given external stress, however, the dislocation attains a final velocity at which the energy losses are equal to the work done by the external forces. This final velocity is clearly an important characteristic of the material investigated and of its dislocations.

The energy losses of moving dislocations may be due to three effects. One is the change of temperature connected with a rapidly changing dilatation. This results in an entropy increase and hence in a mechanical energy loss. The second is generated by the release of high-velocity sound waves by the dislocations, which causes "radiation losses". Finally, the third effect originates from the interaction between the lattice vibrations and the moving dislocations. The role and exact mechanism of these effects constitute one of the least understood parts of the dislocation theory, since the latter two effects depend strongly upon the structure of the investigated crystal. It is rather difficult to take this into consideration [41, 45–47].

2.11. General continuum theory of lattice defects

2.11.1. Introduction

In this section results and principal considerations are repeated which are connected with the foundation of the general continuum theory of lattice defects. Since the treatment involves a rather complicated mathematical apparatus no completeness can be aimed at.

It has already been shown that the continuum model of dislocations is very valuable and leads to results which are in good agreement with practical observation. Our considerations are now extended to defects of any type and defects of both continuous and discrete distribution will be considered. This general treatment, however, cannot be carried out within the frame of classical elasticity theory, since this describes deformations and stresses which in a single continuous body are due to external forces. This type of deformation has no inner sources, and it can be characterized with a continuous and single-valued displacement vector $\mathbf{u}(\mathbf{r})$ everywhere in the material. [If this were not so the definition of the strain tensor as given in (B.4) would have no sense.] Let us investigate the condition necessary for $\mathbf{u}$ to be single-valued and continuous. The field of elastic displacements is not free of curls, i.e. the vector $\boldsymbol{\omega} = \text{curl}\,\mathbf{u} = \nabla \times \mathbf{u}$ is usually not zero [∇ denotes the symbolic vector operations $\nabla = (\partial/\partial x, \partial/\partial y, \partial/\partial z)$], and consequently $\mathbf{u}$ is single-valued if the following condition holds:

$$\oint_L d\boldsymbol{\omega} = \oint_L \frac{d\boldsymbol{\omega}}{d\mathbf{r}}\, d\mathbf{r} = \oint_L \left[\frac{d}{d\mathbf{r}}(\nabla \times \mathbf{u})\right] d\mathbf{r} =$$

$$= \oint_L \left[\nabla \times \frac{d\mathbf{u}}{d\mathbf{r}}\right] d\mathbf{r} = \oint_L (\nabla \times \boldsymbol{\epsilon})\, d\mathbf{r} = 0\,, \qquad (2.90)$$

where L represents any closed curve within the body. The line integral along a closed curve of a non-scalar function, however, is only zero if its curl is identically zero [48]. With this, the condition of applicability of classical elasticity may be written in the following form

$$-\nabla \times (\nabla \times \boldsymbol{\epsilon}) = \nabla \times \boldsymbol{\epsilon} \times \nabla = 0. \qquad (2.91)$$

This equation represents the St. Venant compatibility condition. In order to lay the foundations of the continuum theory of lattice defects the classical theory of elasticity must also be generalized, since the lattice defects are internal sources of elastic deformations which cannot be described everywhere within the body with a single-valued, continuous displacement vector $\mathbf{u}(\mathbf{r})$. Consequently, these deformations do not fulfil the condition of compatibility (2.91). As a result, internal stresses are generated in the material which cannot be produced by external forces. In other words, external forces cannot create an elastic deformation of the body in which the generated stresses correspond with the stress-field of a given lattice defect. In the case of lattice defects the deviation from the compatibility condition, i.e. the incompatibility of the deformations, can be considered as the sources of stresses.

These conclusions are not in contradiction with the properties of dislocations determined in the previous sections by the theory of classical elasticity. It was always tacitly assumed that the cut surface F of the dislocation is carefully eliminated from the investigated space by a surface F'. In the remaining space the displacement vector is everywhere single-valued and continuous so that the classical equations may be used, and finally, with the limit $F' \to F$, the results obtained may be extended over the entire body (similarly to the calculation of the energy of the dislocation by integrating over the cut surface). However, it is apparent that this method cannot be applied, for example, to dislocations of continuous distribution, and thus for these cases the classical equations should be generalized.

2.11.2. The general deformation tensor

A general theory has been worked out by Kröner [49] to describe the internal stresses generated by lattice defects. He also developed a very well-usable analogy between the stress-field of an elastic continuum and the magnetostatic field. The main results are briefly reviewed as follows.

On the analogy of the operators "grad" and "curl", well known from vector analysis, Kröner introduced the operators "Def" (deformation) and "Inc" (incompatibility) with the definitions

$$\boldsymbol{\epsilon} = \text{Def}\,\mathbf{u} = \frac{1}{2}(\nabla\mathbf{u} + \mathbf{u}\nabla), \tag{2.92}$$

$$\boldsymbol{\eta} = \text{Inc}\;\boldsymbol{\epsilon} = \nabla\times\boldsymbol{\epsilon}\times\nabla, \tag{2.93}$$

where $\mathbf{u}\nabla$ denotes the transpose of the tensor $\nabla\mathbf{u}$. $\boldsymbol{\eta}$ is the incompatibility tensor, which is zero if the displacement vector is continuous and single-valued. If lattice defects exist in a body this condition is not satisfied. Thus $\boldsymbol{\eta}$ is not zero and its value characterizes the deviation of the deformations from the compatibility condition. It can be seen from (2.91) that for a continuous and single-valued vector $\mathbf{u}$

$$\text{Inc Def}\,\mathbf{u} = 0. \tag{2.94}$$

It can further be shown that

$$\text{Div Inc}\,\boldsymbol{\epsilon} = 0 \tag{2.95}$$

in perfect analogy with the equations of electrodynamics curl grad $U = 0$, and div curl $\mathbf{A} = 0$. The deformations relating to $\boldsymbol{\eta} \neq 0$ may be defined in the following way. Let the difference of the displacement of points of distance $d\mathbf{r}$ be $d\mathbf{u}$. In the general case this is not a perfect differential, or in other words the distortion-tensor $\boldsymbol{\beta}$ in the equation

$$d\mathbf{u} = d\mathbf{r}\,\boldsymbol{\beta} \tag{2.96}$$

is generally not symmetrical. The generalized strain tensor may be defined by the following equation:

$$\boldsymbol{\epsilon} = \frac{1}{2}(\boldsymbol{\beta} + \tilde{\boldsymbol{\beta}}) \tag{2.97}$$

where $\tilde{\boldsymbol{\beta}}$ is the transpose of the $\boldsymbol{\beta}$ tensor. It is easy to see that (2.97) corresponds with the classical definition if the compatibility condition is satisfied.

2.11.3. Analogy between the elastic and magnetostatic fields

Let us place a homogeneous, isotropic body of permeability μ^* (this notation departs from custom: it has been introduced to distinguish it from the elastic modulus μ), into a magnetic field. No electric current flows in the body. The external magnetic forces generate in the body a magnetic field $\mathbf{H}$ which because of the previous condition can be produced as the gradient of a scalar function $\psi(\mathbf{r})$ which is continuous and single-valued everywhere in the body, i.e.

$$\mathbf{H} = \text{grad } \psi. \tag{2.98}$$

The magnetic field is free of curls, i.e. curl $\mathbf{H} = 0$. The magnetic vector $\mathbf{H}$ is related to the induction vector $\mathbf{B}$ by the following equation

$$\mathbf{B} = \mu^*\mathbf{H}. \tag{2.99}$$

Since no magnetic charge exists

$$\text{div } \mathbf{B} = 0. \tag{2.100}$$

The above three equations unambiguously describe the magnetic field under the given conditions.

Let an external force act on the elastic body. A deformation field $\boldsymbol{\epsilon}$ develops in the body, which, as has been seen, may be written as the deformation of an everywhere continuous and single-valued vector function $\mathbf{u}(\mathbf{r})$

$$\boldsymbol{\epsilon} = \text{Def } \mathbf{u}. \tag{2.101}$$

That is, the incompatibility of the deformation field is zero, so Inc $\boldsymbol{\epsilon} = 0$. The stress-field $\boldsymbol{\sigma}$ is related to the strain tensor $\boldsymbol{\epsilon}$ by the equation

$$\boldsymbol{\sigma} = \mathbf{C}\,\boldsymbol{\epsilon} \tag{2.102}$$

(generalized Hooke law) where $\mathbf{C}$ represents a tensor of rank four determined by the elastic constants. For isotropic materials [50]

$$C_{iklm} = \lambda\delta_{ik}\delta_{lm} + \mu(\delta_{il}\delta_{km} + \delta_{im}\delta_{kl}). \tag{2.103}$$

The condition of equilibrium is

$$\text{Div}\ \boldsymbol{\sigma} = (-\mathbf{f}), \tag{2.104}$$

where **f** is the density of the body forces. It can be seen that the classical elastic field is the perfect analogy of the magnetic field. This analogy may be applied quite successfully to solve the basic equations of the general continuum theory. For the sake of completeness, let us review the relations of the static magnetic field for the more general case when a stationary current flows in the body. In this case the equations describing the magnetic field are

$$\text{div}\ \mathbf{B} = 0\,, \tag{2.105}$$

$$\text{curl}\ \mathbf{H} = \frac{4\pi}{c}\,\mathbf{j}\,, \tag{2.106}$$

where **j** is the current density vector. It is known from electrodynamics that the solution of the above system of equations is reduced to the determination of the vector potential **A**, which is defined by the equation

$$\mathbf{B} = \text{curl}\ \mathbf{A}. \tag{2.107}$$

Thus (2.105) is automatically fulfilled. Substituting (2.107) into (2.106) (using the identity curl curl = grad div $-\Delta$) one obtains the equation

$$\text{grad div}\ \mathbf{A} - \Delta\mathbf{A} = \frac{4\pi}{c}\,\mu^*\mathbf{j}\,. \tag{2.108}$$

However, it is known that (2.107) does not fix the value of **A** absolutely, because the relation remains invariably valid if the gradient of any single-valued scalar function $\varphi(\mathbf{r})$ is added to **A**. This uncertainty may be cancelled by a suitable secondary condition so selected that (2.108) becomes simpler. The well-known secondary condition

$$\text{div}\ \mathbf{A} = 0 \tag{2.109}$$

reduces the determination of the magnetic field to the solution of the Poisson equation

$$\Delta\mathbf{A} = -\frac{4\pi}{c}\,\mu^*\mathbf{j}\,. \tag{2.110}$$

The solution satisfying the secondary condition (2.109) is

$$\mathbf{A}(\mathbf{r}) = \frac{\mu^*}{c} \iiint \frac{\mathbf{j}(\mathbf{r}')}{|\mathbf{r} - \mathbf{r}'|} dV' . \tag{2.111}$$

2.11.4. The solution of the basic equations of the general continuum theory (isotropic case)

Let us follow the train of thought applied to the magnetic field in the previous section. The basic equations describing the internal stresses are

$$\text{Div}\, \boldsymbol{\sigma} = 0, \tag{2.112}$$

$$\text{Inc}\, \boldsymbol{\epsilon} = \boldsymbol{\eta}, \tag{2.113}$$

where $\boldsymbol{\epsilon} = \text{Def}\, \mathbf{u}$ and $\boldsymbol{\sigma} = \mathbf{C}\boldsymbol{\epsilon}$. Omitting the detailed calculations, only the more essential steps are considered. By analogy with the vector potential (as a generalization of the Airy stress function) a "stress-potential" $\boldsymbol{\chi}$ is introduced. $\boldsymbol{\chi}$ is defined by the equation

$$\boldsymbol{\sigma} = \text{Inc}\, \boldsymbol{\chi}. \tag{2.114}$$

Because of the identity (2.95), the condition (2.112) is automatically satisfied. Using the relation (2.102) and substituting (2.114) into (2.113) one obtains for $\boldsymbol{\chi}$ a fourth-order partial differential equation which also contains the incompatibility tensor $\boldsymbol{\eta}$. It is quite clear, however, that because of the identity (2.94) the relation (2.114) does not fix the value of $\boldsymbol{\chi}$ unambiguously, since by adding the deformation of any single-valued vector function to it, the relation remains unaltered. This uncertainty makes possible in this case too the introduction of a secondary condition so selected that the differential equation for $\boldsymbol{\chi}$ becomes simpler. For this purpose the following condition should be prescribed:

$$\text{Div}\, \boldsymbol{\chi}' = 0 , \tag{2.115}$$

where

$$2\mu\boldsymbol{\chi}' = \boldsymbol{\chi} - \frac{\nu\chi_1}{1 + 2\nu}\mathbf{I} .$$

Here $\chi_1 = \chi_{xx} + \chi_{yy} + \chi_{zz}$ denotes the first scalar invariant of the tensor χ and $\mathbf{I}$ is the unit tensor. The equation determining the function χ' with secondary condition (2.115) is

$$\Delta^2\chi' = 0. \tag{2.116}$$

The solution satisfying the secondary condition (2.115) for the case of an infinite medium is

$$\chi'(\mathbf{r}) = -\frac{1}{8\pi}\iiint \eta(\mathbf{r}')\,|\,\mathbf{r} - \mathbf{r}'\,|\,dV'. \tag{2.117}$$

In order to calculate the integral the incompatibility tensor must be known as a function of the locus over the whole body.

2.11.5. The dislocation density tensor

From the definition (2.1) of the Burgers vector, by using (2.96)

$$\mathbf{b} = \oint_L d\mathbf{u} = \oint_L (d\mathbf{r}\,\boldsymbol{\beta}). \tag{2.118}$$

Applying the Stokes theorem

$$\mathbf{b} = \int_F (d\mathbf{F}\,\boldsymbol{\alpha}) \tag{2.119}$$

where

$$\boldsymbol{\alpha} = \mathrm{Curl}\,\boldsymbol{\beta}. \tag{2.120}$$

$\boldsymbol{\alpha}$ is the dislocation density tensor. If this tensor is multiplied by a surface element vector $d\mathbf{F}$ of the material, one obtains the resultant Burgers vector of the dislocations of infinitesimal Burgers vectors (see Chapter 4) passing through this surface element:

$$d\mathbf{b} = d\mathbf{F}\,\boldsymbol{\alpha}. \tag{2.121}$$

The relation (2.120) is equivalent to (2.113). If $\boldsymbol{\alpha}$ is known, the deformation field can be determined. It follows from (2.97) that a well-defined relation exists between $\boldsymbol{\alpha}$ and $\boldsymbol{\eta}$.

One may prove that if $\boldsymbol{\beta}$ is a tensor function which can be differentiated twice (this is now, of course, assumed), then Div Curl $\boldsymbol{\beta} = 0$.

Using this equation one obtains from (2.120) that the divergence of the tensor of the dislocation density vanishes

$$\operatorname{Div} \boldsymbol{\alpha} = 0. \tag{2.122}$$

It follows from the concept of divergence that this expresses the act that the dislocations cannot terminate inside the material.

2.11.6. An example of the incompatibility tensor [51]

In order to describe the internal stresses generated by dislocations the dislocation density tensor must be known as a function of the locus. The incompatibility tensor may then be determined in the following way. Write the relation (2.120) for transposed tensors too and add it to the previous one. One obtains the equation

$$\operatorname{sym}\{\boldsymbol{\alpha}\} = \frac{1}{2}(\boldsymbol{\alpha} + \tilde{\boldsymbol{\alpha}}) = \frac{1}{2}\operatorname{Curl}(\boldsymbol{\beta} + \tilde{\boldsymbol{\beta}}) = \operatorname{Curl} \boldsymbol{\epsilon},$$

where $\operatorname{sym}\{\boldsymbol{\alpha}\}$ denotes the symmetrical part of the tensor $\boldsymbol{\alpha}$. Using (2.113) one obtains for the incompatibility tensor

$$\boldsymbol{\eta} = -\operatorname{sym}\{\operatorname{Curl} \boldsymbol{\alpha}\}. \tag{2.123}$$

Consider as an example for the value of $\boldsymbol{\eta}$ a single dislocation of any form with a Burgers vector **b**. Let one axis of the coordinate system chosen point in the direction of a tangential unit vector **t** in every point of the dislocation line, while the other two coordinate axes (q_1, q_2) are perpendicular to the dislocation line. In this case the dislocation density tensor which satisfies the definition (2.121) is

$$\boldsymbol{\alpha} = \mathbf{t}_\circ \mathbf{b}\, \delta(q_1)\, \delta(q_2), \tag{2.124}$$

where $(\mathbf{t}_\circ \mathbf{b})$ is the tensor product of the vectors **t** and **b**, and $\delta(q_1)$ and $\delta(q_2)$ denote the Dirac δ-function.

Using (2.124), (2.117) may be transformed for easier calculation

$$\chi'_{ij} = \frac{1}{8\pi}\operatorname{sym}\left\{\epsilon_{jkl}\, b_l\, \nabla_k \oint |\mathbf{r} - \mathbf{r}'|\, dl'_i\right\}. \tag{2.125}$$

dl'_i is an element of the dislocation line and the integration must be extended over the entire dislocation line. ϵ_{jkl} is the permutation tensor, whose non-zero components are

$$\epsilon_{123} = \epsilon_{231} = \epsilon_{312} = 1,$$
$$\epsilon_{132} = \epsilon_{321} = \epsilon_{213} = -1.$$

For straight-edge dislocations χ_{zz} is proportional to the Airy stress function as calculated in Appendix C. Thus this constitutes a special case of (2.125).

2.11.7. *Summary of the continuum theory*

In this chapter it has been endeavoured to give an insight into the problems of the study of dislocations by the continuum model. It has been shown that the classical elasticity theory may yield quite valuable results provided that it is applied with certain restrictions. However, it also turned out that without a proper generalization of our concepts almost every problem requires an *ad hoc* solution. This deficiency of the classical theory creates the need for a generally applicable theory, a need which is supported by considerations of principle. It has been seen that a suitable analogy can be developed between the magnetostatic field and the field of the internal stresses. This analogy may presumably be extended to further properties of the magnetic field. The polarization of the elastic field may also be defined [50]; this concept can well be applied to anisotropic crystals, to describe, for instance, the effect of point defects on the elastic behaviour of the crystal, or to characterize the properties of an infinitesimal dislocation loop.

Finally, a few remaining questions should be pointed out. All considerations are based on two assumptions. One of them is the isotropy of the investigated materials, and the other the linear relationship between the stresses and strains. Though both assumptions are primarily mathematical simplifications, they have also another significance. As has been seen, the basic equations of the stress-field have general solutions for the isotropic case; no solution, however, is known for an arbitrary anisotropic case. Assuming the validity of Hooke's law one obtains only a first approximation, which does not explain the higher-order effects. Thus, for instance, the local volume change developing around a screw dislocation can be determined only by considering a non-linear stress–strain relationship [52].

Chapter 3

The properties of point defects

3.1. Point defects in crystals

Besides dislocations, other defects too exist in real crystals. In this chapter those defects which develop in a finite volume of the crystal and which are surrounded solely by good material are discussed.

This type of defect may be generated in various ways. For example, during the growth of the crystal, in some volume element, another crystalline structure, another phase, develops consisting of the same atoms as the matrix. In this case the material of the volume element has other properties than its surroundings. A quite similar situation arises if foreign atoms segregate somewhere in the crystal. The volume defects so generated are called inclusions or elastic inhomogeneities. The point defects, vacancies and interstitial atoms already mentioned in the first chapter may also be regarded as volume defects, though they are usually called zero-dimensional defects as well. Our investigations are mainly directed towards the properties of these point defects, though some of the results may also be applied to more general volume defects. Similarly to other defects, point defects create elastic deformations in the material surrounding them and consequently energy is stored in the stress-field generated. Their presence, however, greatly increases the configurational entropy of the crystal (the formation of one single point defect in one mole of perfect material increases the number of possible micro-states from one to 6×10^{23}). For this reason the thermodynamic equilibrium at a given temperature without any external force requires the presence of a certain number of point defects in the crystal. The equilibrium of any thermodynamic system at constant temperature and constant pressure is determined by the minimum of the free enthalpy which is usually given by $\Delta G = \Delta U + p\Delta V - T\Delta S$. ($U$ denotes the internal energy, T the temperature, S the entropy, p the pressure and V the volume.) It can be seen from this expression that if the entropy becomes large enough during the formation of the point defects, the free enthalpy may decrease ($\Delta G < 0$) in the real crystal as compared to the perfect crystal. This actually occurs in the present case. It can be shown by statistical considerations that at temperature T the equilibrium concentration c of point defects in a crystal is given by the following expression:

$$c = c_0 e^{-E_F/kT}, \tag{3.1}$$

where E_F is the energy of formation of the defect and k is Boltzmann's constant. It can be seen from the above formula that the point defect concentration changes very rapidly with temperature.

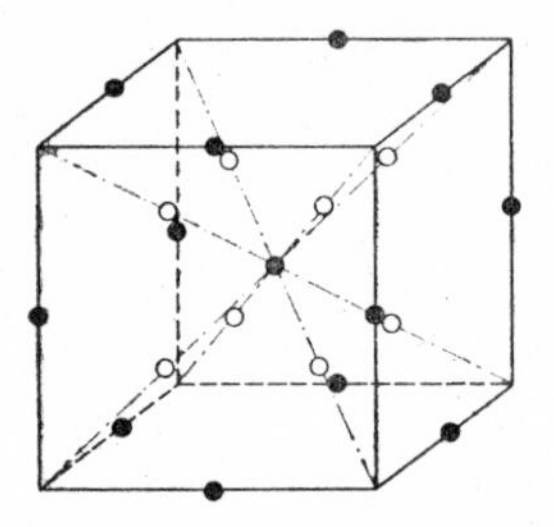

FIG. 3.1. Interstitial positions in a fcc lattice. The empty circles indicate tetrahedral, and the full circles octahedral positions

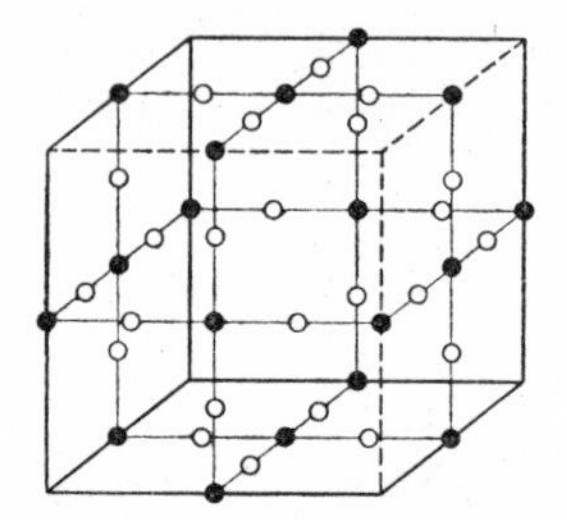

FIG. 3.2. Interstitial positions in a bcc lattice. The smaller positions are indicated by full circles, and the large ones by empty circles

In any crystal, only those point defects can be formed which do not disturb the electrical equilibrium of the crystal, i.e. the crystal remains outwardly neutral. For metals this condition does not mean any restriction, since a vacancy (which is the absence of a positive ion) always means too the absence of a valency electron; the interstitial atom, on the other hand, always creates a valency electron surplus. With ionic crystals the situation is quite different. If, for instance, a vacancy were formed by the removal of a negative ion the electrical equilibrium would be upset. Consequently, in an ionic crystal positive and negative vacancies are always formed together at the same time (Schottky defect). Another possibility consists of the formation of a vacancy–interstitial pair of the same type (Frenkel defect). If the crystal also contains impurities whose ions have an electrical charge different from the charge of the base material, the vacancy concentration (e.g. the number of vacancies created at positive and negative lattice sites) may deviate from that of the pure crystal.

In principle vacancies may be created in any crystal. Interstitial atoms, on the other hand, can be situated only in cavities of appropriate size. This depends upon the three-dimensional atomic structure. In face-centred cubic metals, for example, the possible positions are the ($\frac{1}{4}$, $\frac{1}{4}$, $\frac{1}{4}$) tetrahedral and the ($\frac{1}{2}$, $\frac{1}{2}$, $\frac{1}{2}$) octahedral sites (Fig. 3.1). In body-centred cubic crystals the possible interstitial sites are defined by the coordinates ($\frac{1}{2}$, $\frac{1}{2}$, 0), ($\frac{1}{2}$, $\frac{1}{4}$, 0), and ($\frac{1}{2}$, 0, 0) (Fig. 3.2). The mean volume of cavities is larger for the fcc than for the body-centred cubic (bcc) case, which, however, contains nine possible sites per atom, whereas with the face-centred crystals there are only three.

3.2. The energy of point defects

The energy necessary to generate the lattice defects is the formation energy. The formation energy of one vacancy for instance is equal to the work necessary to remove an atom or an ion from the inside of the crystal to the surface. The formation energy of a vacancy in metals arises from three effects: from the elastic deformation, and the quantum-mechanical and electrical interactions originating from the removal of the ion core and the valency electron. The effect of the elastic deformation is very small. We shall see later that the magnitude of elastic displacements is inversely proportional to the third power of the distance. Consequently, it is enough to take into account only the displacement of the nearest neighbours. The magnitude of the energy originated by the elastic deformation is of the order of a few tenth electron volts. The larger part of the energy of formation comes from the electrical and quantum-mechanical interactions. The theoretical determination of their values is rather difficult, since when calculating the energy change of the valency electrons many effects must be considered. The presence of a vacancy alters the charge distribution in the material. For this reason the potential and kinetic energies of the electron gas change and further additional interactions develop between the electron gas and the positive cavity formed at the site of the vacancy. Moreover, the exchange and correlation energies and the Fermi energy of the crystal also change. Finally, the energy change due to ionic effects must be considered. This is caused by the change of charge distribution within the ions and by a repulsive force between the ions.

Thus the energy of formation is the algebraic sum of many terms of different signs and relatively large absolute values, which means that a considerable uncertainty exists in the final value if the estimation of the individual terms was not exact enough. Huntington and Seitz calculated the value of the formation energy for copper to be 1.8 eV [1]. Later Brooks showed [2] that Huntington considered the deformation of the ions twice. After correcting this error the value 1.0 eV was obtained. By calculating the eigenvalue of the electrostatic valency energies and the Fermi level of the crystals more exactly Amar [3] obtained 2.0 eV. For other fcc metals (silver, gold) no similar calculations exist, because the repulsive potentials of the ions of these metals are not known.

Gold and silver, on the other hand, are experimentally much more easily manageable than copper (see Chapter 9). For this reason reliable experimental data are available first of all for silver and gold [4–7]. The formation energy of a vacancy in silver is 1.1 eV, and in gold 0.97 eV. For copper 1.17 eV has been obtained experimentally [8]. All these values are in good agreement with the value calculated for copper. The comparison is based on the fact that no substantial deviations are expected for the above-mentioned metals.

According to experimental observations the following relationship exists between the formation energy of a vacancy and the Debye temperature, atomic mass number (M) and atomic volume (Ω) [9]:

$$E_F = \alpha\, M\, \Theta_D^2 \Omega^{2/3}\,,$$

where α is constant. Figure 3.3 shows the validity of this relation for various metals.

The determination of the energy of formation of interstitial atoms is even more difficult. It is certain, however, that a large part of the activation energy is due to repulsive forces between ions and to the lattice deformation, and that the interaction with the electron gas is much less decisive. For this reason the energy of formation of an interstitial atom is approximately 4–5 eV [10], though this value depends upon the lattice site of the atom. From the large energy of formation it follows from (3.1) that for a fcc crystal in thermal equilibrium the concentration of interstitial atoms is many orders of magnitude smaller than the vacancy concentration.

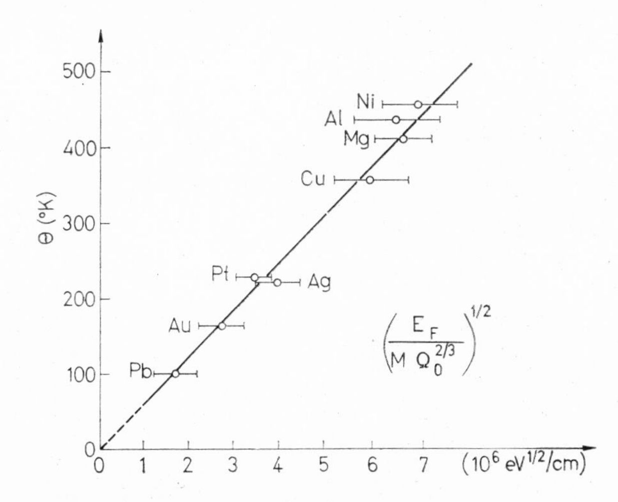

FIG. 3.3. Semi-empirical relation between the energy of formation of a vacancy (E_F), the Debye temperature (Θ_D), the atomic mass number (M) and the atomic volume (Ω)

In fcc metals the self-diffusion and the diffusion of impurity atoms take place by vacancy movement. The motion of a vacancy is produced by the jumping of an adjacent atom into the vacancy site. If the vacancy comes to surface of the crystal, the crystal-defect is annihilated. The recombination of the vacancies and interstitial atoms annihilates both defects. The number of defects produced and annihilated by the above processes are in average equal to one another at thermal equilibrium. The activation of the motion of vacancies depends upon the energy necessary to move an adjacent atom into the vacancy site. This is the activation energy of the motion of the vacancy. If the vacancy concentration of the crystal is larger than necessary for the thermodynamic equilibrium, the rate of vacancy annihilation at constant temperature is proportional to the surplus concentration and to an exponential factor similar to (3.1), but in this case the activation energy of motion E_M is inserted into the expression instead of E_F. Its value for interstitial atoms is approximately 0.5 eV [11], and for vacancies 0.8–1.0 eV [4–7].

3.3. The continuum model of volume defects

The quantitative properties of point defects are, in the same way as dislocations, described most simply by the theory of linear elasticity [12]. These investigations do not yield exact results, because the deformations and stresses are concentrated in the surroundings of the defects. It is not known with certitude to what extent the theorems of linear elasticity are valid in these regions, but nevertheless they give an acceptable possibility of estimating the elastic properties of point defects (at least their order of magnitude). It is worth while to survey the methods applied in the field, all the more so because the results obtained are frequently quoted in the literature.

The interstitial atom is wedged into a volume smaller than itself and for this reason it deforms its surroundings. In an isotropic medium the deformation has spherical symmetry which may be regarded as the result of a force acting centro-symmetrically from one point. A model of such properties can easily be given. For the sake of simplicity let us first consider only spherical defects in infinite bodies. Imagine that a small, but finite rigid body of volume ΔV has been forced into one point of the continuum. With this so-called dilatation centre the same effects as considered above may be produced. Apart from this point no volume change takes place in the material; thus the displacement vector must satisfy the following relation:

$$\operatorname{div} \mathbf{u} = \Theta = c\,\delta(\mathbf{r}), \tag{3.2}$$

where c is a temporarily unknown proportionality factor representing the strength of the dilatation centre, and $\delta(\mathbf{r}) = \delta(x)\,\delta(y)\,\delta(z)$ is the Dirac δ function.

The value of c can be determined from the condition that the entire change of volume of the body is ΔV; thus by the definition of the δ function

$$\Delta V = \int_{-\infty}^{\infty} \operatorname{div} \mathbf{u} \, dV = c \int_{-\infty}^{\infty} \delta(\mathbf{r}) \, dV = c. \tag{3.3}$$

It is known from the potential theory that

$$\operatorname{div} \operatorname{grad} \left(\frac{1}{r}\right) = \Delta \left(\frac{1}{r}\right) = -4\pi\delta(\mathbf{r}).$$

Applying this relation, the displacement vector satisfying the condition (3.3) is

$$\mathbf{u} = -\frac{c}{4\pi} \operatorname{grad} \left(\frac{1}{r}\right) = \frac{c}{4\pi} \frac{\mathbf{r}}{r^3}. \tag{3.4}$$

Let us examine the force which must operate so that the displacement (3.4) satisfies the equilibrium conditions (B.26). Let us produce the expressions in the latter with the help of (3.2) and (3.4). From (3.2)

$$\operatorname{grad} \operatorname{div} \mathbf{u} = c \operatorname{grad} \delta(\mathbf{r}).$$

Further, using (3.4)

$$\Delta\mathbf{u} = -\frac{c}{4\pi} \operatorname{grad} \Delta \left(\frac{1}{r}\right)\Bigg] = c \operatorname{grad} \delta(\mathbf{r}).$$

On multiplying the first relation by $(\lambda + \mu)$ and the second by μ, one obtains, after addition, the equation

$$\mu\Delta\mathbf{u} + (\lambda + \mu) \operatorname{grad} \operatorname{div} \mathbf{u} = c(\lambda + 2\mu) \operatorname{grad} \delta(\mathbf{r}).$$

This shows that the displacements defined by (3.4) are produced if a point force

$$\mathbf{f}_p = -c(\lambda + 2\mu) \operatorname{grad} \delta(\mathbf{r}) \tag{3.5}$$

exists. If the displacements are known the deformations and stresses may be written in the following form:

$$\sigma_{ik} = 2\mu\varepsilon_{ik}, \qquad \varepsilon_{ik} = \frac{c}{4\pi r^3}\left[\delta_{ik} - 3\frac{x_i x_k}{r^3}\right]. \tag{3.6}$$

If the sphere of volume ΔV forced into the material in the point $r = 0$ is a perfect rigid body, only the surrounding material, the matrix, becomes deformed. (Generally, if defects or inhomogeneities are embedded in some homogeneous base material, this latter is the matrix.) For this reason the elastic energy—the self-energy of the defect as defined—can easily be calculated from (3.4) and (3.6). Accordingly, the energy density at a distance r from the defect is obtained from (2.2), (3.4) and (3.6) as

$$u = \frac{1}{2} \sum_{i,k=1}^{3} \sigma_{ik}\, \varepsilon_{ik} = \frac{3\mu c^2}{8\pi^2 r^6} \,. \tag{3.7}$$

Integrating over the entire volume of the matrix, the self-energy of the defect is obtained:

$$U_s = \frac{2}{3}\, \mu \Delta V \,. \tag{3.8}$$

The real situation is better approximated in a somewhat more complicated way if the volume defect is described by the following model:

Let the elastic constants of the matrix be μ and K. Assume further that a spherical part of the matrix with volume V_0 is removed and replaced by a homogeneous sphere of elastic constants μ', K' which has a volume $V = V_0 + \Delta V$, when it is free of any external influence. After relaxation both materials become deformed and consequently elastic energy is stored in both. Such a defect is called *inhomogeneous inclusion.*

If the volume of the inclusion so formed is V_h, it is apparent that $V_0 < V_h < V$. To determine the stress-field and self-energy of the defect, the deformation of the inclusion must also be considered. Denote the volume change of the matrix by ΔV_m and that of the defect, by ΔV_h. Together, they are equal to the surplus volume ΔV. In this case, the force as defined by (3.5) acts in the matrix (here $c = \Delta V_m$). Consequently, the stresses in the matrix are given by (3.6) with $r > r_h$, where r_h denotes the radius of the defect in equilibrium. The matrix exerts a uniform hydrostatic pressure on the defect and for this reason the volume of the sphere changes uniformly. The displacement vector of this deformation is

$$\mathbf{u}' = -A\mathbf{r}, \tag{3.9}$$

where A is a temporarily unknown proportionality factor. Applying the relation $K' = \lambda' + \frac{2}{3}\mu'$, the stresses within the inclusion are

$$\sigma'_{ik} = 2\mu' \varepsilon'_{ik} + \lambda' \Theta' \delta_{ik} = -3AK' \delta_{ik}. \tag{3.10}$$

The value of A may be determined using the condition that at equilibrium the resultant pressure along the surface of the sphere is zero and, consequently

$$\boldsymbol{\sigma}\mathbf{r} = -\boldsymbol{\sigma}'\mathbf{r}, \qquad \text{if} \quad r = r_h.$$

By applying the relations (3.6) and (3.10) one obtains the following result:

$$A = -\frac{4\mu c}{9K'V_h}, \tag{3.11}$$

since the value of the stress in (3.10) is

$$\sigma'_{ik} = -\frac{4\mu c}{3V_h}\delta_{ik}. \tag{3.12}$$

However, consider that (3.12) may also be expressed according to the definition (B.17) of the pure hydrostatic pressure in the following way:

$$\sigma'_{ik} = -K'\frac{\Delta V_h}{V_h}\delta_{ik}.$$

combining this relation with (3.12), V_h may also be ex pressed by the equation

$$\Delta V_h = \frac{4\mu c}{3K'}.$$

ince, on the other hand, $\Delta V_h + \Delta V_m = \Delta V$,

$$\Delta V_m = c = \frac{3K'}{4\mu + 3K'}\Delta V. \tag{3.13}$$

With this, every datum of the stress-field has been determined. Let us calculate the self-energy of a defect, which is the sum of the elastic energies stored in the inclusion and the matrix. The energy of the matrix may be obtained from (3.7). The result is

$$U_m = \frac{2}{3}\mu \frac{(\Delta V_m)^2}{V_h}. \tag{3.14}$$

The elastic energy within the inclusion is

$$U_h = \frac{V_h}{2}\sum_{\substack{i=1\\k=1}}^{3} \sigma'_{ik}\,\varepsilon'_{ik} = \frac{1}{2}\frac{(4\mu\, c)^2}{9K'V_h}. \tag{3.15}$$

The self-energy of the spherical inhomogeneous inclusion is the sum of the energies stored in the matrix and in the sphere, i.e.

$$U_s = U_m + U_h = \frac{2}{3}\mu \frac{3K'}{4\mu + 3K'} \frac{(\Delta V)^2}{V_h}. \tag{3.16}$$

This expression is valid for any spherical inclusion. One can see that the result does not depend upon the bulk modulus of the matrix. Therefore, if for the defect in question $\mu' = \mu$ but $K' \neq K$, instead of (3.16) one may write

$$U_s = \frac{2}{3}\mu \frac{2\mu + 3\lambda'}{6\mu + 3\lambda'} \frac{(\Delta V)^2}{V_h} = \frac{2}{9}\mu \frac{1 + \nu'}{1 - \nu'} \frac{(\Delta V)^2}{V_h}, \tag{3.17}$$

where ν' is the Poisson number for the material of the defect.

In this case the self-energy of the defect does not depend upon the properties of the matrix. The expression (3.17) must also be applied in the case when the material of the defect is the same as that of the matrix (homogeneous inclusion). The results obtained permit the numerical estimation of the elastic self-energy of a point defect. Calculating with the values $\mu = 4 \times 10^{11}$ dyne/cm^2, $\nu = \frac{1}{3}$ and $V_h = 12 \times 10^{-24}$ cm^3, one obtains $U_s = 1.4 \times (\Delta V/V_h)^2$ eV. For substitutional impurity atoms with $\Delta V/V_h \cong 0.1$ the elastic self-energy is 0.014 eV, and this may clearly be neglected. For an interstitial atom, however, one may take $\Delta V/V_h \cong \frac{1}{2}$ and thus $U_s \cong 0.3$ eV, which must be considered in the calculations.

In order to demonstrate the validity and good quality of the continuum model, it is practical to compare our results with those obtained by considering the atomic structure. So let us apply our results to impurity atoms present in fcc metals. Let the lattice parameter of the crystal be a_0, the radius of a normal atom r_0 and let us assume that the radius of the impurity atom is $r_0(1+\varepsilon)$ where $\varepsilon \ll 1$. It is easy to see that in our case the unit cell contains four atoms, and consequently

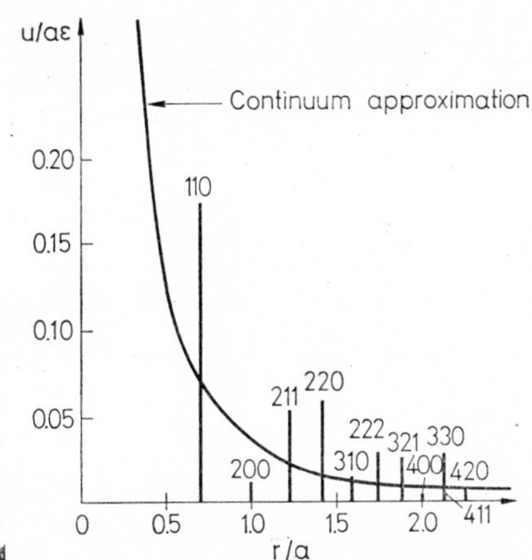

FIG. 3.4. The displacement of the atoms around an impurity atom as a function of their position in a fcc lattice. The discrete vertical lines indicate the displacement of the atoms characterized by the number triples written above the lines

the atomic volume is $a_0^3/4$. It follows from the proportionality of a_0 and r_0 that the surplus volume of the impurity atom is $3a_0^3\,\varepsilon/4$. If no essential difference exists between the elastic constants, by applying (3.17) the equation

$$U_s = \frac{1}{2}\mu\frac{1+\nu}{1-\nu}a_0^3\,\varepsilon^2 \tag{3.18}$$

is obtained (here the elastic constants refer to the base material).

It has been mentioned already that any calculation using the discrete, atomic structure involves serious mathematical as well as physical difficulties. Calculations have been carried out by Flinn and Maradudin [14] for impurity atoms in fcc crystals by this method. According to their results a considerable anisotropy exists in the displacement of the immediate neighbours around the impurity atoms. For the elastic energy these authors obtained instead of (3.18) the equation

$$U_s = 0.271\,\mu\frac{1+\nu}{1-\nu}a_0^3\,\varepsilon^2 \tag{3.19}$$

which is about half the value obtained by the previous method. This may be regarded as good agreement if one considers that only an estimation of the order of magnitude may be expected from the continuum model. Figure 3.4 depicts the position dependence of the displacement of the atom adjacent to the impurity atom. The single, discrete lines are displacements of the atoms around the point defect. The three numbers written above the lines denote the coordinates of the respective atoms in a coordinate system starting from the point defect. It can be seen that the displacements of the individual atoms differ from those obtained by the continuum model. Further, it can also be shown that the displacements are radial only for atoms lying in the $\langle 100 \rangle$, $\langle 110 \rangle$ and $\langle 111 \rangle$ directions. Summing up, one can say that valuable information may be obtained on the properties of lattice defects with the continuum theory. One must not forget, however, that the results obtained in this way are only approximations.

3.4. Volume defects in bodies with free surfaces

Up to now only the volume defects in infinite media have been investigated. In this section the results are extended to the cases of finite bodies, i.e. to bodies with free surfaces. In these cases the respective boundary conditions too have to be considered. In equilibrium the stresses of the resultant stress-field which are normal to the surface of the body must vanish along the boundary. Consequently, from the sequence of ideas developed in Section 2.9, not only the (3.4) displacements but image displacements $\mathbf{u}^k$ too must be considered which have no source (singularity) in the body. With these image displacements, however, the boundary conditions may be satisfied.

The complete determination of the boundary conditions can be attained only in the simplest cases, but the accompanying volume change ΔV is generally determinable.

The image stresses have no internal sources, and consequently everywhere in the body

$$\operatorname{Div} \boldsymbol{\sigma}^k = 0. \tag{3.20}$$

Multiply this equation by the vector $\mathbf{r}$, and integrate over the entire volume of the body. By applying the identity $\mathbf{r} \operatorname{Div} \boldsymbol{\sigma}^k = \operatorname{div} (\boldsymbol{\sigma}^k \mathbf{r}) -$

$- 3K\,\Theta^k$, the following relation is obtained

$$\Delta V_k = \int_V \Theta^k dV = \frac{1}{3K}\int_V \operatorname{div}\,(\boldsymbol{\sigma}^k\mathbf{r})dV = \frac{1}{3K}\int_F (\boldsymbol{\sigma}^k\mathbf{r})d\mathbf{F} =$$

$$= \frac{1}{3K}\int_F \mathbf{r}\boldsymbol{\sigma}^k\mathbf{n}\; dF\,,$$

where the final integrals refer to the surface of the body, and $\mathbf{n}$ is a unit vector normal to the surface. According to (2.78), however, along the surface of the body $\boldsymbol{\sigma}^k\mathbf{n} = -\boldsymbol{\sigma}\,\mathbf{n}$, where $\boldsymbol{\sigma}$ denotes the stress occurring in the infinite body for the same defect. In our case the stress is given by the relation (3.6). From (3.4) and (3.6) it is easy to see that $\boldsymbol{\sigma}\,\mathbf{r} = -4\mu\,\mathbf{u}$, and therefore

$$\Delta V_k = -\frac{1}{3K}\int_F (\boldsymbol{\sigma}\mathbf{r})\; d\mathbf{F} = \frac{4\mu}{3K}\int_F \mathbf{u}d\mathbf{F} = \frac{4\mu}{3K}\int_V \operatorname{div}\mathbf{u}\,dV.$$

By taking into account (3.3)

$$\Delta V_k = \frac{4\mu c}{3K} = (\gamma - 1)c. \tag{3.21}$$

The strength of a dilatation centre in a body of finite dimensions or with free surfaces is

$$\Delta V = \Delta V^\infty + \Delta V_k = \gamma c, \tag{3.22}$$

with

$$\gamma = \frac{3K + 4\mu}{3K} = 3\,\frac{1 - \nu}{1 + \nu}\,.$$

The results show that the effect of the free surface contributes considerably to the volume change of the body. This means that in practice equation (3.22) must be used.

For vacancies in an infinite medium (3.22) is also valid because the normal component of the stress along the surface of the cavity constituting the vacancy is zero. Let us determine the change of volume of a body containing a vacancy taking into consideration the diffusion of an atom to the surface whenever a vacancy is created. If no other change took place in the lattice, every vacancy would increase the volume of the body by one atomic volume. However, because of the image displacement the volume of the crystal decreases by $\Delta V_k = (\gamma - 1)\,\Omega$. ($\Omega$ is the atomic volume.) The resultant change of volume in a body containing a single vacancy is consequently

$$\Delta V = \Omega - (\gamma - 1)\Omega = \frac{5\nu - 1}{\nu + 1}\,\Omega. \tag{3.23}$$

Taking $\nu = \frac{1}{3}$ one obtains $\Delta V = 0.5\,\Omega$, which is in very good agreement with the values obtained for gold experimentally (0.57 ± 0.05) [6], and (0.45 ± 0.1) [15].

3.5. The elastic interactions of point defects with other stress-fields

The investigation of dislocations has already shown that whenever a stress-field exists which was generated simultaneously by two different sources, besides the self-energies belonging to each source, an interaction energy also appears. This is equal to the work which must be done to develop an elastic displacement field of one source in the stress-field of another source. In this way, the point or volume defects must necessarily interact with other stress-fields.

This interaction causes a considerable effect in particular on the plastic behaviour of the material, since the volume defects considerably impede the motion of the dislocations. This was first recognized by Cottrell [16] and Nabarro [17] from studies of the yield points of carbon steels. These authors concluded that because of the interaction of the carbon atoms and dislocations, a region enriched with carbon atoms develops around the dislocations (Cottrell atmosphere); consequently, at least at the beginning of the deformation the motion of dislocations also requires the displacement of the Cottrell atmosphere with the consequence that the stress necessary for the displacement, i.e. the yield stress, increases considerably compared with the pure material. This problem is discussed in some detail in Chapter 7.

As regards the nature of the interaction, Cottrell and Nabarro assumed an elastic effect originating from the difference in size of the matrix and the solute atoms (size effect). This type of interaction was first investigated mathematically by Cottrell and Bilby [18], and later by Bilby [19].

The effect of the dissolved atoms, or more generally of the foreign phase, does not depend only upon the difference in size. Since the elastic properties of the solute atom or foreign phase generally differ from those of the matrix, it is expected that these effects are present even if there is no dimensional difference. This so-called modulus effect is commensurable with the size effect [20].

The electrostatic field around the dislocations [21], stacking faults connected with partial dislocations [22], the effects of the local order [23], etc., all produce interactions. The magnitudes of these contributions, however, are considerably smaller than those of the elastic effects. In the following sections only the elastic interactions are dealt with.

3.5.1. The interaction of the dilatation centre with other stress-fields; size effect

Let us first investigate the interactions of the dilatation centre defined by (3.2) with a stress-field generated by an arbitrary external or internal source. Denote the stresses of this field by $\boldsymbol{\sigma}^A$ and the respective displacement by $\mathbf{u}^A$. Assume first that only the dilatation centre is present in the body. In this case force (3.5) is exerted on every volume unit. The work done on the elementary volume dV during the displacement $\mathbf{u}^A$ against the force is $-\mathbf{f}_p \mathbf{u}^A \, dV$. The interaction energy originating from a dilatation centre and a source A is the sum of these primary works

$$U_i = -\int_V \mathbf{f}_p \mathbf{u}^A dV. \tag{3.24}$$

Using the expression (3.22), $\mathbf{f}_p$ can be written as

$$\mathbf{f}_p = -\Delta V K \operatorname{grad} \delta(\mathbf{r}). \tag{3.25}$$

Substituting into (3.24), and using the identity $\operatorname{div}[\mathbf{u}^A \delta(\mathbf{r})] = \delta(\mathbf{r}) \operatorname{div} \mathbf{u}^A + \mathbf{u}^A \operatorname{grad} \delta(\mathbf{r})$ and the definition of $\delta(\mathbf{r})$, the interaction energy may be expressed in the following form:

$$U_i = -\Delta V K \int_V \delta(\mathbf{r}) \operatorname{div} \mathbf{u}^A \, dV = -\left\{K[\operatorname{div} \mathbf{u}^A]_{\mathbf{r}=0}\right\} \Delta V.$$

The term in brackets is the hydrostatic stress originating from the source A at the site of the dilatation centre, and therefore

$$U_i = -\frac{\Delta V}{3}(\sigma_{xx} + \sigma_{yy} + \sigma_{zz}). \tag{3.26}$$

Our results are also valid for small spherical volume defects whose elastic constants agree with those of the matrix, and within which the changes of the stress-field $\boldsymbol{\sigma}^A$ may be neglected. If this latter condition is satisfied and the elastic constants of the spherical defect differ from those of the matrix, (3.26) is modified in the following way [25]:

$$U_i = -\frac{\Delta V}{3} \frac{K'}{\alpha(K' - K) + K}(\sigma_{xx} + \sigma_{yy} + \sigma_{zz}), \tag{3.27}$$

where

$$\alpha = \frac{1}{\gamma} = \frac{1 + \nu}{3(1 - \nu)}.$$

A dilatation centre can interact only with hydrostatic stresses. From this it follows that interactions between two dilatation centres exist only in finite bodies, because their dilatation in an infinite body is zero everywhere except at their centre.

The interaction of volume defects was investigated in detail by Eshelby. According to his results [24], the interaction energy of defects which may be regarded as perfectly rigid spheres in an isotropic medium changes according to the sixth power of their distance. The sign is always positive and consequently they repel one another. It is interesting to note, however, that the anisotropy increases the effective length of interaction.

For two volume defects at a distance r from each other in a cubic crystal Eshelby obtained the following energy of interaction [12]:

$$U_i = -\frac{15}{8\pi\,\gamma^2}\alpha_0 \frac{\Delta V_1 \Delta V_2}{r^3}, \tag{3.28}$$

where ΔV_1 and ΔV_2 are dilatation strengths originating from th defects, and α_0 is a proportionality factor depending upon the elasti constants and the directional cosines of the vector $\mathbf{r}$. The energy (3.28 for metals is approximately 0.01 eV. This is about one order of magnitude smaller than the energy originating from the electrical interactions. For this reason the elastic interactions of the point and volume defects may generally be neglected.

3.5.2. The interaction of inhomogeneities with other stress-fields; modulus effect

If an isotropic body of elastic constants μ and K contains a region of volume V whose elastic constants, μ' and K', differ from μ and K but which fits into the body without any dimensional difficulty, the region in question is an inhomogeneity. If no external stress acts on the body, the entire body is free of stresses. In the case of an external stress, however (because of the differences of the elastic constants of inclusion and the matrix), the defect interacts with the stress-field. This is the modulus effect. According to the investigations of Eshelby, the interaction energy between a small spherical volume defect and the stress-field $\boldsymbol{\sigma}^A$ is [26]

$$U_{im} = -\frac{V_h}{2}\left\{\frac{A}{9K}(\sigma^A)^2 + \frac{B}{2\mu}\sum {}'\sigma_{ij}^A\,{}'\sigma_{ij}^A\right\}, \tag{3.29}$$

where

$$\sigma^A = \sigma_{xx}^A + \sigma_{yy}^A + \sigma_{zz}^A, \quad {}'\sigma_{ij}^A = \sigma_{ij}^A - \frac{1}{3}\sigma^A\,\delta_{ij},$$

$$A = \frac{K' - K}{\alpha(K - K') - K}, \qquad B = \frac{\mu' - \mu}{\beta(\mu - \mu') - \mu}$$

and

$$\beta = \frac{2}{15}\left(\frac{4 - 5\nu}{1 - \nu}\right).$$

In the following section the results are applied to dislocations and point defects.

3.6. The interaction of dislocations and point defects

Dislocations and atoms whose properties differ from those of the matrix are always found in crystalline solids. Their interactions change the properties of the crystal considerably. This applies especially to alloys in which the atoms of the alloying material are present in great quantities either as dissolved atoms or in the form of precipitations. A knowledge of the nature of the interaction between the dislocations and the foreign atoms is of basic importance for an understanding of the properties of the solids.

The solute atom is an inhomogeneous inclusion because its dimensions and elastic properties generally differ from those of the base material. Their interaction with a stress-field may thus be calculated in two steps. With spherical defects the calculations may be carried out from the size effect with (3.27); the interaction energy due to the modulus effect, on the other hand, is calculated with (3.29). The total interaction energy is equal to the sum of the energies due to the size and modulus effects, since a simultaneous treatment of both effects yields only second-order corrections [20].

3.6.1. Straight edge dislocations

(a) *Size effect.* The hydrostatic stresses around an edge dislocation are given by (2.19). If they are substituted into (3.27) the interaction energy originating from the size effect of the edge dislocation and a small, spherical defect (e.g. dissolved atom) is obtained:

$$U_i^{\text{edge}} = \frac{\mu b(1+\nu)V_0}{3\pi(1-\nu)} \frac{K'}{\alpha(K'-K)+K} \frac{\Delta V}{V_0} \frac{\sin\vartheta}{r}. \tag{3.30}$$

From this expression if $K' = K$ the result of Cottrell is obtained [16]. It can be seen that if $\Delta V > 0$, i.e. the dimension of the solute atom is larger than that of the atoms of the matrix, the interaction energy in the compressed zone ($\sin\vartheta > 0$, Fig. 3.5) is positive. If $\Delta V < 0$ the situation is reversed. All this indicates that an attractive interaction always exists between the edge dislocation and the solute atoms. This explains why impurity atoms are always enriched along a dislocation. The interaction energy is inversely proportional to the distance; consequently, only those impurity atoms with which the interaction energy is larger than the thermal energy of the defect, become attached to the dislocation. The radius within which this is accomplished is defined by the equation $U_i^{\text{edge}} \approx kT$. With $\sin \vartheta = 1$, and with the usual values, the radius at about room temperature is approximately 80–100 Å. Thus the radius of interaction is small

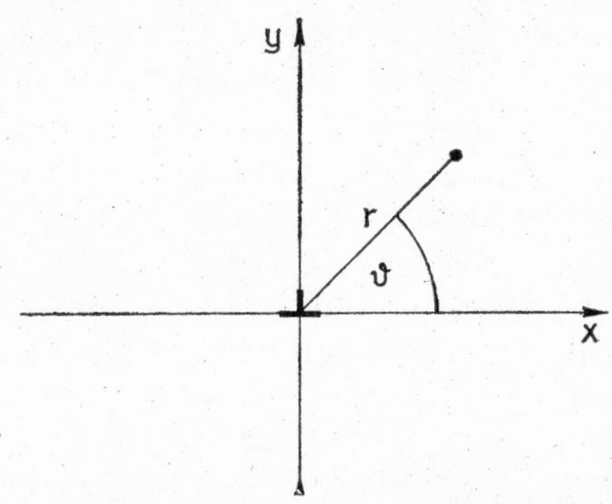

FIG. 3.5. Point defect in the stress-field of an edge dislocation

and it further decreases with increasing temperature. If the temperature is not too low, as a result of the attractive interaction some of the solute atoms are situated in the immediate neighbourhood of the dislocation line. The maximum interaction energy developed in this way ($\sin \vartheta = 1, r = b$) is the binding energy (U_b) between the impurity atom and dislocation.

Let us restrict ourselves to the calculation of the binding energy in a few special cases. Only the size effect is considered, i.e. $K' = K$. For the binding energy one obtains

$$U_b^{\text{edge}} = \frac{\mu V_0}{3\pi} \frac{1+\nu}{1-\nu} \frac{\Delta V}{V_0}. \tag{3.31}$$

According to observations in solid solution, at least to a limited concentration, the average volume V relating to one atom is a linear function of the concentration c of the solute atoms [27]:

$$V(c) = (1-c)\, V_0 + cV^*,$$

where V_0 is the atomic volume in the case of pure material and V^* is a constant characteristic of the solute atoms. This volume change is caused by the size effect of the solute atoms. If their number is N_s, and the total number of atoms is N, $c = N_s/N$ and the volume change produced by one solute atom is given by the equation

$$\Delta V' = \frac{N(V - V_0)}{N_s} = c(V^* - V_0)\frac{N}{N_s} = V^* - V_0 .$$

From this the relative volume change per atom, the volume size factor, is

$$\eta_V = \frac{\Delta V'}{V_0} = \frac{V^* - V_0}{V_0} = \frac{1}{V_0}\frac{\partial V}{\partial c} . \tag{3.32}$$

It has to be considered that in (3.27) $\Delta V/V_0$ denotes the relative difference of the "free" volume of the solute atom as related to the cavity, whereas η_V refers to the deformed volume of the solute atom (i.e. its volume in the matrix). According to equation (3.22) the two quantities are related by the following equation:

$$\frac{\Delta V}{V_0} = 3\frac{1 - \nu}{1 + \nu}\eta_V .$$

If one considers that in the case of a fcc crystal the volume of one atom is

$$V_0 = \frac{4\pi}{3}\left(\frac{b}{2}\right)^3 ,$$

one obtains as the binding energy between the edge dislocation and a solute atom the equation

$$U_b^{\text{edge}} = \frac{1}{6}\mu b^3 \eta_V . \tag{3.33}$$

The value of η_V can be determined experimentally. Table 3.1 summarizes binding energies characterizing the size effect of copper- and aluminium-based solid solutions as obtained from literature data [27].

(b) *Modulus effect.* The interaction of the edge dislocation with a solute atom regarded as an inhomogeneity is characterized by (3.29).

TABLE 3.1. *Binding energies between solute atoms and edge dislocations in Al and Cu base alloys*

Al: $\mu = 2.7\times 10^{11}$ dyne/cm^2, $b = 2.86\times 10^{-8}$ cm

Cu: $\mu = 4\times 10^{11}$ dyne/cm^2, $b = 2.55\times 10^{-8}$ cm

Solute atom	η_V (%)	E_b edge (eV)	Solute atom	η_V (%)	E_b edge (eV)
Zn	− 5.74	0.037	Zn	+17.10	0.117
Si	−15.78	0.103	Cr	+19.72	0.135
Cu	−37.77	0.247	Al	+19.99	0.138
Ag	+ 0.12	0.001	Ag	+43.32	0.300
Sn	+24.09	0.159	Au	+47.59	0.327
Mg	+40.82	0.267	Sn	+83.40	0.572

Using (2.19) the interaction energy originating from the modulus effect may be expressed by the equation

$$U_{im}^{edge} = -\frac{\mu b^2 V_h}{8\pi^2(1-\nu)^2 r^2}\left\{\left[(1-\nu-2\nu^2)A + (2\nu^2-2\nu-1)B\right]\frac{2\sin^2\vartheta}{3} + B\right\}. \quad (3.34)$$

From this the interaction energy between a vacancy and an edge dislocation can easily be obtained [28]. The vacancy corresponds to an impurity atom for which $K' = 0$ and $\mu' = 0$, that is in the case of a vacancy

$$A = \frac{3}{2}\frac{1-\nu}{1-2\nu}, \qquad B = \frac{15(1-\nu)}{7-5\nu},$$

and therefore

$$U_{im}^{edge} = -\frac{15\mu b^2 V_h}{8\pi^2(1-\nu)(7-5\nu)}\left[1 - \frac{1+6\nu-5\nu^2}{5}\sin^2\vartheta\right]\frac{1}{r^2}. \quad (3.35)$$

The interaction energy is negative for every ϑ, which means that the edge dislocation attracts the vacancy from every direction. For the maximum binding energy ($\vartheta = 0$, and $r = b$) one obtains approximately 0.15 eV.

In order to investigate the modulus effect for solute atoms more closely, it is practical to trace back the quantities A and B, which contain the difference of the elastic constants, to known macroscopic parameters [29].

Consider a homogeneous body with an external stress σ_{ij}^A. Its elastic energy is

$$U_0 = \frac{1}{2} N V_0 \sigma_{ij}^A \varepsilon_{ij}^A = \frac{1}{2} N V_0 \left\{ \frac{1}{9K} (\sigma^A)^2 + \frac{1}{2\mu} \sum {}'\sigma_{ij}^A \, {}'\sigma_{ij}^A \right\},$$

where N is the number of atoms, V_0 the atomic volume and K and μ the macroscopic elastic constants of the body. Exchange, in theory, N_s atoms of the body for foreign (solute) atoms. As a result of these solute atoms, which are inhomogeneities in the base material, the elastic energy is altered. If, for the elastic constants of the inhomogeneities $K' > K$ and $\mu' > \mu$, i.e. in expression (3.29) $A < 0$ and $B < 0$ and thus $U_{im} > 0$, then the elastic energy of the body decreases, compared to U_0, because the same stresses produce smaller deformations, and vice versa. For this reason the elastic energy of a body containing N_s inhomogeneities is

$$U_{el} = U_0 - U_{im} = \frac{N V_0}{2} \left\{ \frac{1}{9K} (\sigma^A)^2 + \frac{1}{2\mu} \sum {}'\sigma_{ij}^A \, {}'\sigma_{ij}^A \right\} + \\ + \frac{N_s V_h}{2} \left\{ \frac{A}{9K} (\sigma^A)^2 + \frac{B}{2\mu} \sum {}'\sigma_{ij}^A \, {}'\sigma_{ij}^A \right\},$$

where V_h denotes the volume of the solute atom. Since $N_s = cN$, one obtains

$$U_{el} = \frac{N V_0}{2} \left\{ \frac{1}{9} \frac{1 + c\dfrac{V_h}{V_0} A}{K} (\sigma^A)^2 + \frac{1}{2} \frac{1 + c\dfrac{V_h}{V_0} B}{\mu} \sum {}'\sigma_{ij}^A \, {}'\sigma_{ij}^A \right\}.$$

From this the macroscopic elastic constants of a body containing inhomogeneities (dilute solid solutions) are

$$K_m = \frac{K}{1 + c\dfrac{V_h}{V_0}A} \cong K\left(1 - c\frac{V_h}{V_0}A\right),$$

$$\mu_m = \frac{\mu}{1 + c\dfrac{V_h}{V_0}B} \cong \mu\left(1 - c\frac{V_h}{V_0}B\right).$$

By differentiation one obtains

$$-V_h A = \frac{1}{K}\frac{\partial K_m}{\partial c}V_0 = \eta_K V_0,$$

$$-V_h B = \frac{1}{\mu}\frac{\partial \mu_m}{\partial c}V_0 = \eta_\mu V_0. \qquad (3.36)$$

K_m and μ_m are measurable quantities. One can see that to calculate the energy (3.34) it is enough to know the volume of the atoms of the base material. Using these latter relations, (3.34) takes the following form:

$$U_{im}^{edge} = \frac{\mu b^2 V_0 \eta_\mu}{8\pi^2(1-\nu)^2 r^2}\left\{1 - \frac{2}{3}\left[(1+2\nu-2\nu^2) - (1-\nu-2\nu^2)\frac{\eta_K}{\eta_\mu}\right]\sin^2\vartheta\right\}.$$

With $\nu = \frac{1}{3}$ and $V_0 = 4\pi/3\,(b/2)^3$

$$U_{im}^{edge} = 0.015\frac{\mu b^5 \eta_\mu}{r^2}\left(\cos^2\vartheta + 0.3\frac{\eta_K}{\eta_\mu}\sin^2\vartheta\right). \qquad (3.37)$$

The results show that if $\eta_\mu > 0$ and $\eta_K > 0$ then $U_{im}^{edge} > 0$ for every ϑ, i.e. the modulus effect is not necessarily an attractive interaction. The experimental values for η_μ and η_K are rather uncertain. Thus, instead of the actual evaluation of the energy (3.37), it is more practical to compare the contributions to the binding energy of the size effect and the modulus effect. Since $\eta_K/\eta_\mu < 3$ [29], by substituting $\vartheta = 0$ and $r = b$ into (3.37) and by using (3.33) one obtains

$$\frac{U_b^{\text{edge}}}{U_{im}^{\text{edge}}} = 11\,\frac{\eta_V}{\eta_\mu}.$$

The value of η_V/η_μ is in most cases approximately $\frac{1}{3}$ [29]; consequently, in the case of edge dislocations, the contribution of the modulus effect to the binding energy is about a quarter of that of the size effect.

3.6.2. Infinite screw dislocations

(a) *Size effect.* According to the expression (3.27) obtained for the size effect, the volume defect of spherical symmetry interacts only with hydrostatic stresses. On the basis of linear elasticity theory no hydrostatic component of the stress-field of a screw dislocation is obtained. The application of second order relations, however, leads to a non-zero volume dilatation in the case of screw dislocations too. According to the calculations of Stehle and Seeger [30]

$$\sum_{i=1}^{3} \varepsilon_{ii}^{\text{screw}} = \frac{\kappa}{4\pi}\left(\frac{b}{r}\right)^2, \tag{3.38}$$

where the constant κ is a quantity characteristic of the forces acting between the atoms, its value being around unity. Applying this relation, one obtains for the energy of interaction originating from the size effect between the screw dislocation and the solute atom ($K' = K$)

$$U_i^{\text{screw}} = \frac{\mu\kappa(1-\nu)b^5}{12(1-2\nu)r^2}\,\eta_V. \tag{3.39}$$

κ is generally negative, and consequently the interaction is attractive if $\eta_V > 0$ and repulsive if $\eta_V < 0$. By substituting the values $\nu = \frac{1}{3}$ and $r = b$, the binding energy is

$$U_b^{\text{screw}} = \frac{\mu\kappa b^3}{6}\,\eta_V. \tag{3.40}$$

Thus the size effect of the screw dislocation may be compared with the size effect of an edge dislocation.

(b) *Modulus effect.* Substituting (2.23) into (3.29) the interaction energy between the screw dislocation and the solute atom is

$$U_{im}^{\text{screw}} = \frac{\mu b^2 V_0}{8\pi^2 r^2}\,\eta_\mu\,. \tag{3.41}$$

The interaction is attractive if $\eta_\mu < 0$, i.e. $\mu' < \mu$, and repulsive if $\eta_\mu > 0$, i.e. $\mu' > \mu$. In the case of attractive interactions, the binding energy resulting from (3.41) is

$$U_{bm}^{\text{screw}} = \frac{\mu b^3}{48\pi}\,\eta_\mu\,. \tag{3.42}$$

For a vacancy

$$V_0\,\eta_\mu = -V_h\,\frac{15(1-\nu)}{7-5\nu}\,,$$

and therefore the energy of interaction is

$$U_{\text{ivac}}^{\text{screw}} = -\frac{15(1-\nu)\,V_h\,\mu b^2}{8\mu^2(7-5\nu)r^2} \tag{3.43}$$

which is always negative. Consequently the screw dislocations attract the vacancies.

Chapter 4

Dislocations in crystals

4.1. The Frenkel model

Though the continuum theory explains quite satisfactorily a great number of the properties of dislocations, it does not deal, for example, with the effect of the core of atomic dimensions. The most conspicuous problem is that of the movement of dislocations. According to the continuum theory a dislocation lying at the centre of a cylindrical body can be displaced by any arbitrarily small shear stress since, in this position, its elastic energy does not change. It is quite obvious, however, that the binding forces acting between the atoms impede the motion of dislocations. A cavity of finite dimensions taken around the core of a dislocation in the continuum theory presents a similar problem, which – at least in the case of small Burgers vectors – seems to be rather arbitrary. No such difficulties arise if the crystal is investigated as an atomic structure. Nevertheless, the consideration of an atomic structure – even if the forces acting between the atoms are known exactly – meets with unsurmountable difficulties. To date, only isolated attempts have been made to interpret the various phenomena solely on the basis of the atomic structure.

It seems to be much more suitable to make a transition between the two extreme cases with appropriate models [1]. One of the most simple possibilities is the Frenkel model. In the one-dimensional case this model consists of a chain of mass points connected by equal springs. The points are located in a periodic potential field whose period corresponds to the equilibrium distance (Fig. 4.1a). An edge dislocation is obtained if an element of the chain is removed from or inserted into the chain, after which the chain is again connected (Fig. 4.1b). Two forces now act on the individual mass points: the force of the adjacent springs and the gradient of the potential field. Accordingly, the equation of motion of every mass point can be determined. In this way a system of non-linear differential equations is obtained whose general solution is not known. Nevertheless, if a potential field consisting of a single sine term, or of the superposition of not more than two sine terms and only a few mass points are investigated, the solution is easily obtained [2]. From this model valuable information concerning the dynamic behaviour of the

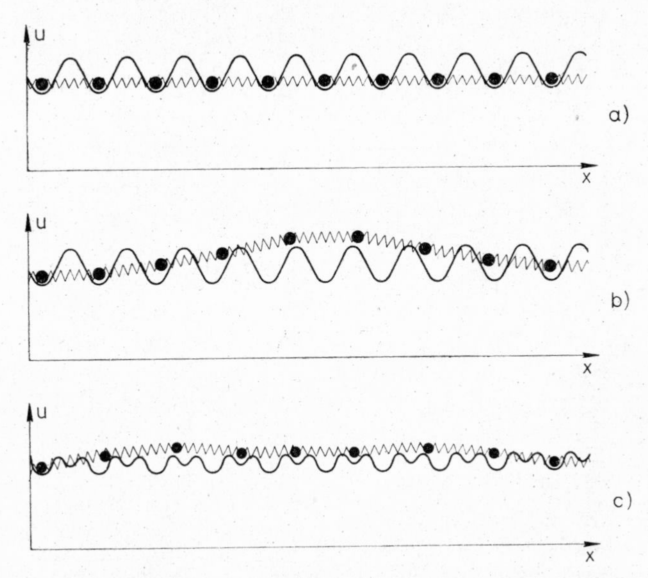

FIG. 4.1. The Frenkel model. Mass points connected by an elastic force in a periodic potential field. (a) perfect lattice, (b) perfect dislocation, (c) partial dislocation

dislocations can be deduced. Thus, it can be shown, for example, that the dislocations cannot move at a speed larger than the velocity of sound, while a dislocation which moves at the velocity of sound has infinite kinetic energy. If, on the other hand, a potential distribution is assumed which has two minima within one period (Fig. 4.1c), according to the solution of the equations, the dislocation splits into two partial dislocations (see Chapter 6).

Since the Frenkel model is today only of historical importance, no detailed discussion of it is given here.

4.2. Peierls model [3]

If a dislocation is in a symmetrical position in an otherwise undistorted crystal (Fig. 4.2b), the lattice does not exert any force on it and the dislocation is in equilibrium (its potential energy has an extreme value). With any deviation from the equilibrium position

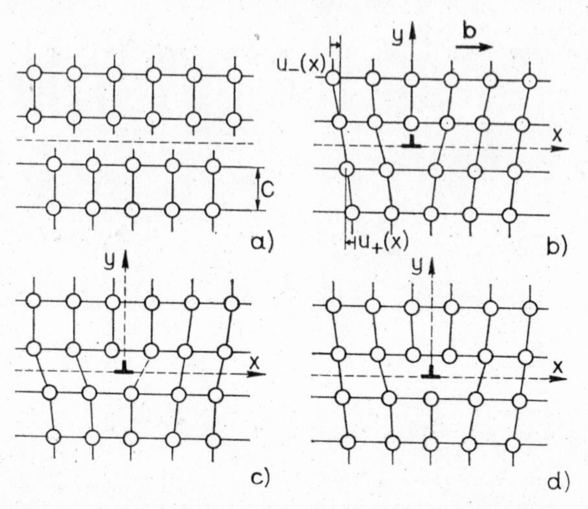

FIG. 4.2. The formation of a dislocation to deduce the Peierls model (a, b) and the displacement of the dislocation [by $\alpha < \frac{1}{2}$ (c) and $\alpha = \frac{1}{2}$ (d), see equation (4.11)]

as a result of the asymmetrical arrangement of the atoms, a force becomes active. Let us now investigate more closely the forces which are active on the atoms lying along the slip plane.

In contrast to the methods in the foregoing sections, a dislocation is created in the following way. Two infinite half-crystals are placed side by side with a displacement $b/2$ (Fig. 4.2a), one side of the contacting planes is stretched, and the other compressed, and finally the two half-crystals are reconnected at some distance from the centre (Fig. 4.2b). In this way the atoms of the upper and lower rows are displaced to an equal extent but in opposite directions by an amouen $u(x)$. (The assumption that $u(x)$ differs only in sign on the two sidst of the x-axis is to some extent arbitrary. Nevertheless, Huntington [4] obtained similar results without this assumption.) The relative displacement of the two sides is thus $2u(x)$. Since the two half-crystals were originally displaced by $b/2$ relative to each other, after their connection, $|u(x)| \to b/4$ if $x \to \infty$.

The displacements $u(x)$ are in relationship with the component σ_{xy} of the stress-field of the dislocation. Because of the lattice structure, σ_{xy} is clearly a periodic function of $u(x)$ with periodic length $b/2$ and for $u(x) \ll b$ satisfies Hooke's law. As a first approximation, let us assume that

$$\sigma_{xy} = \frac{\mu b}{2\pi c} \sin \frac{4\pi u(x)}{b}, \tag{4.1}$$

where c is the distance of the atomic layers gliding on each other. Since $u(x)$ is continuous and approaches $b/4$, one may imagine an infinitesimal dislocation in an interval dx' at any x' whose Burgers vector is

$$db = 2\left[\frac{du(x)}{dx}\right]_{x=x'} dx'. \tag{4.2}$$

The total dislocation is the resultant of these. The results of the continuum theory may safely be applied to the individual elementary dislocations since, because db is infinitesimal, the stress-field has no singularities. Thus from (2.19) the stress at x created by an elementary dislocation lying between x' and $x' + dx'$ is

$$d\sigma_{xy} = \frac{\mu}{2\pi(1-\nu)} 2\frac{du}{dx'} dx' \frac{1}{x-x'}. \tag{4.3}$$

The total stress (excluding the position $x' = x$) is

$$\sigma_{xy}(x) = \lim_{\varepsilon \to 0} \left\{ \int_{-\infty}^{x-\varepsilon} d\sigma_{xy} + \int_{x+\varepsilon}^{\infty} d\sigma_{xy} \right\} =$$

$$= \frac{\mu}{\pi(1-\nu)} \int_{-\infty}^{\infty} \frac{du}{dx'} \frac{1}{x-x'} dx'. \tag{4.4}$$

By comparing (4.1) and (4.4) one obtains

$$\int_{-\infty}^{\infty} \frac{1}{x - x'} \frac{du(x')}{dx'} dx' = \frac{(1 - \nu)b}{2c} \sin \frac{4\pi u(x)}{b}. \tag{4.5}$$

This is the Peierls integral equation. With screw dislocations, the displacements are parallel to the dislocation line (the z-axis) and the integral equation is similar to (4.5).

On looking for a solution of equation (4.5) of the form $u(x) = k \text{ arc tan } Cx$ which satisfies the condition $u(0) = 0$, $|u(\pm\infty)| = b/4$, one obtains

$$u(x) = \pm \frac{b}{2\pi} \text{arc tan} \frac{2(1 - \nu)x}{c}. \tag{4.6}$$

The negative sign clearly refers to the compressed, and the positive sign to the expanded half-space (Fig. 4.2). The two values are equivalent, and thus in the following discussion only the positive sign is used. The interval for which $-b/8 \leqq u(x) \leqq b/8$ is the width of the dislocation; this is an important parameter of the solution. Obviously

$$u\left[\frac{c}{2(1 - \nu)}\right] = -\frac{b}{8},$$

and consequently the width of an edge dislocation is

$$s = \frac{c}{1 - \nu}. \tag{4.7}$$

With most metals $\nu = \frac{1}{3}$, which means that the width is approximately 1.5 atomic distances. (If one does not adhere to a purely sine force law, the width of the dislocation is considerably larger.)

From a knowledge of $u(x)$, σ_{xy} is obtained from (4.1)

$$\sigma_{xy} = \frac{\mu b}{2\pi c} \sin\left(-2 \text{ arc tan} \frac{2x}{s}\right). \tag{4.8}$$

In addition to $u(x)$, this is also depicted in Fig. 4.3, together with the shear stress component as derived in the elastic continuum theory. One may see that, in contrast to the latter, the stress remains finite

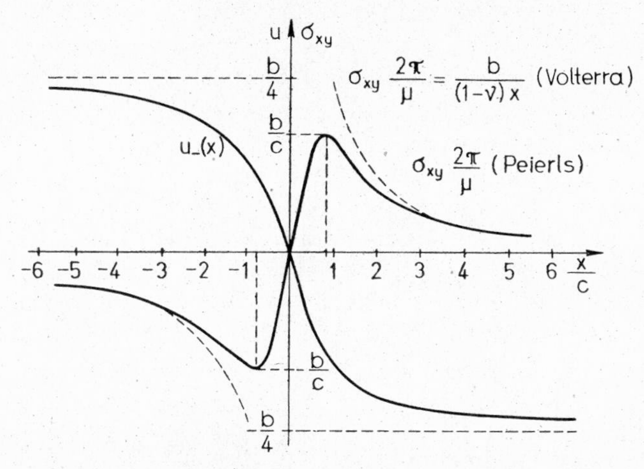

FIG. 4.3. The solution of the Peierls integral equation $[u(x)]$, the stress distribution obtained for $b = c$, and the stress distribution of the Volterra continuum model

in the core of the dislocation, and σ_{xy} $(x = 0) = 0$. If, on the other hand, $x \gg s$, $2 \arctan 2x/s = \pi - \varepsilon$ where $\varepsilon \ll 1$. Thus

$$\sin\left(-2 \arctan \frac{2x}{s}\right) = \sin(\varepsilon - \pi) \approx -\varepsilon =$$

$$= -\pi + 2 \arctan \frac{2x}{s} \approx \frac{s}{x},$$

and so

$$\sigma_{xy} = \frac{\mu b}{2\pi(1 - \nu)x}, \qquad (4.9)$$

which means that the Peierls model leads to the same result as the continuum theory.

With the Peierls model more complex problems can also be treated, such as "dipoles" consisting of two parallel dislocations of opposite Burgers vectors, small-angle grain boundaries, and extended dislocations (Chapter 6) [5, 6].

4.3. The stress necessary to slip a dislocation

We now calculate the change of the energy of a dislocation of unit length for the case of a dislocation leaving its assumed symmetrical position. Since the energy variation is negligibly small compared to the total energy, special care must be taken with the calculations lest by applying approximated formulae this variation is "smoothed out". The aim is rather to determine only this part instead of the total energy. Thus as a first approximation the deformation energy of the two atomic layers lying along the glide plane is calculated. One may assume that the energy accumulated in more distant layers is already independent of the position of the dislocation. It follows from the nature of the model that, for the total energy of the dislocation, a relation similar to (2.20) is obtained, but instead of the core radius, the width of the dislocation now appears in the formula [5].

If the dislocation is created in the above way, during the displacement du' of the atomic row of unit length adjacent to the glide plane and parallel to the dislocation (i.e. by the displacement of material of width b), work $b\,\sigma_{xy}\,du'$ is done by the stress σ_{xy}. Using (4.1), the energy of the row at the end of the deformation is

$$\int_0^u b\,\sigma_{xy}\,du' = \frac{\mu b^3}{8\pi c}\int_0^u \sin\frac{4\pi u'}{b}\,d\left(\frac{4\pi u'}{b}\right) =$$

$$= \frac{\mu b^3}{8\pi^2 c}\left[1 - \cos\frac{4\pi u(x)}{b}\right]. \tag{4.10}$$

This must be summed for the atomic rows on both sides of the slip plane. Let the dislocation now be in an asymmetrical position, and the distance of its central line from the nearest symmetrical position be $\alpha\,b$. The coordinates of the atomic rows on the two sides of the slip plane are

$$x = \left(\alpha + \frac{1}{2} n\right) b, \quad (n = 0, \pm 1, \pm 2, \ldots). \tag{4.11}$$

Substituting into (4.6) and summing (4.10) for every possible n, the so-called misfit energy is obtained

$$U_{\text{misfit}} = \frac{\mu b^3}{8\pi^2 c} \sum_{n=-\infty}^{\infty} \left\{1 - \cos 2\left[\arctan\left(\alpha + \frac{n}{2}\right)\frac{2b}{s}\right]\right\}. \tag{4.12}$$

Nabarro has shown that for $2\pi\, s/b \gg 1$, that part of the equation which depends upon α is to a good approximation [7, 8]

$$U' = \frac{\mu b^2}{2\pi(1 - \nu)} e^{-2\pi s/b} \cos 4\pi\alpha. \tag{4.13}$$

This potential is periodic according to $b/2$, and thus the energies of the symmetrical positions (Figs 4.2b and d) are identical. This follows from the sine force law and the approximation (4.11). If the dislocation moves at a distance $d(\alpha\, b)$, its energy changes. This means that one has to consider a force

$$F = -\frac{dU'}{d(\alpha b)}, \tag{4.14}$$

which has a maximum value if $\sin 4\pi\alpha = 1$. Thus, in accordance with (2.30), the critical shear stress (Peierls stress) necessary to move a dislocation is

$$\tau_p = \frac{F}{b} = \frac{2\mu}{1 - \nu} e^{-2\pi s/b} = \frac{2\mu}{1 - \nu} e^{-2\pi c/b(1-\nu)}. \tag{4.15}$$

If $a = b$, $\nu = 0.30$, then $\tau_p = 3.6 \times 10^{-4}\, \mu$; or if $\nu = 0.35$, $\tau_p = 2 \times 10^{-4}\, \mu$. This is still one order of magnitude larger than the value $4 \times 10^{-5}\, \mu$ measured for Cd and Zn, and the agreement is further reduced if one considers that the motion of dislocations is also impeded by other effects (foreign atoms, stress-fields of other dislocations). The above critical shear stress, on the other hand, is further decreased by many other effects. Of course, no far-reaching quantitative agreement with experiment can be expected from the

Peierls model which is based on numerous approximations. Nevertheless, the model is useful, describes the physical processes correctly at least qualitatively, and yields acceptable results as regards order of magnitude.

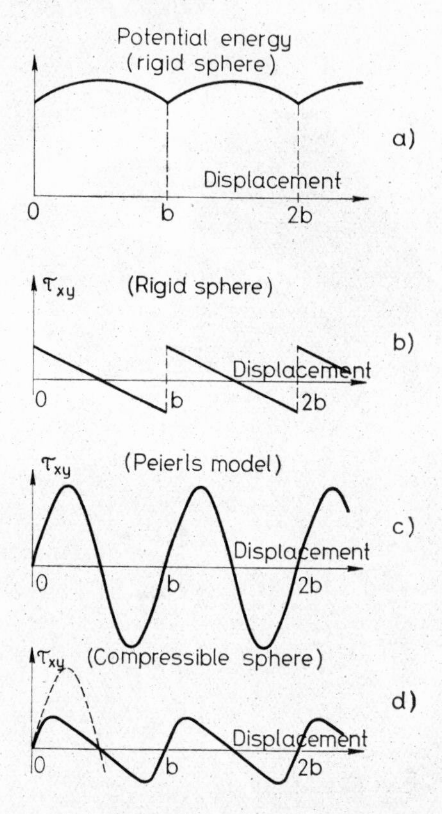

FIG. 4.4. The stress necessary to maintain equilibrium during the slip of the layers for various models

It is remarkable that (4.15) depends strongly upon the ratio c/b, or more exactly upon s/b, and decreases rapidly with the increase of these quantities. This explains why the slip always starts along close-packed crystallographic planes (i.e. along planes which are relatively distant from each other). We shall return to this question later.

Of course in many instances $c < b$, but in the close-packed hexagonal and cubic lattice the elementary slip is only $1/\sqrt{3}$ of the atomic distances (see partial dislocations). Thus $c/b = 3/2$ and calculations with this value reduce the critical shear stress calculated for (4.15) by approximately two orders of magnitude.

It must not be forgotten that the ratio c/b is introduced into (4.15) by applying a pure sine law. This is clearly a rough approximation which can be understood if one considers that the stress necessary to slip a structure consisting of rigid spheres corresponds approximately to Fig. 4.4b, and does not follow Hooke's law. Atoms, however, are not rigid bodies, and thus for small deformations, σ_{xy} follows Hooke's law and the sine relation is almost fulfilled, but for larger values of u, it approaches the curve corresponding to rigid spheres (Fig. 4.4d). A more exact course of the function can be derived from the various atomic potentials. Foreman *et al.* [9] sought a parametric function $u(x)$ which yields the required stress.

They assumed that

$$u(x) = \frac{b}{2\pi}\left(\text{arc tan}\frac{x}{a_0 s} + (a_0 - 1)\frac{sx}{a_0^2 s^2 + x^2}\right), \qquad (4.16)$$

where a_0 is a properly selected constant. If $a_0 = 1$ one obtains the original Peierls solution. According to the investigations of these authors, the stress obtained corresponds best to the theoretically expected function—which was also used by Mackenzie in his estimation of $\tau_{cr} = \mu/30$ as mentioned in Chapter 1—if $2 < a_0 < 4$. This means that the width of the dislocation increases considerably, and hence according to (4.15), leads to the decrease of τ_p. Naturally the new solution somewhat changes the form of the expression of the stress necessary to move a dislocation. However, this may be neglected to a first approximation.

According to the calculations, the width of the dislocation (practically independently of a_0) is now

$$s_F = \frac{\mu}{2\pi}\frac{s}{\tau_{cr}}, \qquad \left(s = \frac{c}{1-\nu}\right), \qquad (4.17)$$

where τ_{cr} is the value of the critical shear stress of a perfect crystal derived from the force law corresponding to the given a_0. According to the various approximations

$$\frac{\mu}{2\pi} \geqq \tau_{cr} \geqq \frac{\mu}{30}$$

and therefore

$$1.5\, s \leqq s_F \leqq 4s.$$

In the above calculations the properties of the actual crystal structure are considered only in a limited way in spite of the fact that the core structure is decisive in the Peierls stress. Unfortunately, our knowledge in this field is extremely imperfect and contradictory. Vitek *et al.* [10] calculated the displacements of the ⟨111⟩ screw dislocations of the body-centred cubic lattice with Johnson potentials fitted to the elastic constants of α-iron, but their results were not applied to determine the Peierls stress. Heinrich *et al.* [11] calculated the Peierls stress for the dislocations of the body-centred lattice from Johnson potentials calculated for α-iron. They also investigated how the result changes if, instead of considering only the interactions between the atoms on both sides of the glide plane, the atomistic model is gradually extended over several layers. According to their results the one-layer atomic model is far from satisfactory, and the Peierls potential for pure screw dislocations is *ca.* 0.15 eV/atomic plane, and in other cases about 0.01–0.03 eV/atomic plane. The relatively large value of the Peierls potential of the screw dislocation is probably due to the fact that the core of the screw dislocation shows an anisotropic broadening due to the decomposition into three partial dislocations [10] (see Chapter 6). Similar results were also obtained by Suzuki [13]. For metals and ionic crystals, equation (4.16) yields good approximations with suitable a_0 values; however, this is not the case with covalent crystals (e.g. Si, SiC) in which the bonding is strongly oriented and, accordingly, the motion of the dislocations requires larger stresses. Nevertheless, it can be generally stated that the Peierls stress is larger for smaller Poisson numbers ν. This is mainly due to the decrease of the width s of the dislocation.

Suzuki tried to determine the Peierls stresses of Si and Ge and, for an edge dislocation, obtained 4.5×10^{10} dyne/cm^2 and 2.5×10^{10} dyne/cm^2 respectively. For the screw dislocation these values were considerably larger. Labusch [14], on the other hand, by fitting the interaction potentials to the spectra of the lattice vibrations for the screw dislocations in Si and Ge, obtained Peierls stresses of 2.3×10^{10} dyne/cm^2 and 2.0×10^{10} dyne/cm^2 respectively, which correspond to potential peaks of 0.26 eV and 0.23 eV per atomic plane.

Stenzel applied the Peierls model [15] to investigate moving dislocations. Numerical results are not known. A critical review of the Peierls mechanism was given by Guyot and Dorn [16].

The effects of lattice vibrations, which even at 0° K cannot be neglected, must also be treated. Kuhlmann-Wilsdorf has shown [17] that because of the vibrations of the individual atoms, some "uncertainty relation" is valid for the position of the core of the dislocation and this causes a further reduction of the shear stress. This effect is largest with metals crystallizing with close-packed and body-centred cubic lattices [18], but it is almost negligible for materials with the diamond structure.

4.4. Kinks on dislocations

In the treatment so far it has been assumed that the dislocation line is parallel to the atomic rows of the lattice, i.e. to valleys of the periodic potential field (U') of the lattice. It was also assumed that during any motion of the dislocation the shear stress lifts the total dislocation simultaneously over the maximum of the potential peak.

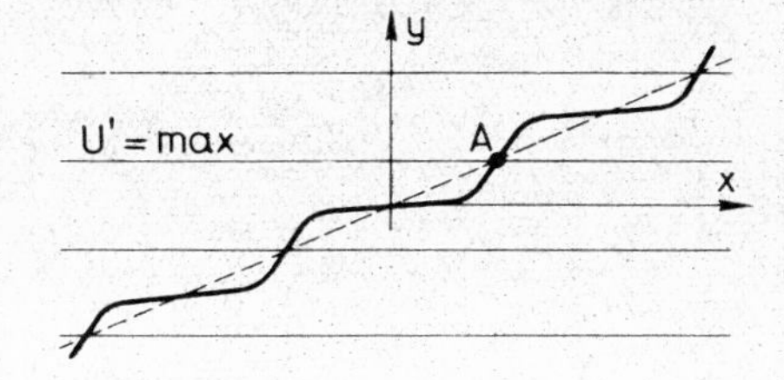

FIG. 4.5. Kinks formed if the dislocation is not parallel to atomic rows

This situation, however, occurs very rarely, because the dislocations usually cross the densest atomic rows and thereby also the maxima of U'. A repulsive force acts on the dislocation segment lying on the sides of the crossing, and as a result the dislocation tends to become wavy, as depicted in Fig. 4.5, in which the segments of minimum energy are relatively long. On the other hand, the wavy shape increases the length of the dislocation, which in turn increases the total elastic energy. These two counteracting effects develop an equilibrium state which depends upon the magnitude of the Peierls potential (and hence upon the width of the dislocation). With face-centred cubic metals – in which the dislocation width is a few atomic distances – the Peierls energy is so small that the elastic energy increment due to the wavy shape fitting the potential valley is many times larger [5].

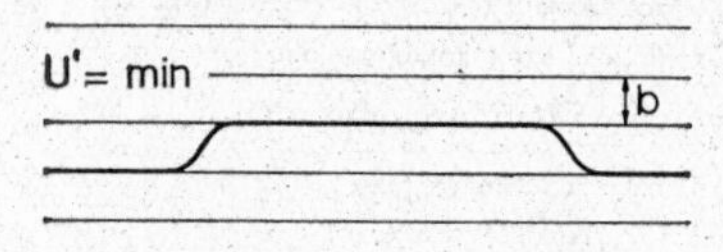

FIG. 4.6. "Kink"-pair in a straight dislocation

Consequently, in this case, the dislocations are smooth curves and their direction is not necessarily parallel to any of the low-index crystallographic directions (i.e. to any close-packed atomic layer). According to the Peierls model too, the energy of this inclined dislocation is to a first approximation independent of the position of the dislocation line and no special energy input is needed for its displacement. Any displacement means only that, for example, point A in Fig. 4.5 (together with the other similar points of intersection) is displaced in the $+z$ or $-z$ direction. Of course, because of the crystalline structure the energy fluctuation exists not only in the x but also in the z direction, and at any instant during the movement this must be overcome by only a short dislocation segment. Consequently, by this mechanism the critical shear stress as obtained by the Peierls model is reduced in the same way as the presence of a dislocation reduces, for example, the tensile strength of an otherwise perfect crystal. For this reason, in fcc metals the Peierls stress may be neglected compared with other effects impeding the movement of dislocations (impurities, the stress-field of other dislocations, and so on). According to Schottky's estimation the activation energy of a kink-motion corresponds to about 0.5 °K [19].

The situation is quite different if the Peierls stress is large. In this case the kink is sharp and the order of magnitude of its energy can approach μb^3. For this reason the dislocations are usually parallel to some low-index crystallographic direction. Silicon is a typical example of this and – as has been shown – in bcc metals the Peierls potential of the ⟨111⟩ screw dislocation is fairly large, in agreement with the relatively frequent occurrence of straight dislocations with this direction.

The motion of straight dislocations lying in the potential valleys starts by the jumping of a short dislocation segment into the next potential valley, as depicted in Fig. 4.6. During this step a pair of kinks is created. This needs an energy-input which comes partly from the work of the shear stress and partly from the thermal energy (thermally activated process). This latter, however, becomes important only at temperatures where kT is comparable with the energy of activation of a kink-pair. (At room temperature $kT \approx 0.025$ eV, and thus it is easy to understand that Si, for which the activation energy of a kink-pair is about 2 eV [14], cannot be deformed at room temperature.)

4.5. Glide systems

Glide systems consist of a glide plane and a glide direction in this plane. These latter are the glide elements.

In order to investigate possible glide systems, let us first study stable Burgers vectors since these determine the glide direction. Up to now it has been said only that the Burgers vector is a lattice vector. However, there are innumerable lattice vectors, and most of them are out of the question since according to experience only a few well-determined glide directions exist in each crystal structure.

The condition of stability is obtained from the energy relation. It has been shown that the energy of a dislocation is proportional to b^2. Any stable dislocation structure must have a minimum deformation energy with constant $\Sigma\, \mathbf{b}$. If, for example, two dislocations (with Burgers vectors $\mathbf{b}_1$ and $\mathbf{b}_2$) meet in identical or intersecting glide planes, and

$$\mathbf{b}_1^2 + \mathbf{b}_2^2 > (\mathbf{b}_1 + \mathbf{b}_2)^2 = \mathbf{b}_3^2, \tag{4.18}$$

then from the two dislocations one single dislocation is produced with the Burgers vector $\mathbf{b}_3$. If, on the other hand,

$$\mathbf{b}_3^2 > \mathbf{b}_1^2 + \mathbf{b}_2^2 \tag{4.19}$$

and $\mathbf{b}_1$ and $\mathbf{b}_2$ are two possible Burgers vectors such that $\mathbf{b}_1 + \mathbf{b}_2 = \mathbf{b}_3$, then the possibly existing dislocation with the Burgers vector $\mathbf{b}_3$ spontaneously splits into two parts according to the above scheme. The above conclusions are in most cases only slightly modified by the anisotropic elastic constants of the crystals. The condition of stability against any splitting may also be written in a different way. According to (4.19), $\mathbf{b}_3$ is stable if $\mathbf{b}_1\mathbf{b}_2 < 0$ ($\mathbf{b}_1$ and $\mathbf{b}_2$ enclose an obtuse angle), and it is unstable if $\mathbf{b}_1\mathbf{b}_2 > 0$ ($\mathbf{b}_1$ and $\mathbf{b}_2$ form an acute angle). If $\mathbf{b}_1\mathbf{b}_2 = 0$, the stability depends upon the stress relations at the site of the dislocation, upon the angle between the Burgers vectors and the dislocation line, and upon the anisotropy effects which were not considered in the energy relation. The stability is deteriorated in every case by the fact that any splitting is connected with a very small entropy increase.

The possible Burgers vectors which are stable against splitting may be investigated in any lattice type (assuming that the Burgers vectors are always lattice vectors). The results of this investigation for some of the more important cases are given in Table 4.1.

TABLE 4.1. *The stable Burgers vectors in some common crystals*

Lattice type	Materials	Stable Burgers vector $\mathbf{b}$	$\|\mathbf{b}\|$	Metastable Burgers vector $\mathbf{b}$	$\|\mathbf{b}\|$
Cubic primitive	CsCl	$\langle 100\rangle$	a	$\langle 110\rangle$ $\langle 111\rangle$	$a\sqrt{2}$ $a\sqrt{3}$
bcc	Fe, W, Nb	1/2 $\langle 111\rangle$ $\langle 110\rangle$	$a\sqrt{3/2}$ $a\sqrt{2}$	–	
fcc	Cu, Al, NaCl, Sphalerite	1/2 $\langle 110\rangle$	$a/\sqrt{2}$	$\langle 100\rangle$	a
hcp	Mg, Wurtzite	1/3 $\langle 2\bar{1}\bar{1}0\rangle$ $\langle 0001\rangle$	a* c**	1/3$\langle 2\bar{1}\bar{1}3\rangle$	$\sqrt{a^2+b^2}$

* "a" dislocation. ** "b" dislocation.

Let us investigate the face-centered cubic lattice. The shortest lattice vector is $\frac{1}{2}\langle 110\rangle$ (pointing from a cube corner to the middle of one of the adjacent cube faces), this cannot be composed of two other vectors satisfying (4.19) and thus this vector is absolutely stable. The $\langle 100 \rangle$ vector can be split in the following way

$$\langle 100 \rangle = \frac{1}{2}\langle 110 \rangle + \frac{1}{2}\langle 1\bar{1}0 \rangle . \tag{4.20}$$

These two components are always perpendicular to each other; consequently, there is no definite stability but, depending upon the effect of other factors, $\langle 100 \rangle$ may be a stable Burgers vector. All other lattice vectors of face-centred cubic lattices may be produced as the sum of the two vectors mentioned, so as to satisfy (4.19). This can easily be seen in the individual cases, but it can also be proved generally.

Let us now deal with the possible glide planes. It is not easy to establish these planes since two different effects which depend upon the anisotropy of the elastic constants must be considered. First, as will be seen in Chapter 6, the glide plane must coincide with the plane of the Burgers vector and the dislocation line. Because of the anisotropy of the elastic constants, however, the energy of the dislocation depends upon its orientation, and primarily dislocations corresponding to positions of lower energy develop during the crystal growth. On the other hand, according to (4.15) the stress necessary to move a dislocation depends strongly upon the ratio s/b, which in the isotropic case was simply $c/b(1 - \nu)$. Thus, the further apart the gliding atomic planes (or the more closely packed the atoms on them), the more easy to slip a dislocation.

However, taking into account the elastic anisotropy the ratio of the s/b values for the different lattice planes may be different from that of the isotropic case. According to the calculations of Seeger [6], for example, for an edge dislocation in copper $(s/b)_{111} = 0.98$ and $(s/b)_{100} = 0.39$; for aluminium, on the other hand, $(s/b)_{111} = 0.65$ and $(s/b)_{100} = 0.53$. For the isotropic case with $\nu = \frac{1}{3}$,

$$\left(\frac{s}{b}\right)_{111} = \frac{0.816}{1 - \nu} = 1.26 \, , \quad \left(\frac{s}{b}\right)_{100} = \frac{0.707}{1 - \nu} = 1.06 \, .$$

The dislocations formed during the crystal growth or developed by heat-treatment from the already existing dislocation network, and the planes determined by the smallest Peierls stress do not always

TABLE 4.2. *Glide systems in some crystals**

Crystal	Lattice type	Glide direction	Glide plane	Remarks
Cu, Ag, Au, Ni, α-CuZn	Al-type fcc	$\langle 110\rangle$	$\{111\}$	
Al	Al-type fcc	$\langle 110\rangle$	$\{111\}$	
		$\langle 110\rangle$	$\{110\}$	above 450 °C
NaCl, KCl, KJ, KBr	Bl-type fcc	$\langle 110\rangle$	$\{110\}$	
PbTe	Bl-type fcc	$\langle 100\rangle$	$\{110\}$	
Si, Ge	A4-type fcc	$\langle 110\rangle$	$\{111\}$	
α-Fe	A2-type bcc	$\langle 111\rangle$	$\{110\}$	
		$\langle 111\rangle$	$\{112\}$	
		$\langle 111\rangle$	$\{123\}$	
α-Fe + 4% Si, Mo, Nb	A2-type bcc	$\langle 111\rangle$	$\{110\}$	
AuZn, AuCd, MgTl, NH_4Cl, TlCl, NH_4Br	B2-type cubic	$\langle 100\rangle$	$\{100\}$	
AgMg	B2-type cubic	$\langle 111\rangle$	$\{321\}$	
β′-CuZn	B2-type cubic	$\langle 111\rangle$	$\{110\}$	
Cd, Zn ZnCd	A3-type hcp c/a = 1.85	$\langle 2\bar{1}\bar{1}0\rangle$	$\{0001\}$	
		$\langle 2\bar{1}\bar{1}0\rangle$	$\{01\bar{1}1\}$	
		$\langle 2\bar{1}\bar{1}3\rangle$	$\{\bar{2}112\}$	
Mg	A3-type hcp c/a = 1.623	$\langle 2\bar{1}\bar{1}0\rangle$	$\{0001\}$	
		$\langle 2\bar{1}\bar{1}0\rangle$	$\{01\bar{1}1\}$	above 225 °C
		$\langle 2\bar{1}\bar{1}0\rangle$	$\{01\bar{1}0\}$	below room temp.
Be	A3-type hcp c/a = 1.568	$\langle 2\bar{1}\bar{1}0\rangle$	$\{0001\}$	
		$\langle 2\bar{1}\bar{1}0\rangle$	$\{01\bar{1}0\}$	
Ti	A3-type hcp c/a = 1.587	$\langle 2\bar{1}\bar{1}0\rangle$	$\{01\bar{1}0\}$	
		$\langle 2\bar{1}\bar{1}0\rangle$	$\{01\bar{1}1\}$	
		$\langle 2\bar{1}\bar{1}0\rangle$	$\{0001\}$	

* According to A. Seeger, *Kristallplastizität*, Hdb. d. Phys. VII/2, p. 26, Springer, Berlin, 1958.

coincide. In such cases the glide can start only in an unfavourable direction. For this reason it is most probable that the "as grown" dislocations already present in the crystal do not participate in the mechanism of deformation, which is realized completely by dislocations created during the deformation itself (see Chapter 5).

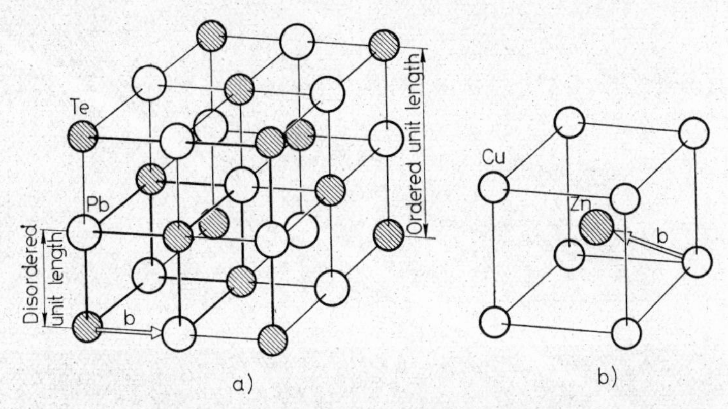

FIG. 4.7. Burgers vectors for NaCl and CsCl type lattices of low ordering energy; (a) the [100] vector of the disordered primitive cell in an ordered face-centred lattice is $\frac{1}{2}$[100] (PbTe); (b) the $\frac{1}{2}$[111] vector of the disordered body-centred lattice is no longer a lattice vector of the ordered primitive cubic lattice (AgMg, β'-CuZn)

Table 4.2 contains the experimentally established glide systems of some crystals. The following remarks are made in connection with this table. In the hexagonal close-packed (hcp) metals the glide on the (0001) basal plane loses its absolute character with the decrease of c/a, and its importance becomes gradually less. In the three ordered alloys (PbTe, AlMg, β'-CuZn) the slip direction does not coincide with the direction of the shortest lattice vector. This can be explained by considering that during ordering the unit cell of the lattice increases and the Burgers vector which was so far a lattice vector in the disordered state loses this character (see Chapter 6), but the slip direction remains unchanged, corresponding to the disordered lattice (Fig. 4.7).

The behaviour of ionic crystals of otherwise similar structure differs from that of these ordered alloys in so far as, because of the

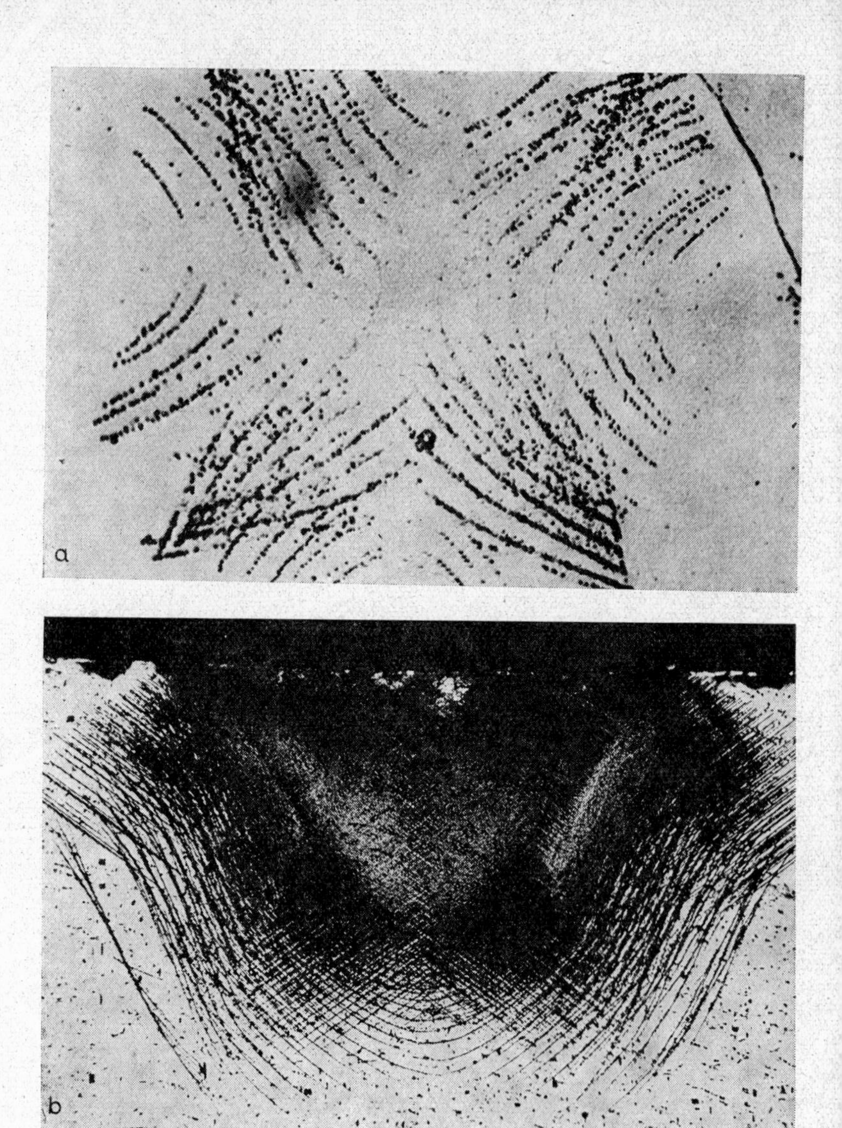

Fig. 4.8. Curved slip lines on the surface of a Fe–3% Si crystal; etch figure obtained in the case of an inhomogeneous stress distribution by a deformation induced by pressing a sphere onto the (100) surface. (a) Section parallel to the pressed (100) surface. (b) Section normal to the previous one in the (011) plane. The traces of four $\frac{1}{2}\langle 111 \rangle$ glide systems can be observed (B. Šesták)

Coulomb interaction, glide directions in which ions of the same charge would become immediate neighbours are not realized (e.g. $\langle 100 \rangle$ in NaCl).

In the case of strongly inhomogeneous stress distribution it may occur that the glide starts not along plane surfaces, but along cylindrical surfaces, parallel to the Burgers vector (Fig. 4.8).

4.6. Glide and climb

The movement and the glide of dislocations have so far been mentioned several times. We shall now specify these movements more exactly.

The concept of glide was always understood above as a motion whereby certain parts of the material slip relative to each other along certain planes (or cylindrical surfaces). The slip is not simultaneous; the boundary between the slipped and unslipped region is a dislocation whose Burgers vector points in the direction of the glide, and consequently the Burgers vector is parallel to the slip surface.

However, a quite different type of motion is also conceivable, and for this reason it is necessary to define the concepts more precisely.

If a segment $d\mathbf{l}$ of the dislocation moves a distance $d\mathbf{s}$, it sweeps the surface element $d\mathbf{f} = d\mathbf{s} \times d\mathbf{l}$. The vector $d\mathbf{f}$ is a normal vector of the new surface element (Fig. 2.2) on that side where the material moves in the direction of the Burgers vector. If the Burgers vector (and thus the displacement of the material too) is in the plane of the surface element ($\mathbf{b}\, d\mathbf{f} = 0$), the cut surfaces move parallel to each other. This type of motion is called *glide* or *conservative motion* of a dislocation. In order to start and preserve this motion, only the internal friction forces must be overcome (Peierls potential, effect of impurities). The motions discussed so far were of this type.

If the Burgers vector has a component normal to the slip surface ($\mathbf{b}\, d\mathbf{f} \neq 0$), the atoms originally situated on the two sides of the surface element $d\mathbf{f}$ are also displaced perpendicularly to this surface element. Thus the motion involves a volume change, since depending upon the direction of the displacement it either squeezes originally adjacent volume elements into each other or separates them so that a new empty volume element is created between them. This is the *climb* or *non-conservative* motion of a dislocation. The magnitude of the volume change is determined by the product of the displacement and the surface element. However, if the vector **b** points in the direction of $d\mathbf{f}$ ($\mathbf{b}\ d\mathbf{f} > 0$), the opposite cut surfaces are squeezed into each other and so the volume decreases. Therefore,

$$\Delta V = -\mathbf{b}\ d\mathbf{f} = -\mathbf{b}\ (d\mathbf{s} \times d\mathbf{l}) = \mathbf{b}\ d\mathbf{l}\ d\mathbf{s}. \tag{4.21}$$

Physically, this means the absorption or production of a series of vacancies or interstitial atoms. This process involves a fairly considerable energy input (the activation energy of these point defects is of the order of magnitude of 1 eV), and consequently any non-conservative motion can be initiated by external forces only with difficulty. However, if the concentration of vacancies does not correspond to the thermal equilibrium for some reason, the dislocations (accompanied by non-conservative motion) become vacancy sources or sinks. Such a motion of the dislocations may also be described in that the point defects exert an "osmotic pressure" on the dislocation line; in the case of vacancies the effect of this is given by the force [20]

$$d\mathbf{F} = (\mathbf{b} \times d\mathbf{l}) \frac{kT}{b^3} \ln \frac{c}{c_0}, \tag{4.22}$$

where c and c_0 are the momentary and the equilibrium vacancy concentrations, respectively. This force is normal to the glide plane and corresponds to the stress

$$\sigma_v = \frac{kT}{b^3} \ln \frac{c}{c_0}. \tag{4.23}$$

Since $\ln c/c_0$ may be 10 and $\mu b^3 \approx 10$ eV, and at room temperature $kT \approx 1/40$ eV, $\sigma_v \approx \mu/40$. Consequently, the effective stress originating from a sufficient vacancy supersaturation can exceed the usual external stresses; this means that a non-conservative motion can take place against external forces too. In special cases the vacancies produced during the non-conservative motion of one segment of the dislocation are absorbed by another and the motion of the total dislocation is finally conservative. This is *conservative climb*. The dislocation loops produced during the condensation of vacancies can move in this way (Section 6.9).

It follows from the definition of conservative motion that motion of a pure screw dislocation is always glide, because **b** and $d\mathbf{l}$ are parallel and thus no volume change takes place (in the surroundings

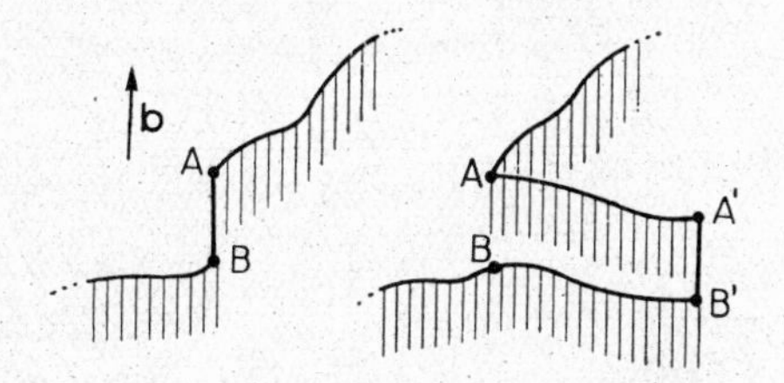

FIG. 4.9. The cross-slip of a pure screw dislocation. The streaks demonstrate the slip plane

of a screw dislocation the deformation is pure shear). In any other case, however, only one surface can be found on which the dislocation moves by glide. The unit vector normal to this surface is given by the equation

$$\mathbf{n} = (\mathbf{b} \times d\mathbf{l}) \, |\mathbf{b} \times d\mathbf{l}|^{-1}. \tag{4.24}$$

Let us finally examine a dislocation which has a pure "screw" segment (Fig. 4.9, section AB). The whole dislocation moves on a slip surface. Nevertheless, under suitable conditions the screw segment may leave this surface and continue to slip on another one while pulling new edge dislocation segments with it (Fig. 4.9, sections AA' and BB'). This is the *cross-slip*.

It has been seen that more than one glide system may exist in a crystal. In a face-centred cubic crystal for example there are twenty-four (four $\{111\}$ planes, and in every plane six $\langle 110 \rangle$ directions), whereas in a close-packed hexagonal lattice there are six glide systems. Of the possible glide systems, that one along which the shear stress has a component larger than the critical value is realized. Naturally, it may be that this condition is satisfied for several glide systems. In this case the slip starts along each glide system (double glide, multiple glide). The fact that fcc metals are easier to deform than hexagonal ones can be explained by multiple glide.

4.7. Jogs on dislocations

It has been pointed out several times that a dislocation may in principle be any three-dimensional curve. However, according to the Peierls model the dislocations formed and slipped during the deformation are generally plane curves lying in the glide planes of closest

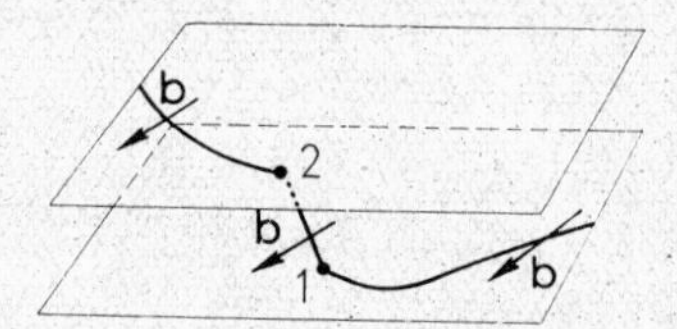

FIG. 4.10. Jog on a dislocation

packing. It frequently occurs that one segment of a dislocation lies in another plane parallel to the original one, and at a distance of only one or a few atomic distances (Fig. 4.10). In this case the two plane curves are connected by a segment with a different orientation but the same Burgers vector. This segment forms a jog. Naturally, the glide plane of the jog differs from that of the original dislocation and for this reason, because of their non-conservative motion, their role, as will be seen later, must be considered separately.

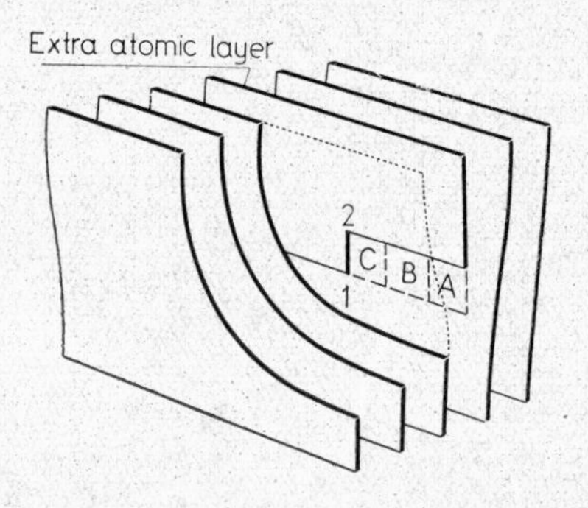

FIG. 4.11. Jog formation with vacancy absorption on an egde dislocation or with the emission of interstitial atoms

Jogs can be created in several ways which differ considerably from each other. As regards the final result of the jog-forming processes, however, there is practically no difference between them.

One of the most important processes is the absorption and emission of point defects (vacancies, interstitial atoms). Consider, for example, a pure edge dislocation. If the point defects arrive from the surroundings of the dislocation to the core by diffusion, they are easily incorpo rated at the edge of the surplus atomic layer (Fig. 4.11, sites ABC . . .) The relevant segment of the dislocation moves in this case by non-conservative motion to a glide plane lying one atomic distance higher and a jog is created in the segment 1–2. The emission of an interstitial atom yields the same result. If vacancies are emitted and/or interstitial atoms are absorbed, the jog created is of opposite direction. Since diffusion processes are rather slow in most materials at room temperature, this jog formation mechanism is important only at high temperatures.

Moving intersecting dislocations may create jogs on each other if their Burgers vectors have components perpendicular to glide planes of the other dislocation. The direction of a jog always corresponds with the direction of the Burgers vector (i.e. its perpendicular component) of the intersecting dislocation. During the slip of the dislocation the two sides of the glide plane are displaced relative to each other by **b**. Consequently, if one dislocation intersects the other, the two parts of the intersected dislocation are displaced relative to each other by the Burgers vector of the intersecting dislocation. Naturally only the displacement perpendicular to the slip plane creates a jog, because the parallel component results only in the overtaking of one part of the dislocation by the other one for a short time during their glide.

In addition to these two mechanisms, the cross-slip of a segment of a pure screw dislocation creates jogs quite frequently, while the spatial (cone) Frank–Read sources (Section 5.2) may also produce jogged dislocations in some circumstances.

Finally, it should be mentioned that according to Kuhlmann-Wilsdorf (as was pointed out in connection with the Peierls model) due to the thermal motion of the atoms, there is some uncertainty in the position of the dislocation line, and hence the position of the glide plane too [21], which may result in a spontaneous jog formation.

Let us now examine the motion of the jogs, i.e. the motion of dislocations containing jogs. Consider a jog on a mixed dislocation as depicted in Fig. 4.12a. It has been mentioned that the glide plane of a jog differs from that of the original dislocation; during slips, the motion of the jog remains conservative only if its displacement is at the angle ϑ with the direction of the dislocation (ϑ is the angle between the Burgers vector and the dislocation line). For this reason it remains in line with the dislocation only if its velocity is $(\sin \vartheta)^{-1}$ times greater than that of the dislocation. If the dislocation has a strong screw character the jog has to move at a very high speed (it tends to infinity in the limit $\vartheta \to 0°$). It is known that the velocity of a dislocation cannot exceed the velocity of sound because the kinetic energy of the dislocation would become infinitely large. On the other hand, the rapidly moving dislocation transfers much energy to its surroundings (see Chapter 2). Thus the presence of jogs always increases the stress necessary to initiate the slip of the dislocation.

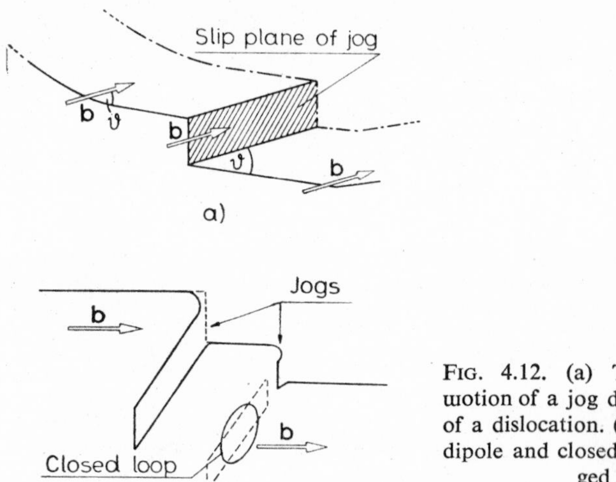

FIG. 4.12. (a) The conservative uιotion of a jog during the motion of a dislocation. (b) Formation of dipole and closed loop by a dragged jog

In the case of pure edge dislocations ($\vartheta = 90°$) there are no such problems, but along the dislocation the jog can move only with non-conservative motion. In contrast, for a pure screw dislocation ($\vartheta = 0°$), any displacement of which is a slip, the jog is an edge dislocation segment which cannot slip together with the dislocation. Here, of course, only motions not parallel to the jog are considered. Otherwise one is concerned with a simple kink. Along the screw dislocation, on the other hand, the jog always moves by slip, even if the dislocation itself is at rest.

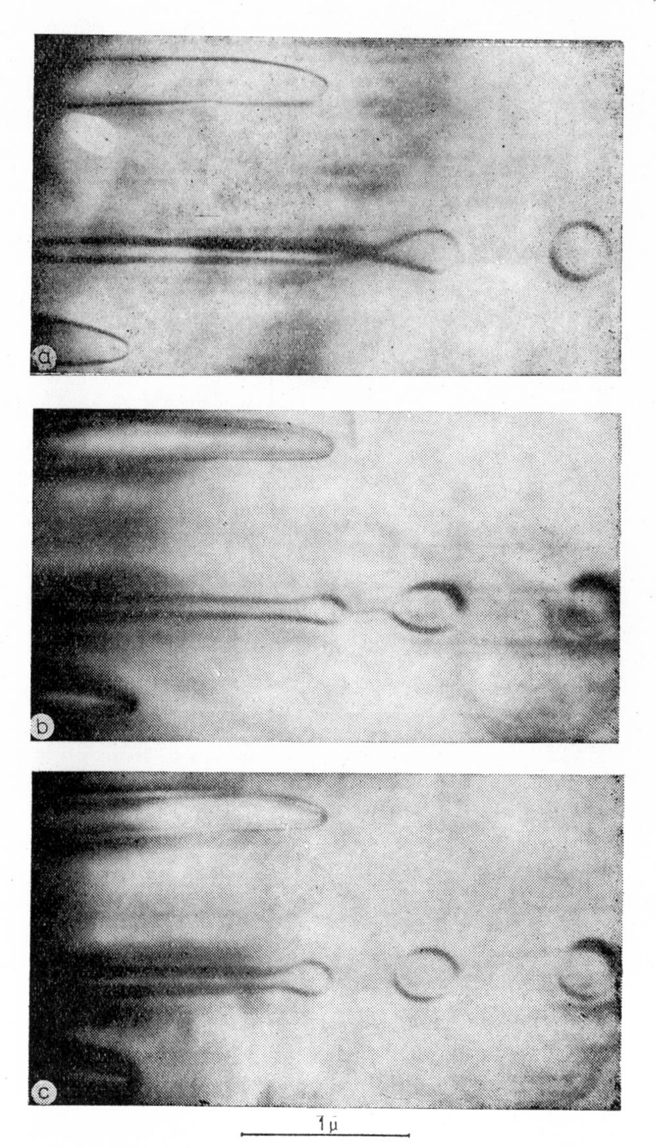

FIG. 4.13. The formation of closed dislocation loops due to a jog dragging in a Zn crystal. Transmission electronmicrogram [Price, *Phil. Mag.* **6** (1961) 499]

Thus, in general during the motion of the dislocation, the jog move by causing volume changes at least in part. Because of the energy needed to create point defects, this motion results in a further increase of the shear stress. Three cases can be distinguished:

1. The force acting on the dislocation is large enough to cover the energy necessary to produce the point defects without the help of thermal activation. In this case the jog moves perpendicularly to the dislocation line, leaving behind an approximately uniform line of point defects. If the height of the jog is equal to b (elementary jog) and the volume of a point defect is assumed to be βb^3 ($\beta \approx 1$), then the average distance between the point defects in the direction of the motion is $\beta b/\cos \vartheta$.

2. If the thermal energy of the lattice also needed to create the point defects, there must be an average "waiting time", which depends on the temperature, for the creation of every single point defect. During this time the jog may move in the permitted glide plane. For a pure screw dislocation the jog may slip along the dislocation in both directions, and the point defects arising from the same jog are situated along a zigzag line and have a rather disordered distribution. Their average distance perpendicular to the dislocation is βb.

3. If the energy needed is very large (in the case of large jogs), the jog cannot move and lags behind the moving dislocation which has predominantly a screw character. During this motion it pulls behind it a tail consisting of two dislocations of opposite signs (dislocation dipole). This dipole later breaks off in the form of a closed loop (Fig. 4.12b, and Figs 4.13a, b, c). During this process the shape of the loop clearly changes as a result of the line tension (Section 2.8), but the enclosed area remains constant.

4.8. The role of dislocations in crystal growth [22]

A detailed theory of crystal growth is not treated in this section; only those problems are touched upon which are closely connected with the concept of dislocations and which were the first experimental proofs of the dislocation theory.

Let us begin with the non-crystalline phase, which may be a solution, a melt or a vapour. In every case there exists a well-defined temperature of transformation T_0. If $T > T_0$ the free energy of the non-crystalline phase is smaller, and if $T < T_0$ the free energy of the crystalline phase is smaller, and consequently this phase is more stable. Of course, this is valid only if the whole material is in either one or the other state. Let us assume that the non-crystalline phase has been overcooled somewhat below T_0. The new phase appears in the form of a smaller or larger crystalline group of atoms. However, this results in a new boundary surface, the creation of which requires a considerable energy input because of the surface tension. It is easy to see that if the crystalline dimensions are small the free energy difference between the crystalline and non-crystalline phases cannot cover the energy necessary to form this new surface. For this reason these formations are not stable below a certain critical dimension, and if small crystals are formed in some way they soon disintegrate. If one succeeds in attaining this critical dimension at which the change of the total free energy with increasing dimensions is not positive, the further growth of the "nucleus" so developed results in energy release, and a very rapid crystal growth sets in.

Naturally, any other (possibly foreign) material introduced from outside (seed) can fulfil the role of the stable nucleus, because this too ensures the new surface. In most cases (because of the impurities) this actually does occur. If no seed exists, a new nucleus can be formed only at the cost of some local energy excess by energy fluctuations. The probability of nucleation depends upon the degree of overcooling; it becomes vanishingly small at a low degree of overcooling, i.e. supersaturation. Nevertheless, small crystalline groups of short life-

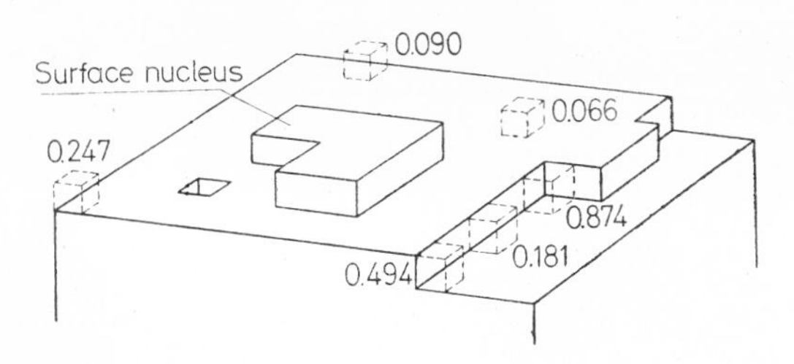

FIG. 4.14. Energy released during the incorporation of individual new ions in NaCl-type ionic crystals, depending on the site of incorporation. The energy released by uniting one singular positive and negative ion is regarded as unity [after Malicskó, L., *Fiz. Szemle* **12** (1962) 172]

time form further, which possibly grow to a stable nucleus before their disintegration.

Kossel and Stranski showed (1929–30) that if a nucleus already exists, further particles are incorporated into it to continue incomplete atomic rows or layers. In this way the energy gain is larger because of the effect of the nearest neighbours. Figure 4.14 depicts a model of possible sites of growth and the energy gain connected with them for NaCl-type ionic crystals. (As a result of the ionic bonding, the energy gain is larger at the corners and along the edges of the crystal.)

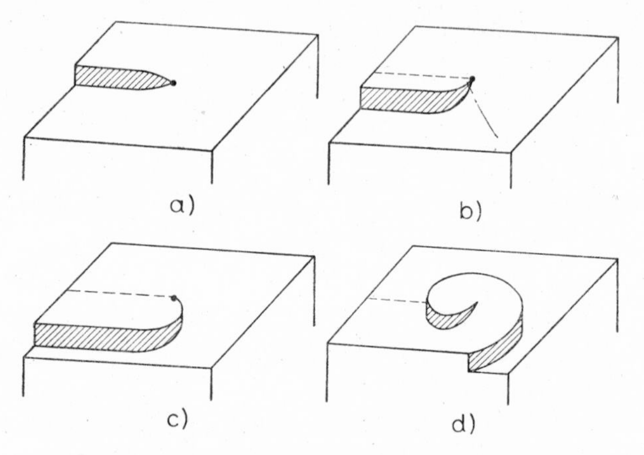

FIG. 4.15. The formation of a growth spiral

The formation of a new layer or a two-dimensional surface nucleus is less favourable than the continued growth of already existing layers of surface nuclei. Burton *et al.* [23] found in 1951 that the formation of a surface nucleus (when the crystal is growing from solution) begins only if a supersaturation of about 25 – 50% has been attained. In practice, however, crystals usually grow quite well at a supersaturation of even 1%. This and some other observations show that in most crystals the steps to which the new material easily adheres are permanent, and grow together with the crystal. If no steps exist the nucleus cannot grow at low supersaturation.

As was shown by Frank, the most simple permanent step is formed by the surroundings of a screw dislocation emerging at the surface

(Fig. 4.15a); here the layer can always be continued. The rate of growth is approximately the same, at every part of the surface and therefore the central part advances whereas the peripheral parts with much smaller angular velocity lag behind (Figs 4.15b, c, d). In this way a microscopically well-observable growth spiral develops on the surface of the crystal (Fig. 4.16). The step-height (which according to other observations can be identified with the absolute value of the Burgers vector) is generally many times as large as the lattice parameter, and may attain a value of a few hundred angstroms.

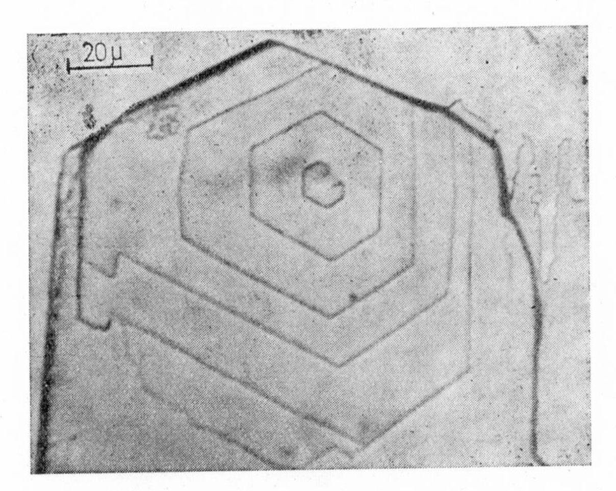

FIG. 4.16. Double growth spiral formed in CdI around two adjacent screw dislocations of opposite sign (Neubauer, I. unpublished)

The formation of dislocations of such large Burgers vectors is still not sufficiently clear (according to the above they cannot be stable). It is generally believed that they are the result of the superposition of several parallel screw dislocations lying close to each other. In some cases such phenomena can undoubtedly be observed [24], but in many instances this is quite clearly not the case (Fig. 4.17). In our opinion the observations can be explained by assuming that the growth does not consist of a series of incorporations of individual atoms (ions), but that the already mentioned unstable groups of atoms, created by fluctuation processes and already containing a certain number of defects, become attached to the nucleus. (This can be well observed on the bubble model of metals.) Even this process does not always result in a perfect fitting, and when the material deposits at the sites of two or more initial nuclei become connected, the lattice planes can meet with a deviation of many atomic distances. The individual elementary dislocations cannot be marked off in the dislocation thus developed. This dislocation structure is stabilized against splitting up by the imperfect fitting of the layers. The moirée

FIG. 4.17. Transmission electronmicrogram of a colloidal gold single crystal obtained by the reduction of $AuCl \cdot 3HCl \cdot 2H_2O$. Both sides show spiral growth and the hexagonal cavity in the centre is easily observable. The streaks are the results of moiré patterns due to a small displacement of the adjacent layers (Suito and Uyeda [25])

patterns of the layer deposited on each other as shown in Fig. 4.17 demonstrate the considerable displacement between them. Impurities clearly play an important role in the mechanism discussed. Only a small number of the nuclei grow further in this way. The observations of Suito and Uyeda [25], according to which only 5–10% of the colloidal gold particles investigated showed a spiral growth, correspond with this finding. The rest of the grains consisted of thin lamellae whose lateral growth stopped after the initial large supersaturation.

Further conclusions can be drawn from the study of the holes formed in the centre of colloidal gold single crystals. In the continuum theory a cavity was cut out from the inside, but it was pointed out that because of the atomic structure such a cavity does not appear in crystals in the case of a Burgers vector of the order of 10^{-8} cm. Now, however, the Burgers vector is much larger, and a measurable hole exists in the core of the dislocation. Let us investigate whether this case agrees with the continuum theory.

The decrease of the deformation energy of a screw dislocation can be achieved by the increase of the volume of the cavity. In this case, because of the surface tension, energy must be fed into the crystal to increase the free surface. Consequently, taking into consideration the surface tension γ, the energy per unit length of a screw dislocation containing a hole of radius r at the centre of a cylinder of radius R is

$$U = \frac{\mu b^2}{4\pi} \ln \left(\frac{R}{r} - 1 \right) + 2\pi r \gamma . \qquad (4.25)$$

If a stable radius r_0 is realized

$$\left(\frac{\partial U}{\partial r} \right)_{r=r_0} = 0 . \qquad (4.26)$$

From this condition one obtains

$$r_0 = \frac{\mu b^2}{8\pi^2 \gamma}, \quad \text{or} \quad b = 2\pi \left(\frac{2 r_0 \gamma}{\mu} \right)^{\frac{1}{2}} . \qquad (4.27)$$

According to the measurements of Suito and Uyeda, the average diameter of the holes is $2r_0 = 1020$ Å, and from the step height

$b = 109$ Å. If $\mu = 2.76 \times 10^{11}$ dyne cm^{-2} and $\gamma = 600$–1000 erg cm^{-2}, one obtains $b = 93$–113 Å which corresponds surprisingly well with the values measured. Thus, to such an extent the linear continuum theory has considerable validity. If $b = 3$ Å is substituted into (4.27) one obtains $2r_0 \approx 0.6$ Å.

Whiskers play an important role in the more recent experimental investigations. Whiskers are single crystals, a few millimetres or centimetres long and 0.01–10 μm thick which contain very few (perhaps only one) dislocations. They may be produced in several ways. They grow from the vapour or the liquid phase at very low supersaturation essentially by the screw dislocation mechanism. Their lateral growth is blocked either by contaminants or for geometrical reasons; e.g. the solution is in contact with the open air through small holes at the edges of which the crystallization begins. In this way, a tube-like formation slowly develops in which the material supply is ensured by capillary forces and surface diffusion. (The inner cavity of a whisker does not always correspond to the cavities of the continuum theory.) In any case, whiskers which have been grown under favourable conditions contain only one or two screw dislocations parallel to the growth direction. For this reason their properties are similar to those of perfect crystals. Many experiments confirm that their tensile strength approaches the value of $\mu/30$ calculated for ideal crystals. Because of their small dislocation density, on the other hand, they are difficult to deform plastically. The tensile strength of thicker samples rapidly decreases with increasing diameter, but the amount of plastic deformation up to fracture increases considerably [26, 27].

Whiskers may also grow from the solid state. The growth mechanism, however, is less clear. It is quite sure that spontaneous growth is initiated in most cases by inner stresses of thermal origin, since the thermal expansion coefficients of metals growing spontaneously from a polycrystalline phase depend upon the orientation. Whiskers have been successfully pressed out of polycrystalline metals with pressures of 10^2–10^3 kp/cm^2 [28]. The growth is presumably connected with the formation of dislocation helices [29]. Let us investigate a mixed but nearly pure screw dislocation fixed at two points (perhaps a screw dislocation with a few jogs). If the concentration of the point defects is 1–2% larger than the equilibrium concentration, the dislocation bows out by climb, and if the "osmotic pressure" (Section 4.6) due to the supersaturation is large enough, the climb continues in spite of the action of the line tension and a helical dislocation results (Figs 4.18 and 1.17). If a helix terminates near to the surface, its

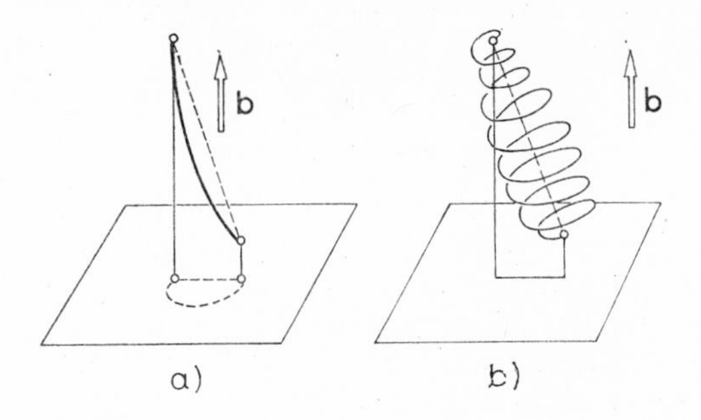

FIG. 4.18. Formation of a dislocation helix

individual threads repel each other, whereas the free surface "attracts" them. Thus, if other factors do not inhibit it, the helix moves towards the surface. Whenever a thread reaches the surface, one atomic layer of material emerges from the crystal. A whisker developed in this way is free of stresses, and consequently in the case of suitably large inner stresses the process is accompanied by energy release. Material transport is ensured by internal diffusion, and if the point defect surplus is maintained, together with the motion new threads are continually generated on the lower section of the helix.

As a result of his experiments with the bubble model of metals, Sines suggests a mechanism based on a cooperative motion of edge dislocations [28]. However, there is as yet no direct experimental evidence regarding this mechanism.

Chapter 5

Multiplication of dislocations

5.1. Sources of dislocations

It follows from what has been said above that the plastic deformation of crystals is brought about by the motion of dislocations. Dislocations are always created during the process of crystal growth. Nevertheless, observations reveal that their movement alone does not account for the deformation.

1. Most well-annealed (easily deformable) crystals do not contain as many dislocations as are needed for a large-scale deformation. Let us assume that 50% shear deformation is produced on a 1 cm cube. If a glide of 2.5 Å takes place for each dislocation, in the case of a homogeneous deformation 2×10^7 (= 0.5 cm/2.5 Å) dislocations must sweep the whole surface of 1 cm^2 (Fig. 5.1). Considering that only a fraction of the dislocations can slip in the proper direction, and even then only a part of the total cross-section is swept, a dislocation density of at least 10^8–10^9 per cm^2 is required for this shear.

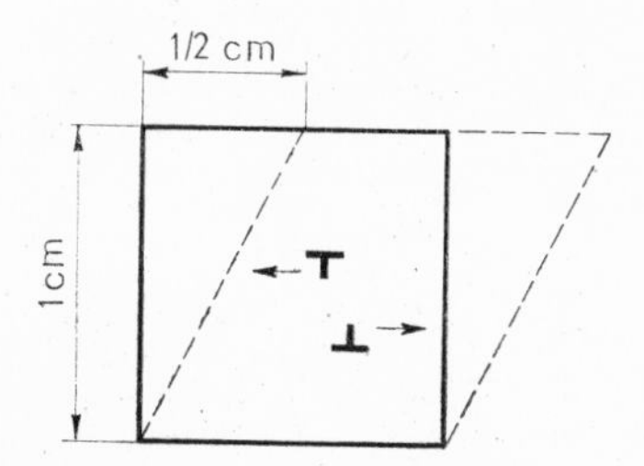

FIG. 5.1. 50% shear deformation of a 1 cm cube

2. The deformation of a single crystal is inhomogeneous, i.e. the glide along some atomic planes is a multiple of the atomic distance, whereas along other planes no glide takes place at all [1].

3. Because of their interaction with impurity atoms (Section 3.5) the "as grown dislocations" developed during crystal growth or annealing can move during deformation with difficulty, or not at all

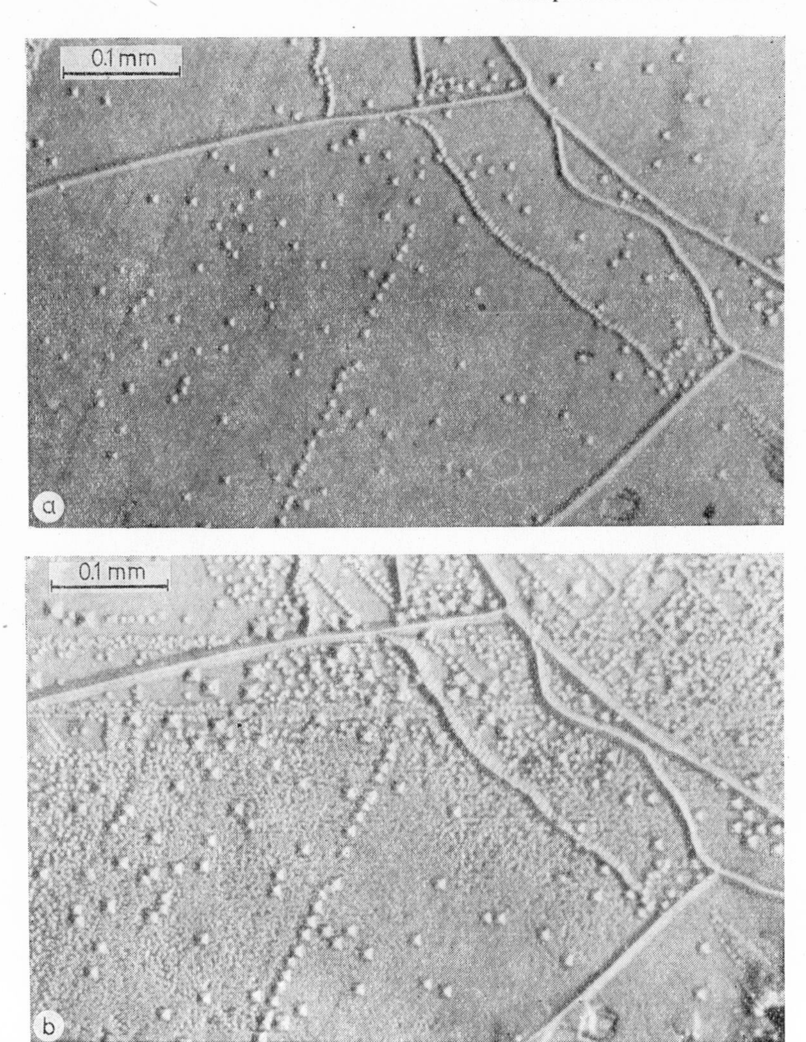

FIG. 5.2. The etched surface of a NaCl crystal (a) before deformation and (b after strong deformation. The pits produced by the first etching were practically all further developed by a second etching (Harta, E., Thesis, ELTE,* 1961)

* ELTE: Eötvös Loránd University, Budapest

(Figs 5.2a and b). New dislocations are created, however, which are also distributed inhomogeneously along the glide planes.

4. With the considerable increase of the number of dislocations an increasingly larger stress is needed to deform the crystal.

All this points to the fact that deformation occurs with the formation of a large number of dislocations, and the observed phenomena are caused primarily by these new dislocations. These, however, do not nucleate spontaneously and homogeneously (by the impact of external stresses or thermal energy fluctuations, see Section 2.7), but with the participation of certain preferred sites, "sources". According to scattered observations, dislocations may be created by stress in the immediate surroundings of microcracks or foreign inclusions with diameters of a few thousand angstroms, but the experimental phenomena cannot fully be explained by these. In 1950 a model to explain the multiplication was proposed by Frank and Read [2], and this has been the basis of all later ideas.

5.2. The geometry of the Frank–Read source

As an example let us investigate a crystal in which a dislocation segment of length l can move in the $[10\bar{1}]$ (101) glide system. Let us further assume that the end-points A and B of the segment are fixed (these fixed points are nodes), or that the direction of the continuing dislocation line is such that the external stress does not produce any force component acting in the slip plane. If a tensile stress in direction x now operates on this crystal, then

$$\boldsymbol{\sigma}\,\mathbf{b} = \left(\sigma_{xx}\frac{b}{\sqrt{2}}, 0, 0\right). \tag{5.1}$$

This vector is parallel to the x-axis, and thus if the line element of the dislocation lies in the xy-plane, the force exerted is normal to the slip plane.

For an edge dislocation in the (101) plane $\Delta\mathbf{l} = (0, \Delta y, 0)$, and thus a force

$$\Delta\mathbf{F} = (\boldsymbol{\sigma}\,\mathbf{b})\times\Delta\mathbf{l} = \left(0, 0, \sigma_{xx}\frac{b}{\sqrt{2}}\,\Delta y\right)$$

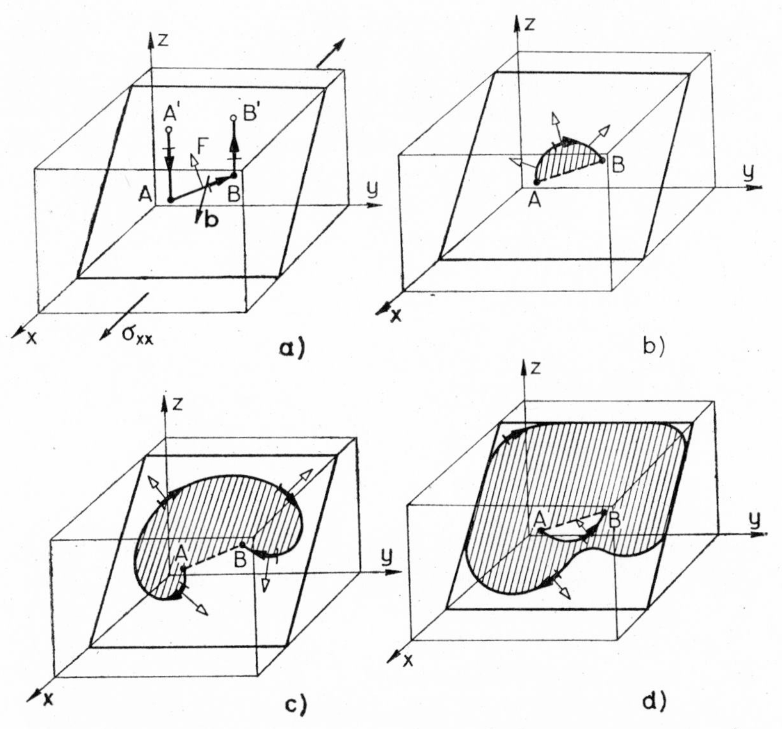

FIG. 5.3. Frank–Read source in the (101) plane. Burgers vector: →, force operating on the dislocation: —▷, line element: —▶. The shaded area indicates the slipped area

is exerted whose component in the slip plane is

$$\Delta F_b = \Delta \mathbf{F}\frac{\mathbf{b}}{b} = -\frac{1}{2}\sigma_{xx}\, b\Delta y\,. \tag{5.2}$$

The negative sign refers to the fact that the direction of the force is opposite to that of the Burgers vector. Similarly, the force on a screw dislocation (perpendicular to $\Delta\mathbf{l}$, i.e. in the y-direction) is

$$\Delta F_y = -\frac{1}{2}\sigma_{xx}\, b\Delta z\,. \tag{5.3}$$

The calculations are similar for a mixed dislocation segment as depicted in Fig. 5.3.

The force causes the AB segment to bow out increasingly, and if the external stress is sufficient to overcome the line tension, a loop gradually develops; this first extends sideways, and later bends backwards on itself (Fig. 5.3). When the two parts of the loop touch at

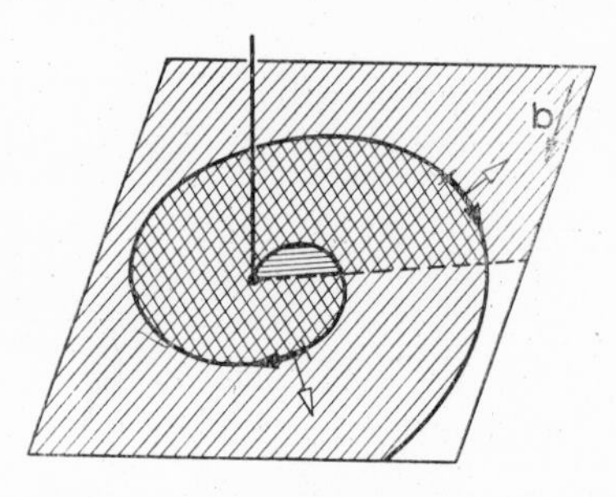

FIG. 5.4. Frank–Read source from L-shaped dislocations

the opposite side of the segment AB, the slipped parts meet. For this reason the dislocation vanishes along the line of contact. In this way a closed dislocation loop is obtained with a dislocation similar to the previous one, between the points A and B, and the whole process starts over again. By this mechanism a series of concentric, continually expanding dislocation loops is produced.

If the dislocation is attached at only one point, dislocation spirals are created by a similar mechanism (Fig. 5.4). Figure 5.5. shows an X-ray topograph of such a spiral source.

If A and B are nodal points, the Burgers vector of the connected dislocations may have, and in most cases actually have, a component normal to the plane of the source. Because of the conservation of the Burgers vector the dislocation network can be resolved, for example, according to Fig. 5.6. When therefore a Frank–Read source begins to operate, and the loop sweeps round the dislocation $A'A''$, according to the sense of the Burgers circuit, after one revolution it arrives at another slip plane displaced by the perpendicular component of b_1. If the perpendicular component of the dislocation $B'B''$ is just equal to the same component of $A'A''$ but is of opposite direction, the displacements of the two lines are the same, and the loops close again. The next cycle begins to operate on another plane parallel to the previous one. This is the spatial (cone) Frank–Read source. In every other case the two parts of the loops bending backwards lie in different planes. As long as the distance between them is not too large,

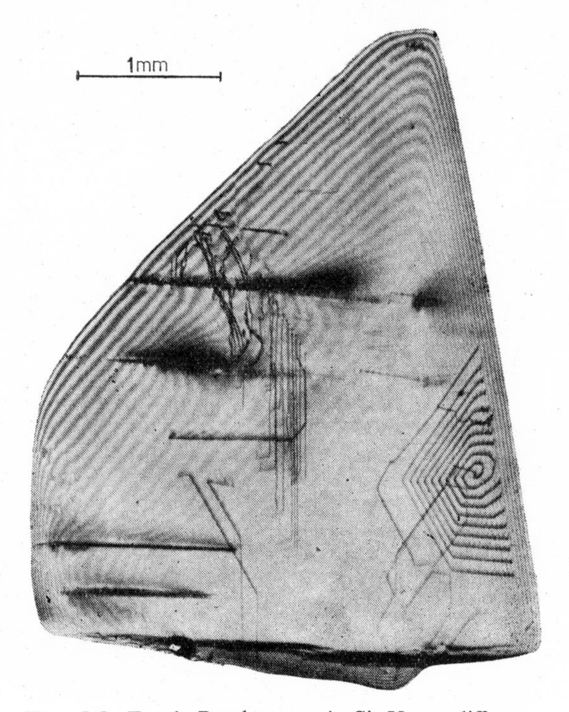

FIG. 5.5. Frank–Read source in Si. X-ray diffraction topogram with AgK α-radiation and the 111 reflection [3]

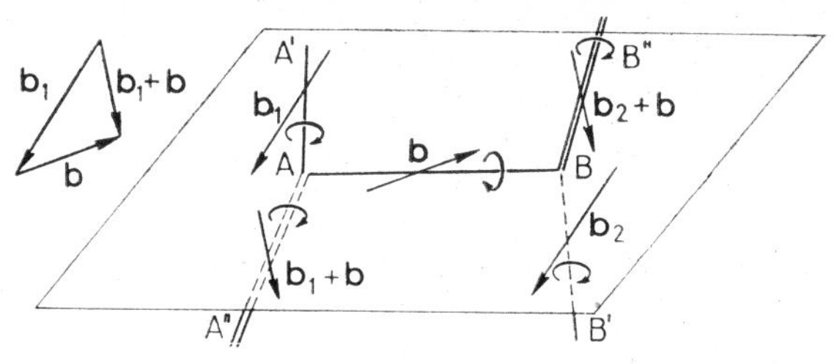

FIG. 5.6. Configuration resulting in a spatial (cone) Frank–Read source

on meeting, the screw components join by cross-slip. Thus, a jog is produced both on the new loop and on the dislocation starting a new cycle. The number of jogs (or their size) naturally increases after each cycle until the two branches are so separated that in practice they operate as separate spiral sources.

Similar results are obtained when closed loops starting from an otherwise plane source intersect a screw dislocation (or parallel screw dislocations).

If it is assumed that the sources are produced mainly from the elements of the dislocation network, it is quite correct to conclude that plane sources are relatively rare. In Section 5.4, however, it is shown that once a slip has started, new plane sources may be created.

5.3. Stress required for the operation of a Frank–Read source

Let us investigate the causes which inhibit the operation of a source, and the stresses which are required to overcome them. The Peierls force always present whenever a dislocation moves, is not taken into consideration here because it is so small that no error is committed by omitting it. The line tension, on the other hand, is extremely important. This has already been discussed to some extent in Section 2.8. According to our results, the smallest stress necessary to overcome the line tension of the source of length l is

$$\tau_F = \frac{2U(\vartheta_1)}{bl \sin(\vartheta_0 + \vartheta_1)}, \tag{5.4}$$

where ϑ_0 is the angle between the initial dislocation segment and the Burgers vector, and ϑ_1 is the angle between the tangent and the Burgers vector at the fixed end points in the case of critical bowing out (Fig. 2.14). $U(\vartheta_1)$ is the energy per unit length of the dislocation at the fixed points. Its value may be taken to a good approximation as equal to the energy per unit length of a circular slip loop of radius $l/2$ at the point where the tangent has direction ϑ_1. By applying the relation (2.42)

$$\tau_F = \frac{\mu b}{2\pi l \sin(\vartheta_0 + \vartheta_1)} \left(\cos^2\vartheta_1 + \frac{\sin^2\vartheta_1}{1-\nu}\right)\left(\ln\frac{4l}{r_0} - 2\right). \tag{5.5}$$

Let the original segment be a pure screw dislocation. In this case $\vartheta_0 = 0$ and $\vartheta_1 = 90°$. With these values

$$\tau_F^{\text{screw}} = \frac{\mu b}{2\pi(1 - \nu)l}\left(\ln\frac{4l}{r_0} - 2\right). \tag{5.6}$$

For a pure edge dislocation $\vartheta_0 = 90°$ and $\vartheta_1 = 0$, therefore

$$\tau_F^{\text{edge}} = \frac{\mu b}{2\pi l}\left(\ln\frac{4l}{r_0} - 2\right). \tag{5.7}$$

One may see that the bowing out of a pure screw dislocation requires a $1/(1 - \nu) \approx 1.5$ times larger stress than for edge dislocation. If the anisotropy is also taken into account this value is altered to a small extent; thus, for instance, in the case of copper the ratio increases to 1.76 [4]. This means that the operation of a source is determined in practice by the bowing out of the pure edge dislocation.

The logarithmic function changes only slowly and consequently no serious error is committed if the quantity $\ln 4l/r_0$ is regarded as constant while $1/l$ changes. Equation (5.7) can then be written in the following form:

$$\tau_F = \alpha\frac{\mu b}{l}, \tag{5.8}$$

where $\alpha \cong 1 - 1.3$, if $l = 10^3 r_0 - 10^4 r_0$. Since $r_0 \approx b$, the stress necessary for the Frank–Read sources to operate, that is the stress needed for a macroscopic glide is obtained from equation (5.8) as $\tau_F \approx 10^{-3}\mu$, which is in good agreement with experiment.

The stress-fields of the loops emitted by an active source inhibit the operation of this source. According to equation (2.49) the stress required to keep a loop of radius R in equilibrium is

$$\tau' = \frac{\mu b(2 - \nu)}{8\pi R(1 - \nu)}\left(\ln\frac{8R}{r_0} - 1\right) \approx \frac{\mu b}{R}. \tag{5.9}$$

[This is just the average of the stresses (5.6) and (5.7) with $R = l/2$.] This stress is clearly equal to the internal stress generated within the loop. After the emission of the first loop with radius $R = pl$, the stress

$$\tau = \tau' + \tau_F \cong \frac{\mu b}{l}\left(1 + \frac{1}{p}\right) \tag{5.10}$$

is needed to start up a new cycle. Assuming that the closed loops are given off by the source in uniform cycles, and so the radii of the successive loops are pl, $2pl$, $3pl$, ... jpl, after the nth loop the further operation of the source is only possible if

$$\tau > \frac{\mu b}{l}\left(1 + \sum_{j=1}^{n} \frac{1}{jp}\right) \approx \frac{\mu b}{l}\left(1 + \frac{1}{p} \ln n\right). \tag{5.11}$$

Hence it can be seen that the number of the loops scarcely influences the operation of the source; if $n = 10^3$ and $p = 1$, the value of the term in brackets is still only about 8.

The time necessary to emit each loop can easily be estimated by taking into consideration that during the emission the dislocation travels a distance of $pl \approx 10^{-4}$ cm. The velocity of the motion depends upon the nature of the crystalline material, the stress and the temperature. With the stresses occurring in practice, however, this velocity varely exceeds 0.01–0.1 times the velocity of sound, i.e. 10^4 cm/s. However, in the case of Ge for example, rates of 10^{-6}–10^{-1} cm/s

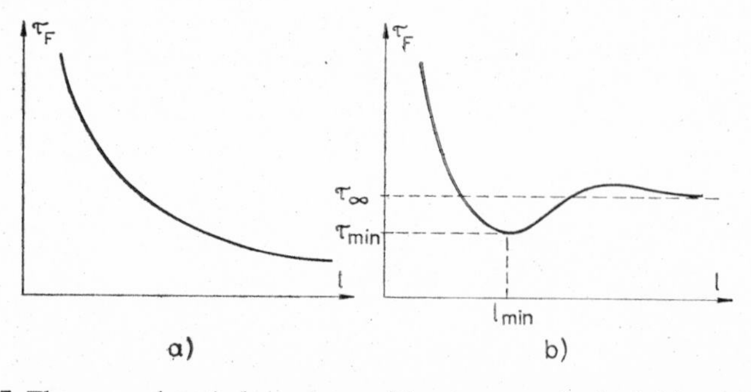

FIG. 5.7. The source-length dependence of the stress required to initiate a Frank-Read source, (a) merely as a result of line tension, (b) considering also the effect of the jogs produced during the intersection of the dislocation network. In this latter case, because of the approximative character of the calculations, the small local maximum of the stress has no physical meaning

have been measured. Such extremely low values occur only rarely. In any case, on calculating with a velocity of 1 cm/s, the time of one cycle is approximately 10^{-4} s. For metals this period is much shorter.

So far no consideration has been given to the fact that the dislocation loop bowed out during its motion intersects the dislocation network, already present in the crystal, and so an increasing number of jogs is produced on the loops. On advancing, these jogs produce point defects on sections predominantly of a screw character. If the jogs could not move along the dislocation, in the case of a sufficiently large loop diameter, the external stress would not be able to supply the energy necessary for the formation of a point defect. (The length of a dislocation increases in proportion to the diameter, and the number of jogs increases as the square of the diameter.) Because of the mobility of the jogs, however, some of them recombine, and after a time their density reaches a saturation value. Orlov [5] has shown that the stress required to initiate the operation of a source depends, in a different way with respect to equation (5.8), upon the source-length (Fig. 5.7). According to calculations for a source of infinite length

$$\tau_F(\infty) = \mu b \sqrt{D\alpha'} \sqrt{\frac{v}{v_j}}. \tag{5.12}$$

where D is the density of the screw dislocations intersecting the slip plane, $\alpha' = U_p/\mu b^3$, and U_p is the average energy of formation of a point defect ($0.1 \lesssim \alpha' \lesssim 1$), v is the velocity of dislocation and v_j the velocity of the jogs on the dislocation.

The stress $\tau_F(l)$ has a minimum if the temperature is so low that

$$v_j < \alpha' v \, \pi/6 \tag{5.13}$$

and then

$$l_{\min} = \frac{8}{\pi D\alpha'}, \tag{5.14}$$

$$\tau_{\min} = \mu b \sqrt{\frac{D\alpha'\pi}{2}}. \tag{5.15}$$

If the velocity of motion of the jogs does not satisfy (5.13), τ_F decreases monotonously to $\tau_F(\infty)$, which also decreases with increasing v_j. Hence, if $\tau_F(\infty) > \tau > \tau_{min}$, only those sources begin to operate whose length falls in a rather narrow interval. Since for $v_j < 0.43\,\alpha' v$ the stress required for the further expansion of the loops already broken away shows a similar course, the sources of the loops and also the operation of the source very quickly stop. The source only really becomes active when the condition

$$\tau > \tau_F(\infty) \gtrsim 1.5\sqrt{\alpha'}\,\frac{\mu b}{\bar{l}} \tag{5.16}$$

is satisfied. This is similar to (5.8) but the source-length is replaced by the average distance of dislocations intersecting the slip plane ($\bar{l} = D^{-1/2}$).

The multiplication so initiated naturally ceases after some time, because the first loop sooner or later runs into some obstacle (e.g. a grain boundary, or a Lomer–Cottrell barrier, see Section 6.6). The consecutive loops thus pile up, and in the case of a large number of loops the internal stress reaction inhibits the operation of the source.

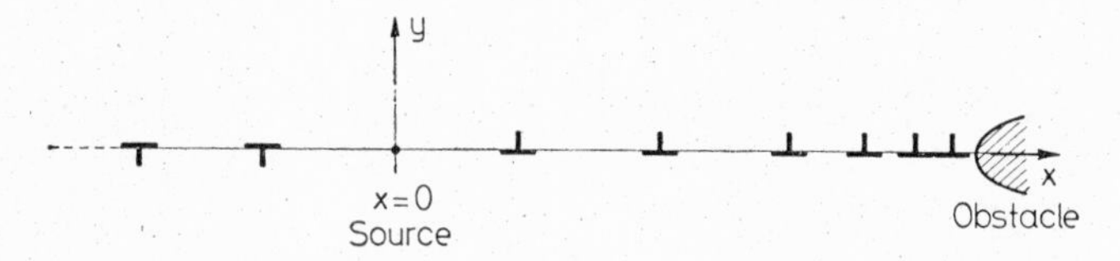

FIG. 5.8. The pile-up of edge dislocations against an obstacle

Since a detailed discussion would be rather lengthy, we merely summarize the results. The simplest case is the "one-dimensional" source, when the dislocations which have moved some distance from the source are regarded not as loops but as straight dislocations. This case occurs in practice in the deformation of thin foils (electron-microscopic investigations). With the obstacle at a distance L from the source (Fig. 5.8), the object of the calculation is to look for a dislocation distribution at which the resultant of the external τ and internal τ' stresses due to the dislocation vanishes at the sites of every dislocation in the interval $(-L, +L)$ around the source (otherwise there could be no equilibrium). Eshelby *et al.* [6] considered a discrete dislocation distribution with dislocation coordinates

$x_1, x_2, \ldots, x_j, \ldots, x_n$. Since the stress-field of every dislocation is of the form $A/(x - x_j)$ [where $A_{edge} = \mu b/2\pi(1 - \nu)$ and $A_{screw} = \mu b/2\pi$], a system of n dislocations is in equilibrium in the presence of an external shear stress τ if

$$\sum_{\substack{j=1 \\ j \neq i}}^{n} \frac{A}{x_i - x_j} + \tau(x_i) = 0\,. \tag{5.17}$$

However, by differentiating the polynomial $f(x) = \prod_{j=1}^{n} (x - x_j)$ (the zero points of which are the values x_j sought), we obtain

$$\frac{1}{f}\frac{\partial f}{\partial x} = \sum_{j=1}^{n} \frac{1}{x - x_j}.$$

Hence the equilibrium condition (5.17) leads to the differential equation relating to the polynomial $f(x)$, and the first derivative of the nth Laguerre polynomial yields $f(x)$. Such a polynomial describing this

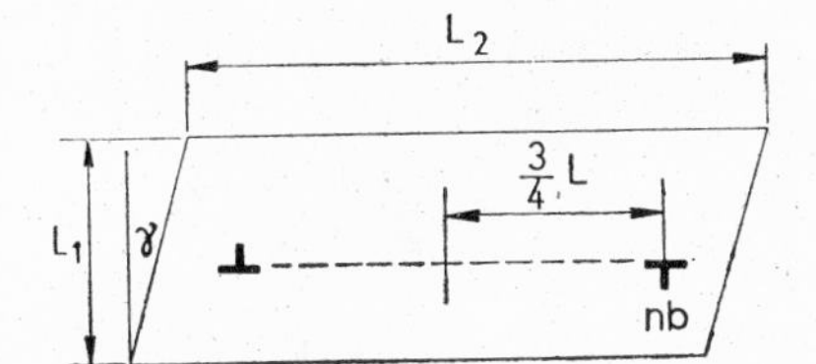

FIG. 5.9. For the derivation of equation (5.19)

distribution of n dislocations is analogous to the polynomial occurring in the radial electron-distribution function of a hydrogen atom in the ns state. According to the above approximation, the number of dislocations formed in the length L by an external stress τ is

$$n = \frac{\pi L \tau k}{\mu b}, \qquad (n \gg 1) \tag{5.18}$$

where $k = 1 - \nu$ for an edge dislocation and $k = 1$ for a screw dislocation. The centre of gravity of the dislocations is at $3L/4$, and hence the slip is as though a single dislocation of the Burgers vector nb had moved a distance $3L/4$ from the source. Consequently, if the dimension of the crystal in the x-direction is L_2, and the distance of the sources perpendicular to x is L_1, the total shear deformation (Fig. 5.9) is

$$\gamma = \frac{3Lnb}{4L_1L_2} = \frac{3\pi}{4}\frac{L^2\tau}{L_1L_2\mu}k\,. \tag{5.19}$$

The above calculations were extended by Mitchell and co-workers to the anisotropic cases [7].

Leibfried [8] characterized the dislocation distribution by a continuous density function $D(x)$. The problem now consists of solving the integral equation

$$\frac{\mu b}{2\pi k}\int_{-L}^{+L}\frac{D(x)}{\xi - x}dx + \tau(\xi) = 0\,, \qquad -L \leqq \xi \leqq L\,.$$

The result is

$$D(x) = -\frac{2\tau k}{\mu b}\frac{x}{\sqrt{L^2 - x^2}}\,. \tag{5.20}$$

The negative sign shows that for a positive τ the Burgers vector points in the $-x$-direction if a line element of positive z-direction has been chosen. The number of dislocations is

$$n = \int_0^L D(x)dx = \frac{2L\tau k}{\mu b}, \tag{5.21}$$

and the shear deformation is

$$\gamma = \frac{1}{L_1L_2}\int_{-L}^{L} bD(x)dx = \frac{\pi}{2}\frac{L^2\tau k}{L_1L_2\mu}\,. \tag{5.22}$$

One can see that the two results scarcely differ. The assumption of straight dislocations naturally means a serious restriction. Leibfried also carried out the calculations for circular loops. The result, similar to the previous one, is

$$D(\rho) = \frac{8}{\pi}\frac{\tau k'}{\mu b}\frac{\rho}{\sqrt{R^2 - \rho^2}}, \tag{5.23}$$

where R and ρ are the radii which replace L and x, and

$$k' = \frac{1 - \nu}{2 - \nu}.$$

Finally, it should be mentioned that in the experimentally observed piled-up rows of dislocations, the distribution does not strictly follow the above relation [9]. The reason for this is clearly the fact that extremely small stresses cannot move the dislocation, partly because of the Peierls stress, and partly because the impurity atoms accumulating in the surroundings of the once stopped dislocations prevent any further movement [10, 11]. The stress-field developing around the piled-up dislocations has been calculated by, among others, Mitchell [12].

5.4. The formation of sources

For several years following the suggestion by Frank and Read, the dislocation network already present in the crystals was regarded as the origin of the sources. With the production of good-quality single crystals, however, many observations were made which contradicted this assumption. It is sufficient merely to refer to the experiments of Johnston and Gilman [13]. These authors found, for instance, that the glide bands of good-quality LiF crystals frequently develop from only one, or a very few dislocation loops which extend from surface to surface, in regions which are perfectly free from any dislocation network. With gradual dissolution of the surface layers the loops were traced back inside the crystal. All the investigations indicated that, in addition to the original loops, many new dislocations with the same Burgers vector as the original one are formed on parallel but not necessarily identical planes during the deformation, and that a glide band is created like an avalanche. Hence the sources of further multiplication are created by the expanding dislocation loops themselves. Similar observations have been made with other materials too [14]. The previously suggested mechanism [15], according to which the sources are developed by the condensation of vacancies, i.e. from dislocation loops of submicroscopic dimensions (R-dislocation, see

Section 6.9), does not seem to play an important part. The formation from a fixed network, on the other hand, is quite possible.

Our present conception as to the formation of the sources can be summarized in the following way. The screw section of a moving dislocation goes over partly or entirely to another parallel plane by a double cross-slip, and thus a smaller or larger jog is created. The cause of the cross-slip is up to now not fully known. With LiF crystals, for instance, it is an experimentally proved fact [13] that the screw dislocations emerging on the surface of the crystal turn with their ends by cross-slip into positions normal to the surface as a result of the surface attraction (Fig. 5.10; see also Section 2.9). Cross-slip

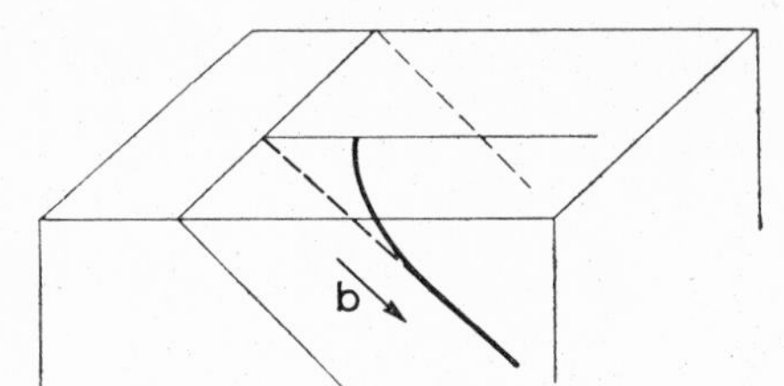

FIG. 5.10. The cross-slip of a screw dislocation near to the free surface

may also occur if the actual stress affecting the dislocation has a component perpendicular to the slip plane, irrespective of whether it originates from an external stress, from internal inhomogeneities or possibly from the stress-field of adjacent dislocations. If the initiating source is not a plane source, some interaction always exists between the stress-fields of dislocations moving on planes which are not too far from each other.

Although the jog developed cannot slip together with the screw dislocation (Section 4.7), the dislocations on the two sides of the jogs move almost independently if the jog is large enough. Thus, the end-points of the jog become fixed points (Fig. 5.11) of single-ended plane sources (Fig. 5.4). Figure 5.12 shows a transmission electron-micrograph of a source of this type [16]. If the jog is not large enough for this, it drags behind itself a dislocation dipole consisting of two parallel edge dislocations with opposite Burgers vectors. Because of the interaction of their stress-fields, these dislocations cannot separate, and in equilibrium their common plane encloses an angle of 45° with the slip plane (Section 2.5.5). According to Fig. 2.7a, the energy required to separate the dipole (if the dislocation is so long that the line tension can be neglected) is

$$\tau \cong \frac{0.25\mu b}{2\pi(1-\nu)y} \approx \frac{\mu b}{20y}, \tag{5.24}$$

where y is the distance of the two dislocations perpendicular to the slip plane. For $y = 50b = 100$–150 Å, from the above relation $\tau \approx$ $\approx 10^{-3}\,\mu$, and for $y = 5\times 10^3 b$, $\tau \approx 10^{-5}\,\mu$. Hence, even 100-Å-wide dipoles become active only as a result of large stresses. For this reason many dipoles perpendicular to the Burgers vector (dislocation frag-

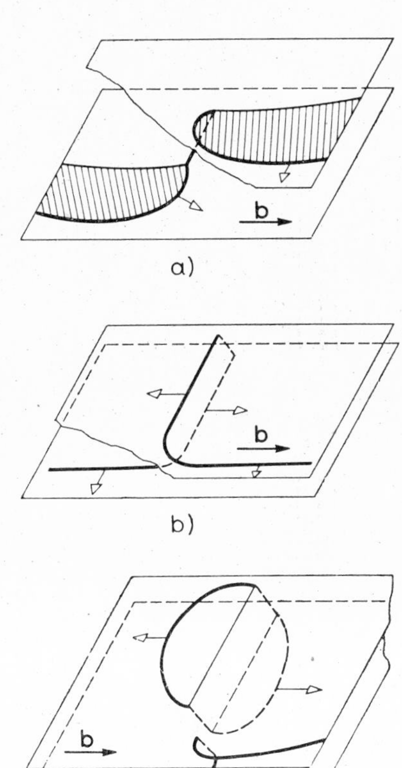

FIG. 5.11. The development of sources in the case of jogs formed on a screw dislocation: (a) large jog, (b) dislocation dipole formed from a smaller jog, (c) disconnected dipole which becomes an active source under the action of a larger stress

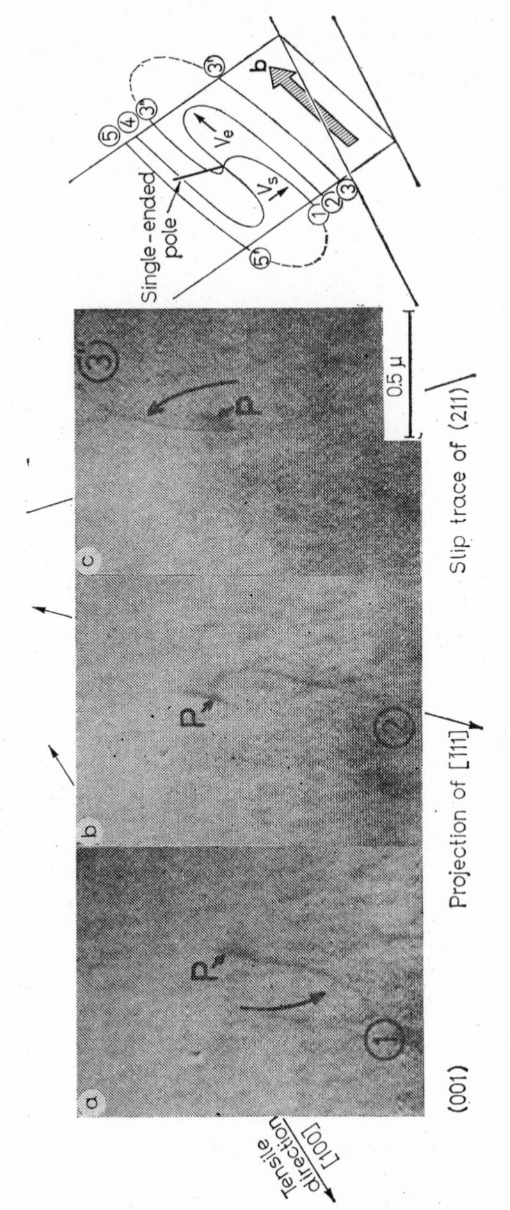

FIG. 5.12. Dislocation multiplication produced by a large jog in iron foil. The pictures were taken with a 500 kV electron-microscope (Saka *et al.* [16])

ments or debris) are left behind the slipped screw dislocation (Fig. 5.13a). This is confirmed in fact by numerous experiments [14, 15]. Figure 5.13b shows the surface of an etched sodium chloride crystal. The acute etch pits were formed at the point of emergence of the dislocations. The large flat-bottomed pits are due to dislocations

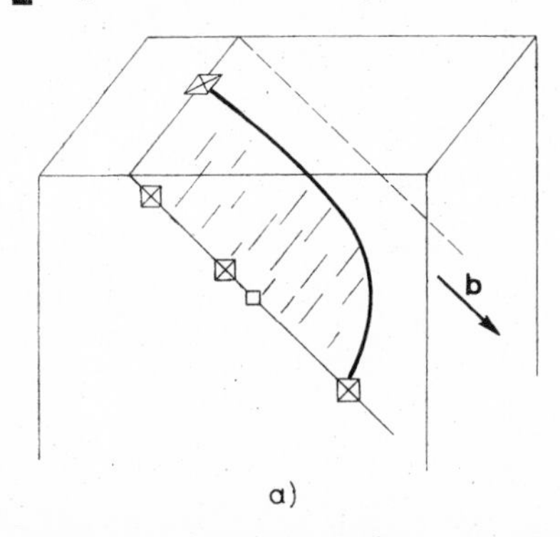

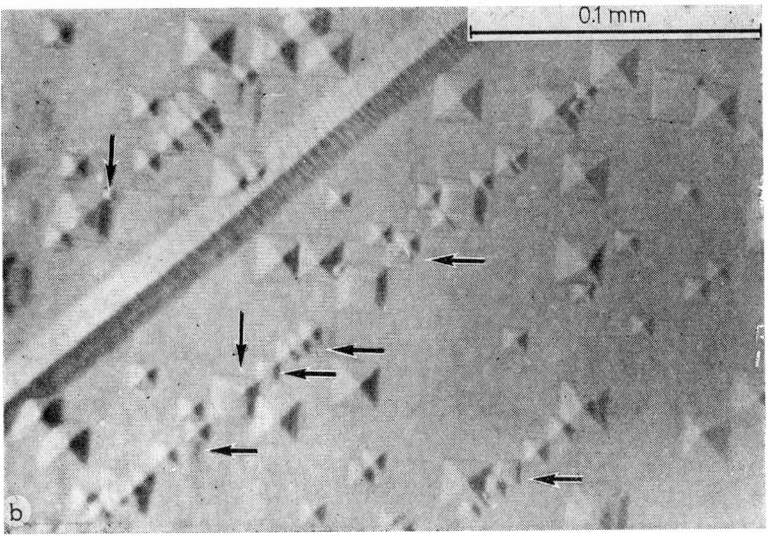

FIG. 5.13. The distribution of "debris" remaining in the wake of a moving screw dislocation (a). Part of the twice deformed and etched surface of a NaCl single crystal (b). The small flat pits indicated by arrows quite certainly originate from fragments (Harta, E: Thesis, ELTE, 1961)

moved away during the deformation between two consecutive etchings. The small flat-etch pits are caused by dipoles which spread to only a very small depth below the surface. The small pits which are visible at the edges of some large ones are also due to dipoles which originally did not reach the surface, but were "developed" by the sideways growth of the large pits.

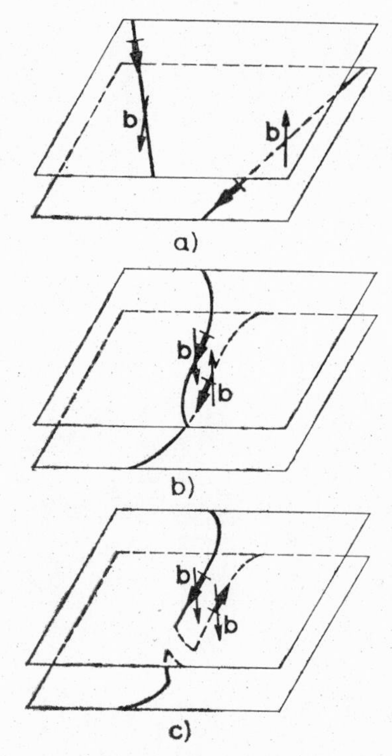

FIG. 5.14. The origin of a dipole according to Tetelman

If the size of the jog is only a few angstroms, instead of dipoles, only a row of point defects is created.

Another interpretation of the formation of dipoles was suggested by Tetelman [17]. In his opinion the cross-slip of a screw dislocation is not probable in the initial stage of the deformation. According to electron-microscopic investigations, screw dislocations were observed in some cases, with tail-like dipoles in both directions. This is difficult to explain with the old concept. Dipoles, however, may also be created if two dislocations of opposite Burgers vectors slipping on two parallel planes approach each other. Depending upon the angle of their inclination, it may be energetically more favourable if two parallel dislocation segments of opposite sign are formed whose stress-fields partially compensate each other (Fig. 5.14b). If one of them has a short pure screw segment, this passes over by cross-slip to another plane as a result of the mutual attraction between the dislocations of opposite sign, and (according to Fig. 5.14c) a dipole and dislocation possessing a jog are produced. There are still not enough experimental observations on this mechanism, but nevertheless it is definitely supported by the fact that for extended dislocations (Section 6.4) it is much more probable than that suggested by Johnston and Gilman.

5.5. Dynamic multiplication

Finally, the concept of dynamic multiplication, introduced by Frank in 1948 [18], is discussed briefly. Let us first investigate a crystal in which the motion of the dislocations is not blocked by obstacles. If the external stress is large enough, the dislocation is accelerated to a high speed and in addition to the energy of the stress-field, acquires considerable kinetic energy; this is the sum of the kinetic energies of the atoms. Every lattice atom transmits its energy by colliding with the next row of atoms. If the dislocation arrives at the edge of the crystal, the last few atomic rows can no longer transmit the energy, and thus they go beyond the original boundary surface dragging a few further rows with them. Hence, these few rows slip one atomic distance further than the rest, producing a slipped area which is again bordered by a dislocation. This dislocation, however, has a sign opposite to that of the first one, and goes backwards under the influence of the shear stress, while the area which has slipped further increases. This process is repeated on the other side of the crystal, and so on.

The process described is, of course, too idealized. Let us imagine, however, that one part of the fast moving dislocation meets an obstacle. The dislocation travels some distance on the two sides of the obstacle, bulges sideways and the two segments meet beyond the obstacle. A similar process occurs on the meeting as that at the edges of the crystal in the previous case. If the kinetic energy of the dislocation is approximately the same as the deformation energy, the volume element which slips last, slips doubly at the junction of the two dislocations. Hence, a closed loop of an opposite Burgers vector is again produced, which expands rapidly and soon reaches the obstacle once more. On the other side of the obstacle the process is repeated. This multiplication mechanism, however, does not seem probable, because according to Section 2.10, a speed approaching the velocity of sound is needed to obtain such large energies. This requires a large free path-length and such a large shear stress ($\tau \gg 10^{-3}\mu$) that the Frank–Read sources begin to operate before this mechanism. For this reason the dynamic multiplication does not occur in general. Only one observation is known where dynamic multiplication presumably occurred [19]. It is probable, however, that this process takes part in the mechanism of fracture.

Chapter 6

Partial dislocations

6.1. Stacking faults in close-packed lattices

For the sake of simplicity let us consider a close-packed lattice made up of spheres of equal size. The spheres can be placed in a close-packed structure in a plane in only one way, whereas a three-dimensional lattice can be constructed in two different ways. If we already

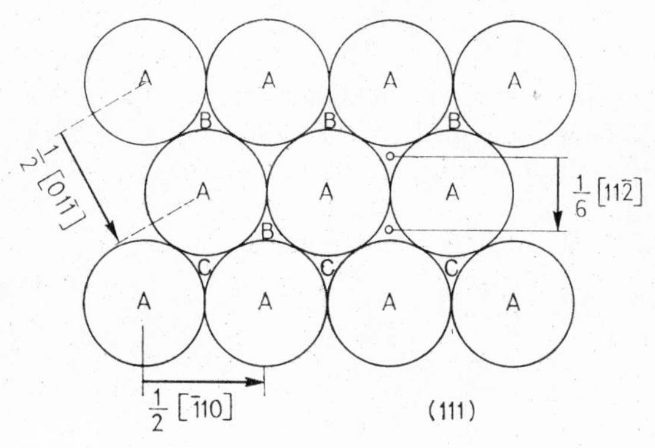

FIG. 6.1. Close-packed layer. (111) plane of the fcc lattice

have a layer—in which the centres of the spheres are denoted by *A* —a further layer can be placed above either of two sets of hollows in the first layer (situations *B* or *C* in Fig. 6.1). If the layers follow each other so that the centres of the spheres are in order over points *A*, *B*, *C*, *A*, *B*, *C* . . . , then a close-packed cubic (face-centred) lattice is obtained, the close-packed layers forming its {111} planes (see Appendix A). However, if the layers are repeated in the stacking sequence *ABABAB* . . . (*BCBCB* . . . *ACACA* . . .), a hexagonal close-packed lattice is formed, and the layers lie in the (0001) plane.

In both lattices, which thus differ from one another only in the sequence of the close-packed layers, every sphere (atom) has twelve immediate neighbours. In the cubic lattice the twelve immediate neighbours are equivalent, whereas in the hexagonal lattice only two groups of six spheres are in equivalent positions.

It is quite obvious that a great energy difference between the two types of stacking order is not possible. This is borne out by the fact that Co, Ni and some other metals, for example, can crystallize in both modifications. Thus it may easily occur that as a result of some effect the otherwise regular stacking sequence becomes disordered, and a *stacking fault* is created, for instance in the following way:

$$\dots ABCAB\overset{\downarrow}{\,}ABCABC\dots$$

At the position denoted by the arrow there is a stacking fault.

Since two layers of identical character cannot follow each other, a more concise notation is obtained by introducing the "stacking operator" which is defined in the following way. If the sequence of two consecutive layers is AB, BC or CA, the symbol describing the transition is Δ, whereas the sequence AC, CB or BA is expressed by the symbol ∇. With these symbols the fcc lattice is $\dots\Delta\Delta\Delta\Delta\Delta\dots$ or $\dots\nabla\nabla\nabla\nabla\nabla\dots$ (these two lattices differ from one another only by an orientation difference of 180° around the $\langle 111\rangle$ axis). The hexagonal close-packed (hcp) stacking is described by the symbol $\dots\Delta\nabla\Delta\nabla\Delta\nabla\Delta\nabla\dots$

What kind of stacking faults are possible in a fcc lattice? If a close-packed plane is removed from the lattice, or a part of the lattice is displaced for instance by vector $\frac{1}{6}\{112\}$ along some $\{111\}$ plane (from situation B to C), then we obtain the sequence

$$\dots\Delta\Delta\Delta\nabla\Delta\Delta\Delta\dots,\ (\dots \overline{ABCA}\,\overline{CABC}\dots) \tag{6.1}$$

in which the stacking order is faulty. On the other hand, if a close-packed plane is inserted into the lattice along a $\{111\}$ plane, the formation obtained is

$$\dots\Delta\Delta\Delta\nabla\nabla\Delta\Delta\Delta\dots,\ (\dots \overline{ABCA}C\overline{BCAB}\dots) \tag{6.2}$$

Here the stacking is faulty twice in succession. The fault is perpendicular to the surface and is a finite extension (with a thickness of one layer), since in contrast to the sequence (6.1) it contains a layer which is not coherently connected with either of the half-crystals. Such stacking faults are generally called *extrinsic* faults, whereas those which do not contain a faulty layer of finite thickness are *intrinsic* faults.

Another type of stacking fault is

$$\ldots\Delta\Delta\Delta\Delta\nabla\nabla\nabla\nabla\ldots \quad (\ldots ABCABACBA\ldots) \tag{6.3}$$

in which there is a common, so-called "coherent" boundary of two twin crystals whose orientations differ by 180°. The fault (6.1) can be regarded to a first approximation as a single-layer coherent twin.

The following stacking faults occur in the hexagonal system (for convenience the symbols describing a faultless lattice are first presented):

$$\ldots\Delta\nabla\Delta\nabla\Delta\nabla\Delta\nabla\ldots \quad (\ldots ABABABAB\ldots) \tag{6.4}$$

$$\ldots\Delta\nabla\Delta\nabla\nabla\Delta\nabla\Delta\ldots \quad (\ldots ABABACACAC\ldots) \tag{6.5}$$

$$\ldots\Delta\nabla\Delta\nabla\nabla\nabla\Delta\nabla\Delta\ldots \quad (\ldots ABABACBCBC\ldots) \tag{6.6}$$

$$\ldots\Delta\nabla\Delta\nabla\nabla\nabla\nabla\Delta\nabla\Delta\ldots \quad (\ldots ABABACBABA\ldots) \tag{6.7}$$

The sequence (6.6) is obtained when a displacement *AB*, *BC* or *CA* occurs along the basal plane between two consecutive layers. If a layer is removed (or inserted) and simultaneously one of the above displacements is made, then we obtain the sequence (6.5). (The displacement at the same time as the removal of a layer is absolutely necessary, because otherwise two layers of the same character would join and this is practically impossible.) Finally, if a further layer is inserted into the lattice without any displacement the sequence (6.7) is obtained. Of the faults listed, the first and the second are intrinsic, while the third is extrinsic.

Of course more complicated faults and system of faults may also be conceived, but these can be built up from the above basic types, which, however, are essentially different from one another. In addition, stacking faults can occur in other lattices too, e.g. in the bcc lattice (see Section 6.5), but naturally along other crystallographic planes. Stacking faults are created whenever there is more than one possibility for the arrangement of the atomic layers. The more complicated the structure, the more numerous are the possible stacking orders. In ordered alloys the boundaries of the anti-phase domains are also stacking faults (see Section 6.9).

6.2. Generalization of the Burgers circuit

Stacking faults have so far been discussed as if they extended from boundary surface to boundary surface. Generally, however, this is not the case: if a two-dimensional stacking fault ends within the crystal its boundary is a dislocation, called a *partial dislocation*. Let us consider, for instance, a close-packed cubic lattice in part of which is a fault of the type (6.1) made by cutting the lattice halfway along a $\{111\}$ plane, displacing the cut surfaces relative to each other for instance by the vector $\frac{1}{6}\langle 112\rangle$ (see Fig. 6.1), and finally reconnecting the cut parts. We thus have a stacking fault bordered within the crystal by a dislocation. The vector of displacement – i.e. the Burgers vector too – is in this case not a lattice vector. There is a misfit between the cut surfaces (in the subsequent treatment these will be important), and consequently the concepts used so far must be generalized.

The first difficulty arises in the interpretation of the Burgers vector, since it is possible to go round the dislocation bordering the stacking fault only by touching the fault too, and this is not "good material". Although the coordination number is the same in planes of faulty stacking as in the good material, the adjacent atomic sites relative to one another are different, and so no unequivocal correspondence can be made between all the points in the good and in the bad lattices.

This difficulty can be overcome by defining (in contrast to the usual definition of the Burgers circuit) a closed circuit in the real crystal such that

(a) the circuit starts from the stacking fault, and everywhere else lies in good material only (Fig. 6.2);

(b) in the case of extrinsic faults, besides this, the circuit intersects the fault perpendicularly (Fig. 6.3).

The image of this closed circuit is constructed in the perfect lattice so that if the original circuit passes through an extrinsic fault this transition is represented in the image-circuit by a parallel section of

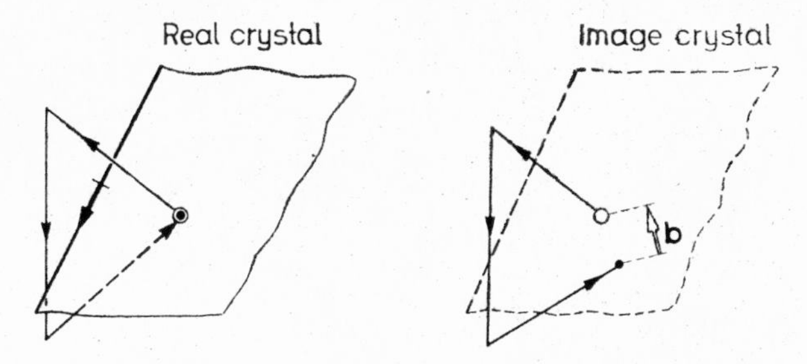

FIG. 6.2. Burgers circuit in the case of a partial dislocation bordering an intrinsic stacking fault. (For the sake of clarity, only the plane of the intrinsic fault is drawn)

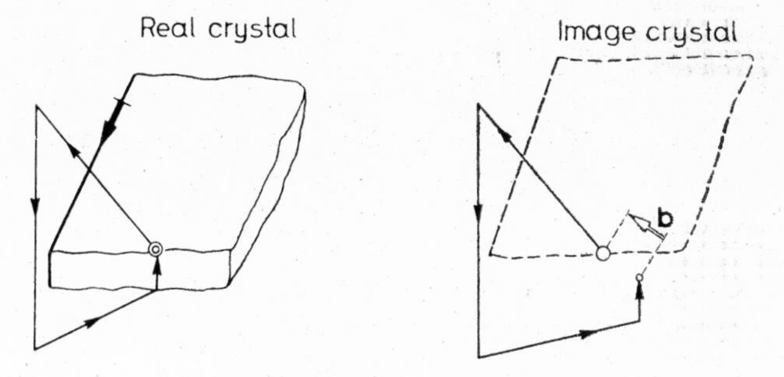

FIG. 6.3. Burgers circuit in the case of a partial dislocation bordering an extrinsic stacking fault

equal length (Fig. 6.3). The closure-failure then occurring is the Burgers vector. Naturally the rotation of the Burgers circuit in this case is opposite to that of the definition given in Chapter 1.

The case when the dislocation runs along the common boundary of two connected stacking faults must be mentioned separately.

The Burgers circuit can still touch the faulty material only once, and for this reason two circuits are taken (Fig. 6.4), only one of which (B_2) surrounds the dislocation considered. The Burgers vector is the difference between those of the two circuits B_1 and B_2. The conservation of the Burgers vector also holds for the partial dislocations. This

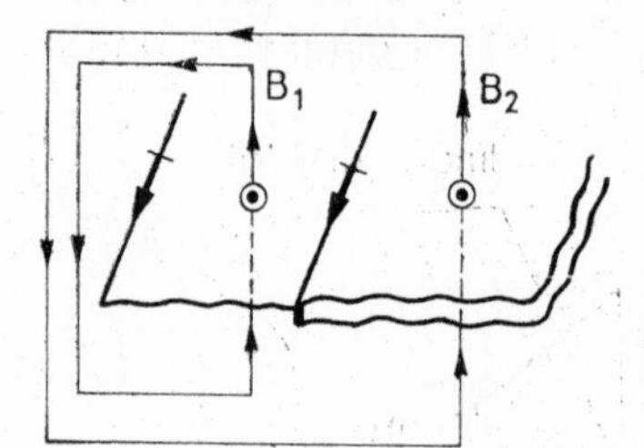

FIG. 6.4. Burgers circuit around a partial dislocation forming a common border between two stacking faults

can be seen from considerations similar to those for the perfect dislocations. The same refers also to the other consequences of the theorem; it should only be added here that a stacking fault ending within the crystal is always bordered by a partial dislocation.

6.3. Partial dislocations in fcc lattices

It has already been mentioned that stacking faults in a fcc lattice can be created by three types of operation. Let us consider what dislocations are obtained if these operations are applied to only one part of the cross-section of the crystal. In this case, however, attention should be drawn to the fact that not only the orientation of the Burgers vector relative to the dislocation line must be considered, but also *its relation to the plane of individual stacking faults.* Consequently, several essentially different dislocations must be discussed.

(a) If the layers along one part of a (111) plane are displaced relative to each other by a shear along $\frac{1}{6}\langle 11\bar{2}\rangle$, then the dislocation of Fig. 6.5 is obtained. Here the Burgers vector is parallel to the stacking fault plane. The diagram gives the position corresponding to an edge dislocation. Of course, such partial dislocations do not usually have pure edge (or screw) character. As an exercise it is worth while drawing the equivalent of Fig. 1.11b in this case too.

This dislocation type was introduced in 1948 by Heidenreich and Shockley [1, s been called a *Shockley partial dislocation.* If follows from the definition that both the Burgers vector and the dislocation line always lie in the plane of the intrinsic stacking fault associated with the dislocation. Consequently the dislocation line

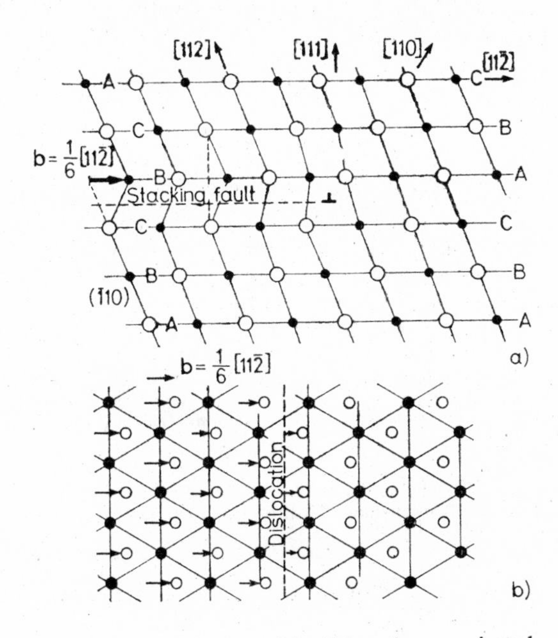

FIG. 6.5. Shockley partial dislocation projected onto planes (a) perpendicular to and (b) parallel to the plane of the stacking fault. The full circles indicate the atoms in the plane, and the empty circles those below (or above) the plane

is always a plane curve and never a three-dimensional curve and the dislocation can move in its own plane by glide. During its motion the area of the stacking fault either increases or decreases.

(b) By removing one part of a close-packed layer a stacking fault . . . *ABCBCA*. . . , and a dislocation bordering this fault are again obtained. The Burgers vector ($b = \frac{1}{3}\langle 111\rangle$) and the glide plane too are perpendicular to the plane of the fault (Fig. 6.6). However, the stacking fault cannot jump from one plane to another, and so this dislocation cannot glide, and for this reason it is called a *sessile* dislocation. The removal of the stacking fault requires approximately the theoretical critical shear stress ($\mu/30$). If the temperature is high enough for self-diffusion to become considerable, the dislocations may be displaced in the plane of the fault with point-defect emission or absorption. Such a dislocation was first considered by Frank in 1949 [2].

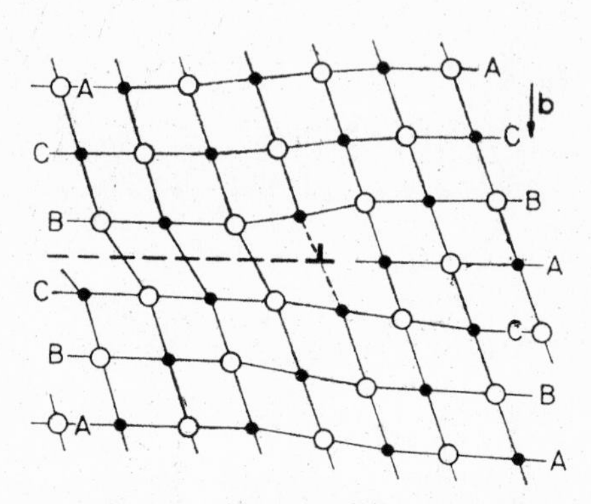

FIG. 6.6. A negative Frank partial dislocation projected onto the $(1\bar{1}0)$ plane

(c) Let us insert an extra plane into the material cut off along close-packed planes. In this way an *extrinsic* fault of type . . . *ABCBABC* . . . is obtained, which is again bordered by a partial dislocation whose Burgers vector is normal to the plane (Fig. 6.7a). In the same way as the above, this dislocation cannot move by glide, and others of their properties are similar to each other. An essential difference, however, is that although both are edge dislocations of Burgers vector $b = \frac{1}{3}\langle 111 \rangle$, they belong to stacking faults of different types.

It can be seen from Figs 6.6 and 6.7 that as a result of the two kinds of stacking fault, Burgers vectors of antiparallel orientations are produced. As we have seen, for perfect dislocations the orientation of the Burgers vector is merely a question of convention. Here, however, rather more is involved, and therefore it is essential to distinguish between the two types. A negative Frank partial dislocation is a partial dislocation associated with an intrinsic stacking fault . . . ΔΔ∇ΔΔ . . . , whereas a positive partial is associated with an extrinsic stacking fault . . . ΔΔ∇∇ΔΔ . . . . In contrast to the Shockley dislocations, these are produced not as a result of glide, but because of the condensation of lattice vacancies or interstitial atoms in the

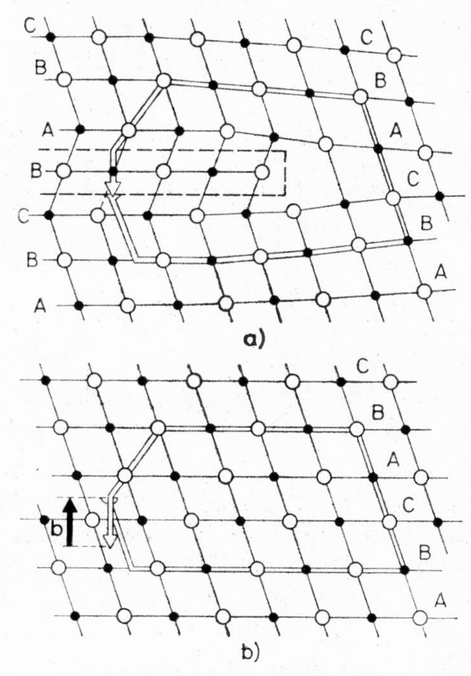

FIG. 6.7. A positive Frank partial dislocation with (a) the Burgers circuit, and (b) the corresponding circuit in a perfect crystal

form of discs, especially if the material is suddenly quenched from high temperature or subjected to high energy radiation (this will be discussed in detail in Section 6.9). Once they are created they cannot be removed by glide and at low temperature therefore they constitute serious obstacles to moving dislocations.

So far, only the simplest dislocations which border only a single stacking fault have been considered. If the common boundary of two stacking faults, perhaps with different orientations, is studied, other partial dislocations are also found. These will be dealt with together with the reactions of the partial dislocations in Section 6.6.

6.4. Extended dislocations in fcc lattices

It has been seen that the energy of dislocations is proportional to b^2. Accordingly, in Section 4.5 a study was made of which lattice vectors can form stable Burgers vectors. It is already known that a Burgers vector is not necessarily a lattice vector. Consequently, the question arises whether the perfect dislocations which have so far

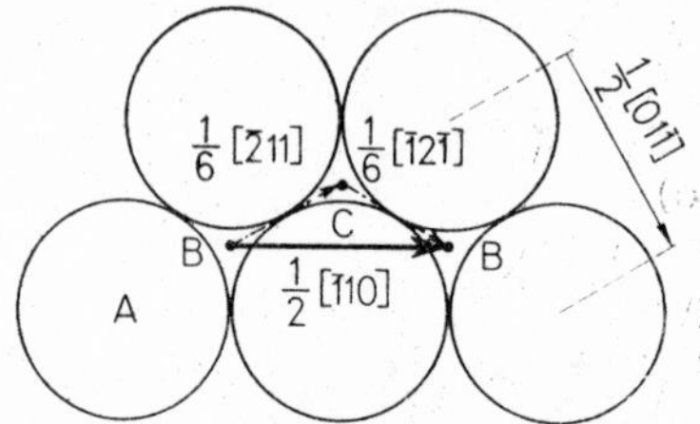

FIG. 6.8. The dissolution of the Burgers vector of a perfect dislocation in a fcc lattice

been regarded as stable can split up spontaneously into partial dislocations. Thus, for instance, the $B \rightarrow B$ glide (Fig. 6.8) in the $\frac{1}{2}\langle\bar{1}10\rangle$ direction in the cubic close-packed lattice can be accomplished more easily in two steps according to the scheme $B \rightarrow C \rightarrow B$. If the dislocation splits into two parts by the spatial separation of the partial glides, the Burgers vector splits according to the relation

$$\frac{1}{2}\langle\bar{1}10\rangle = \frac{1}{6}\langle\bar{2}11\rangle + \frac{1}{6}\langle\bar{1}2\bar{1}\rangle. \qquad (6.8)$$

The components on the right side enclose an acute angle, and this obviously means an energy gain. In contrast, as the two partial dislocations move away from one another, of necessity an intrinsic stacking fault forms between them (Fig. 6.9).

In order to produce a stacking fault work must be performed. This work is proportional to the area of the fault and is supplied from the previously mentioned energy gain. Consequently the stacking faults endeavour to draw together the partial dislocations, which otherwise repel each other. When one of the dislocations is removed from its position by some force, the other one follows it. This configuration is called an *extended dislocation.*

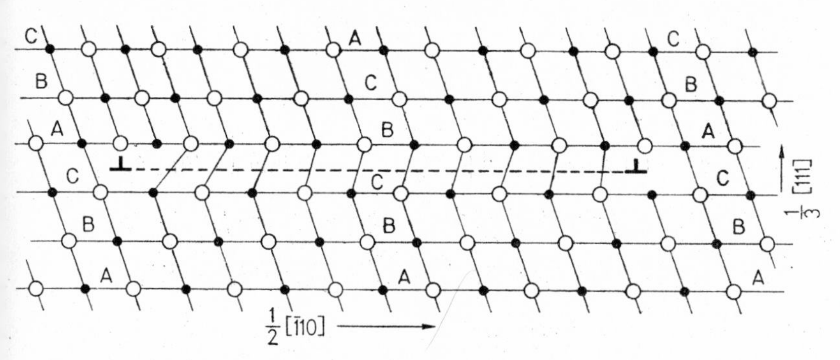

FIG. 6.9. Extended dislocation projected onto the (1$\bar{1}$0) plane in the fcc lattice

Let us calculate the extent of the separation from the continuum model. For simplicity the material is considered as isotropic, of infinite extent, and free from other defects. We further restrict ourselves to cases where the Burgers vectors of the partial dislocations enclose an angle of 60°, and are of equal size. (Other rarely occurring cases can be treated in a similar way.) If the specific energy (related to unit surface) of the stacking fault is γ, the work done along unit length on a change Δd of the distance d is $\gamma \Delta d$, and hence the contracting force of the stacking fault is exactly γ; the equilibrium is maintained by the repulsion between the partial dislocations. This repulsion force can be calculated from the relations discussed in Chapter 2. According to equation (2.27) the force of interaction per unit length is:

$$\mathbf{F} = (\boldsymbol{\sigma}^A \, \mathbf{b}^B) \times \mathbf{t}^B$$

where $\boldsymbol{\sigma}^A$ is the stress tensor of the first partial dislocation, $\mathbf{b}^B$ is the Burgers vector of the second partial dislocation, and $\mathbf{t}^B = (0, 0, 1)$ is its tangential unit vector. Since the glide planes are identical, the stress tensor must be taken in the plane $y = 0$, and hence some of its components vanish. In the end, only a force in the x-direction is obtained:

$$F_x = \sigma_{xy}^A b_x^B + \sigma_{yz}^A b_z^B,$$

where

$$\sigma_{xy}^A = \frac{\mu b_x^A}{2\pi(1-\nu)d},$$

$$\sigma_{yz}^A = \frac{\mu b_z^A}{2\pi d}.$$

With the notation $b = |\mathbf{b}^A| = |\mathbf{b}^B|$, according to Fig. 6.10,

$$b_x^A = b\sin(\phi - 30°), \qquad b_z^A = b\cos(\phi - 30°),$$

$$b_x^B = b\sin(\phi + 30°), \qquad b_z^B = b\cos(\phi + 30°).$$

After suitable rearrangement:

$$F_x = \frac{\mu b^2(2-\nu)}{8\pi d(1-\nu)}\left(1 - \frac{2\nu}{2-\nu}\cos 2\phi\right) \tag{6.9}$$

and, because $\gamma = F_x$,

$$d = \frac{\mu b^2(2-\nu)}{8\pi\gamma(1-\nu)}\left(1 - \frac{2\nu}{2-\nu}\cos 2\phi\right). \tag{6.10}$$

Let us take, for instance, $\mu = 5 \times 10^{11}$ dyne cm^{-2}, $\nu = 0.35$, $b_0 =$ $= 2.5 \times 10^{-8}$ cm $(b = b_0/\sqrt{3})$, $\gamma = 50$ erg/cm^{-2}; then for $\phi = 0$, $d \approx$ ≈ 12 Å, for $\phi = 45°$, $d \approx 20$ Å, and for $\phi = 90°$, $d \approx 29$ Å. The continuum theory is still applicable for such distances.

Although the above refers to isotropic materials, it can be applied in some simple cases (e.g. for dislocations in the basal plane of an hexagonal crystal) to anisotropic crystals too, using apparent ν and μ values calculated from anisotropic constants [3, 4]. A more general solution for an anisotropic crystal which is finite in one direction has been worked out by Spence [5]; he assumed that if the z-axis lies in the direction of the dislocation, only the constants C_{11}, C_{12}, C_{13}, C_{22}, C_{33}, C_{44}, C_{55}, C_{66} are non-zero. According to his calculations, the relations derived for an infinite crystal can be well applied to approximately 10^3 Å thick foils (if the dislocation is approximately in the centre of the foil). Baker *et al.* [6] also carried out calculations on anisotropic crystals. Seeger [4] calculated the width of extended dislocations from the Peierls model. For small separations only this gives acceptable results.

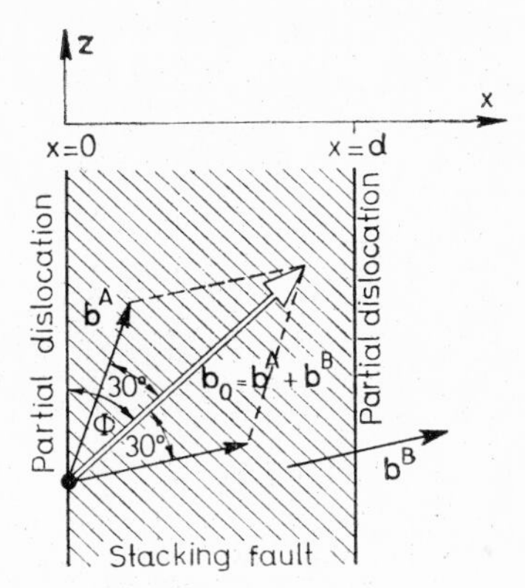

FIG. 6.10. For the derivation of equation (6.10)

It will be seen later that the extended dislocations are very important in the mechanism of plastic deformation. This appears primarily in the motion of screw dislocations; in these too the plane of glide is determined unambiguously by the non-parallel Burgers vectors of the separated partials. Cross-slip becomes possible only if (with a suitable energy input) the extended dislocation first contracts into perfect dislocation ($d \approx$ atomic distance). However, there is a small probability that, for example, the first partial dislocation encounters an obstacle and splits again (naturally also with some energy input) according to the equation

$$\frac{1}{6}[\bar{2}11] \rightarrow \frac{1}{6}[\bar{2}1\bar{1}] + \frac{1}{3}[001]. \qquad (6.11)$$

The new Shockley partial glides further on the $(11\bar{1})$ plane. When the second partial reaches the sessile dislocation with Burgers vector $\frac{1}{3}[001]$, left behind where the change of direction took place, after the reaction

$$\frac{1}{3}[001] + \frac{1}{6}[\bar{1}2\bar{1}] \rightarrow \frac{1}{6}[\bar{1}21] \qquad (6.12)$$

the cross-slip of the whole extended dislocation begins (Fig. 6.11). Naturally, if both processes occur simultaneously, nothing distinguishes them from simple shrinkage cross-slip. These processes have been investigated theoretically by Wolf [7].

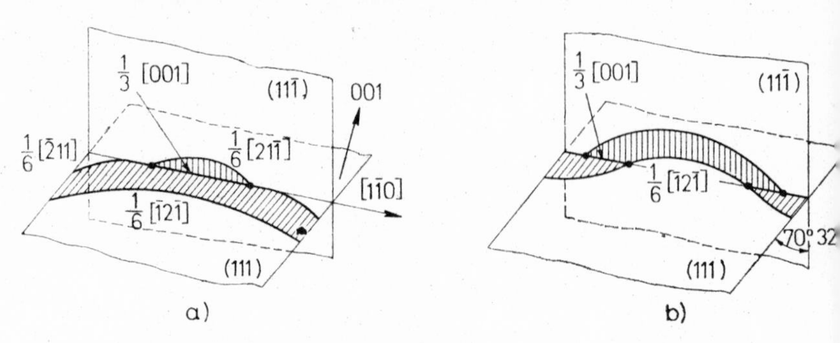

FIG. 6.11. Cross-slip of extended dislocations. The vectors given beside the dislocations are the Burgers vectors of the partial dislocations

Similar problems occur in the case of intersection of dislocations, since with extended dislocations a jog must also be formed in the stacking fault. In most cases this is associated with a large disturbance of the lattice, and so a considerable energy input is required. The jog created may be constricted but does not necessarily remain in a constricted state; instead, an extended jog, perhaps containing several stacking faults of various orientations, may be produced (Section 6.7).

The stacking fault also impedes the climb of dislocations due to the diffusion of point defects, since a stacking fault cannot be transferred from one plane to another.

Extended dislocations and stacking faults occurring in materials with the diamond structure or the very similar sphalerite structure must be dealt with separately. The diamond lattice is shown schematically in Appendix A (Fig. A.10). It differs from the close-packed fcc lattice in that the layers are repeated in pairs in the $\{111\}$ directions according to the sequence

$$\ldots AA'BB'CC'AA' \ldots \tag{6.13}$$

(the prime denotes the fact that the two layers are not symmetrically equivalent). For this reason the stacking faults are situated between layers of different positions as depicted in Figs 6.12a, b. The coherent twin boundaries, on the other hand, run between planes of similar positions (Fig. 6.12c). In practice, both intrinsic and extrinsic faults frequently occur in Si and Ge, and the formation of growth and deformation twins has also been observed.

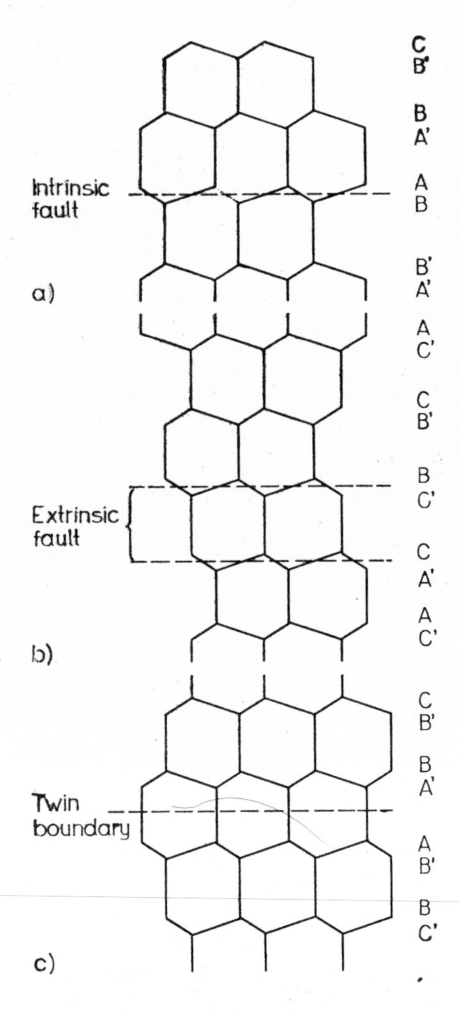

FIG. 6.12. Stacking faults and coherent twin boundary projected onto the $(1\bar{1}0)$ plane in the diamond lattice (see Fig. A.10b)

Before discussing partial dislocations in the diamond structure let us first outline the main properties of perfect dislocations. Because of the oriented covalent bonds, the structure of the dislocation-core

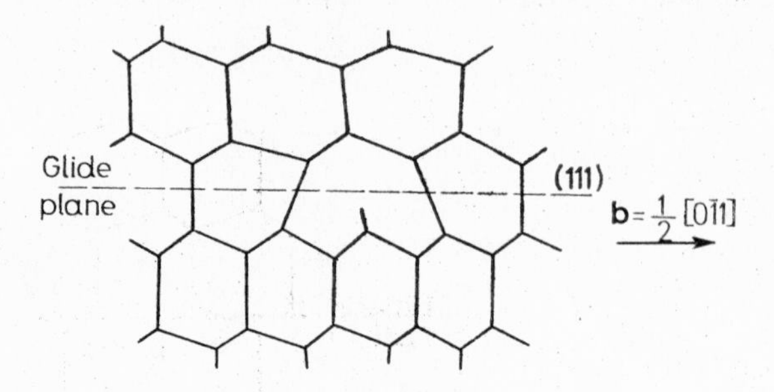

FIG. 6.13. 60°-dislocation projected onto the (1$\bar{1}$0) plane in the diamond lattice

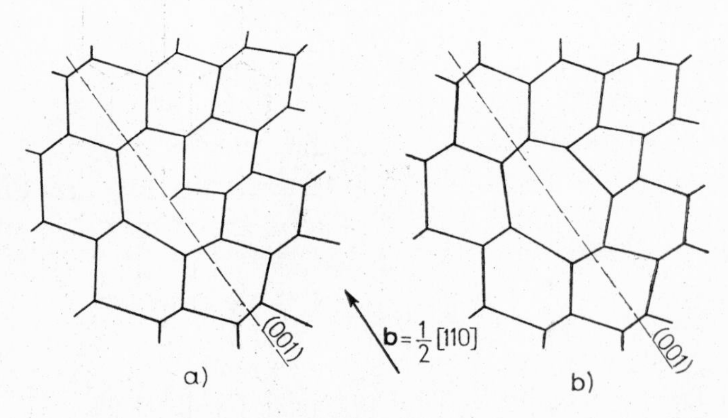

FIG. 6.14. Pure edge dislocation projected onto the (1$\bar{1}$0) plane in the diamond lattice

is of fundamental importance as regards the properties of the dislocations. The various possibilities have been treated in detail by Hornstra [8]. The shortest lattice vector is still $\frac{1}{2}\langle 110\rangle$, and since the Peierls potential is relatively large, the dislocations (in agreement with experience), to a first approximation, are parallel to the ⟨110⟩ direction. In practice, therefore, three dislocation types occur: screw, 60° and edge dislocations. The energy of the dislocation consists mainly of the energy of deformation (approximately 7 eV per atomic layer in Ge), though the energy of broken bonds cannot be neglected (approximately 1 eV per bond). The broken, uncompensated bonds strongly influence the electrical properties of the crystals. In principle they act as acceptors since they take up electrons, but then the dislocations become electrically charged. For a more profound study of the electrical properties the reader is referred to the works by Rhodes, and by Alexander and Haasen [9]. We mention only that the localized levels developing around a dislocation usually lie deep in the forbidden band.

One of the most frequently occurring types is the 60° dislocation depicted in Fig. 6.13; during the glide (in the {111} plane) similar layers are displaced with respect to one another.

The pure edge dislocation can occur in two types of configuration. In the case outlined in Fig. 6.14a, there are two uncompensated bonds per atomic plane. If, however, an atomic row is removed from the end of the extra half-plane and the remaining bonds are connected with one another, the stronger and more stable configuration depicted in Fig. 6.14b is obtained, which presumably forms only at high temperatures [8]. A pure edge dislocation can glide only in the {001} plane. Such a glide system, however, has not so far been observed.

In the case of partial dislocations, further varieties can be observed as a result of the different crystalline features compared with the close-packed structures. Here only the splitting of 60°-dislocations is discussed (Fig. 6.15). The extended dislocations forming still split into two Shockley partials according to equation (6.8). This process, however, can be achieved in two different ways. According to Hornstra [8] the first step of the separation consists of the rotation of the two atomic rows near to the dislocation core as in one or other of Figs 6.15a and b. In this way the atomic configurations corresponding to the two partial dislocations are developed, after which they merely separate from one another (c, d). In one case the glide plane is completely on one side of the stacking fault, whereas in the other case the two partial dislocations glide on both sides of the stacking fault. The splitting may also not take place parallel to the glide plane. A pure edge dislocation, however, can split into partials only in planes non-parallel to the glide plane which is of the form {100}.

Because of the unequal lattice constants of the joining materials, "misfit"-dislocations are formed along semiconductor heterojunctions. The properties of these misfits are well treated by Holt [10], who in another paper [11] deals with the defect structure of the sphalerite lattice too.

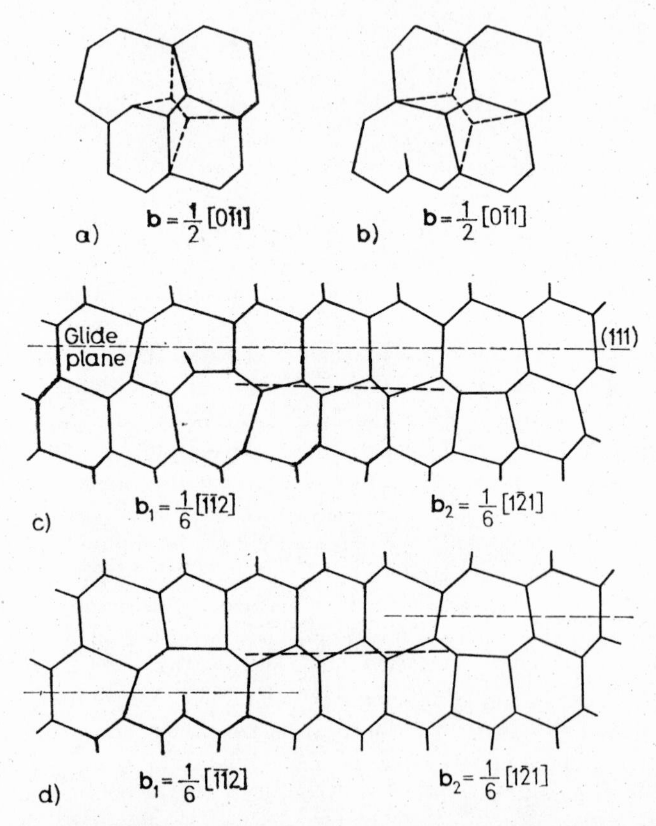

FIG. 6.15. Two possible initial stages (a, b) of the decomposition of a 60° dislocation, and the resulting extended dislocations (c, d) in the diamond lattice. The Burgers vector of the first partial dislocation is parallel with the plane of the drawing. The thick broken line represents the stacking fault

Finally let us consider briefly dislocations lying in the basal plane of a hcp crystal. The geometrical relations here are quite similar to the case of the cubic close-packed (fcc) structure. The splitting occurs according to the equation:

$$\frac{1}{3}[2\bar{1}\bar{1}0] \rightarrow \frac{1}{3}[1\bar{1}00] + \frac{1}{3}[1010]. \qquad (6.14)$$

6.5. Stacking faults and extended dislocations in bcc crystals

The shortest lattice vector in the bcc lattice is the $\frac{1}{2}\langle 111\rangle$ vector, and according to experience the glides almost always take this direction. The glide plane, however, is far from so clear-cut. The glide takes place only at very low temperatures and at large deformation rates, exclusively along the $\{110\}$ planes; otherwise, macroscopically the glide does not occur in a well-defined crystallographic plane, but (depending upon the orientation of the stress) mainly parallel to the

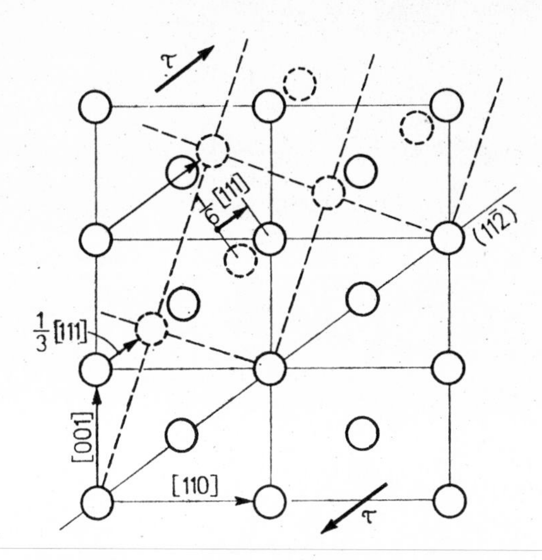

Fig. 6.16. Shear stress acting in the direction of twinning, projected onto the (1$\bar{1}$0) plane in the bcc lattice. (The lattice corresponding to the twinned region is indicated by a dotted line)

Burgers vector along that plane for which the shear stress is a maximum. This *non-crystallographic glide* is the result of a composite glide along the {110} and {211} planes [12, 13]. Apart from this, for the glide along the {211} planes, the critical resolved shear stress is smaller if the shear tends towards twin formation (Fig. 6.16) and not in the opposite direction. It is well known that deformation twinning

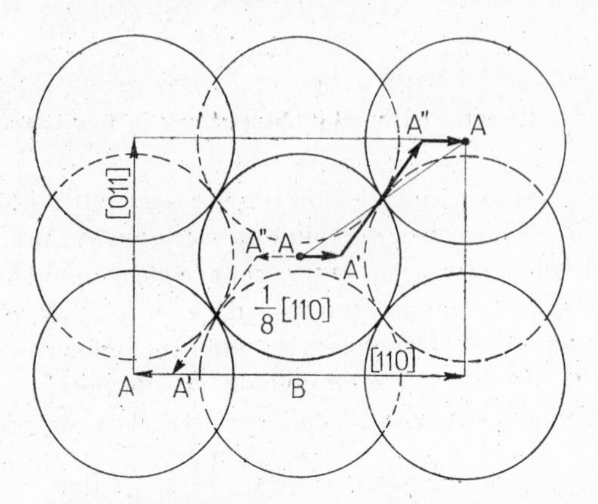

FIG. 6.17. The splitting of the $\frac{1}{2}$[111] Burgers vector in the (110) plane of the bcc lattice

occurs relatively frequently in these crystals. The observations can be explained by the presence of extended dislocations. In pure bcc metals, however, stacking faults have so far not been directly observed. This means that the energy of the stacking faults is fairly large, and that the splitting of the partials is insignificant. It has been pointed out in Chapter 4 that according to the model calculation of Vitek *et al.* the core of a $\frac{1}{2}\langle 111\rangle$ screw dislocation shows a characteristic threefold anisotropic broadening [14]; this can be explained by the symmetrical splitting of the screw dislocation into an extended dislocation consisting of partial dislocations and stacking faults. The width of the stacking faults is accordingly only about $2b$. The partials in this model cannot be isolated and this is presumably the case with real crystals too. Nevertheless, it is a good approximation to speak simply of extended dislocations. The splitting can occur in either the {110} or in the {112} planes.

6.5.1. *Splitting in the* $\{110\}$ *plane*

The bcc lattice can be constructed according to Fig. 6.17 from a set of (110) layers with positions A and B in the sequence . . . $ABAB$. . . When two such layers of rigid spheres are displaced relative to one another parallel to the $\frac{1}{2}[111]$ vector, the spheres denoted by A move in the steps $A \rightarrow A' \rightarrow A'' \rightarrow A$ (or $A \rightarrow A'' \rightarrow A' \rightarrow A$) according to the splitting:

$$\frac{1}{2}[111] \rightarrow \frac{1}{8}[110] + \frac{1}{4}[112] + \frac{1}{8}[110]. \tag{6.15}$$

The $\frac{1}{8}[110]$ glide naturally results in a stacking fault, and the resulting configuration is the extended dislocation shown in Fig. 6.18a. According to Vitek's calculations [14, 15], in the case of a pure screw

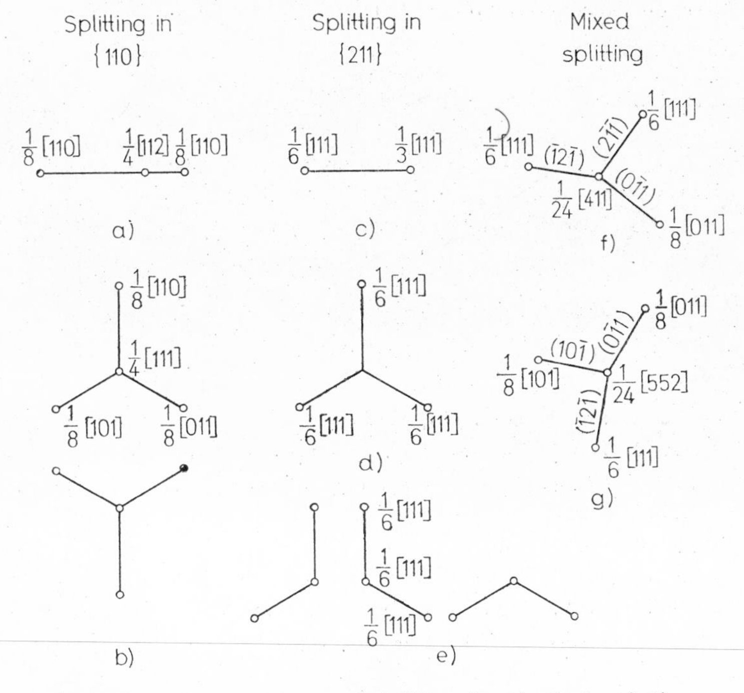

FIG. 6.18. The types of extended dislocations in the bcc lattice (according to Hirsch [13])

dislocation, the splitting

$$\frac{1}{2}[111] \rightarrow \frac{1}{8}[110] + \frac{1}{8}[101] + \frac{1}{8}[011] + \frac{1}{4}[111] \qquad (6.16)$$

may also be expected. The three $\frac{1}{8}\langle 110\rangle$ partial dislocations are displaced symmetrically from the central $\frac{1}{4}[111]$ dislocation along the $\{110\}$ planes (Fig. 6.18b). Wasilewski has pointed out [16] that an alternative splitting

$$\frac{1}{2}[111] \rightarrow \frac{1}{6}[111] + \frac{1}{6}[111] + \frac{1}{6}[111] \qquad (6.17)$$

is also possible. As a result of external stress these symmetrical extended dislocations deform and partly recombine.

6.5.2. Splitting in the $\{112\}$ *plane*

It has already been mentioned that a $\frac{1}{6}\langle 111\rangle$ displacement in the twin direction in the $\{112\}$ plane disturbs the lattice only slightly. It may be assumed, therefore, that the energy of the stacking fault so created is moderate, and the dislocation can split according to the relation:

$$\frac{1}{2}[111] \rightarrow \frac{1}{6}[111] + \frac{1}{3}[111]. \qquad (6.18)$$

As a result an extended dislocation is created in which the $\frac{1}{6}[111]$ partial leads the way (Fig. 6.18c). In the deformation of opposite sign, however, the $\frac{1}{3}[111]$ displacement must take place first, since $\frac{1}{6}[111]$ movement in the opposite direction to the twin formation causes a stronger lattice disturbance, and the corresponding stacking fault is presumably not stable. For this reason the splitting in one plane according to the relation (6.17) is not probable. In the case of a pure screw dislocation, on the other hand, the symmetrical splitting

$$\frac{1}{2}[111] \rightarrow \frac{1}{6}[111] + \frac{1}{6}[111] + \frac{1}{6}[111] \qquad (6.19)$$

may take place along three {112} planes (Appendix, Fig. A.6b) without a central partial dislocation (Fig. 6.18d), though Sleeswyk [17] has shown that this symmetrical arrangement is not stable and spontaneously transforms into one of the configurations depicted in Fig. 6.18e by the jump of one partial dislocation into the centre.

In addition to the above cases, the screw dislocations may also split into partials in a combined way according to Figs 6.18f and g [12].

The distance of the partial dislocations is determined as above. Using the anisotropic elastic constants Vitek [15] made calculations on the splitting in the {110} plane, while Teutonico [18] and Hartley [19], on the other hand, made similar calculations on the {112} splitting. Unfortunately, not much is known of the energy of the stacking faults. With the aid of Morse potentials, Hartley [19, 20] estimated the energy of stacking faults formed in the {112} plane. He obtained, however, the same value for the energies of the stacking faults produced by the displacements of the $\frac{1}{6}\langle 111\rangle$ and $\frac{1}{3}\langle 111\rangle$ partial dislocations; this is in contradiction with the asymmetry of the observed glide.

Although the extensions of the dislocations are small, these influence their motion considerably; because of the three-dimensional extension of the screw dislocations, their glide in any direction can begin only after the thermally activated contraction of the dislocation. For this reason the mobilities of screw dislocations are much smaller than those of edge dislocations. Their cross-slips, on the other hand, may occur at practically the same stresses as glides in any other direction. Thus "non-crystallographic" glide becomes possible.

Experimental data relating to the asymmetry of the shear stress show that the energies of the stacking faults, and accordingly the activation energy of contraction, are smaller on the {110} planes. Consequently, with low temperatures and fast deformations the glide takes place only in the {110} planes.

Šesták and Zárubová [21] calculated the activation energies as a function of the stacking fault energy and the stress orientation. When their results are applied to experimental data on Fe–3%Si single crystals, the values $\gamma_{110} = 120$ erg/cm^{-2} and $\gamma_{112} = 270$ erg/cm^{-2} are obtained [13]. With increasing Si content the stacking fault energy decreases.

6.6. Reactions of dislocations

In this section only the more important interactions between dislocations moving in the $\{111\}$ planes of cubic close-packed lattices are treated. In addition, the calculations refer to the isotropic case: with most cubic close-packed crystals this yields satisfactory results.

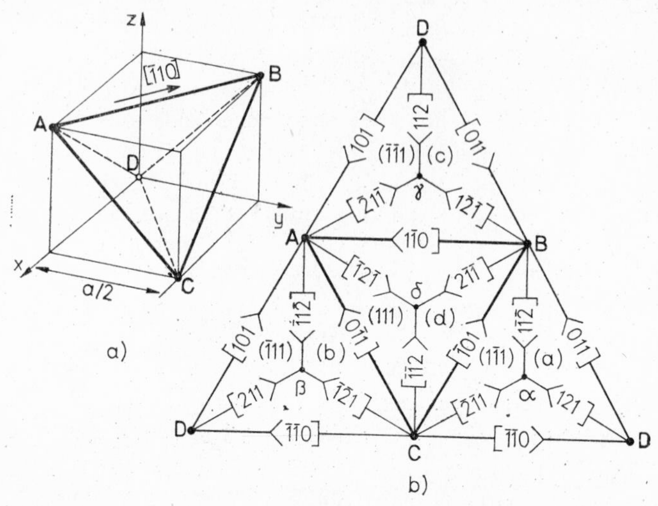

FIG. 6.19. The Thompson tetrahedron (a) in the unit cell and (b) spread out, with indices which can be attached to the individual directions. The asymmetrical bracket [〉 refers to the direction of the vector, consequently [〉 = −〈]

First of all the geometrical situation must be reviewed. For this reason the individual glide systems are characterized by the *Thompson tetrahedron* [22] (Fig. 6.19). The possible glide planes are determined by the faces, and the glide directions by the edges of the tetrahedron. The dislocation can be illustrated graphically by drawing the Burgers vector on the spread out tetrahedron. With the aid of the Thompson tetrahedron it is easy to decide which directions and planes can together make up a glide system. Therefore, instead of the indices used so far, the corresponding symbols of the tetrahedron are often used: first the Burgers vector is given, and then in brackets the glide plane (if it is one of the $\{111\}$ planes). The individual faces of the tetrahedron are denoted by the letters *a*, *b*, *c* and *d*, and their centres by α, β, γ and δ. Thus, for example, **AB**(*d*) means a dislocation with the Burgers vector $\frac{1}{2}[\bar{1}10]$ gliding in the (111) plane. On the other hand, it may happen that the Burgers vector is not a vector connecting the corners or the face centres. According to Thompson's notation in this case, $\delta\alpha/\mathbf{BC}$, for instance, means twice the vector connecting the mid-points of the distances $\delta\alpha$ and **BC**, that is the vector $\frac{1}{3}[010]$. (The distances here are measured on the tetrahedron and not on its spread-out form.)

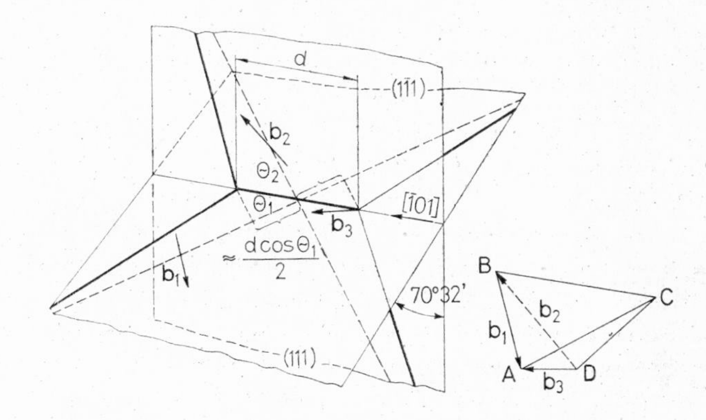

FIG. 6.20. Reaction between perfect dislocations

Let us consider what happens if two perfect dislocations gliding in different planes intersect one another in a point *P*. (One possibility has already been mentioned: after a jog is produced in the intersection the dislocations again separate and continue their glide.) For this, however, an energy input is needed. It is possible that the dislocations react with each other along a segment of length *d*, and thus the energy of deformation decreases (Fig. 6.20). The course of the reaction is conceived as follows: a resultant dislocation of a few atomic diameters is first produced at the point of intersection; this immediately widens and continues to do so while the process results in an energy reduction. If it is taken into account that a new dislocation of length *d* is formed by the reaction, and that at the same time the original dislocations become reduced in length by $d \cos \Theta_i$ (where Θ_i is the angle between the ith dislocation line and the line of intersection of the two planes), then using relations (2.37) referring to the energy per unit length, the energy change per unit length of the new dislocation segment is

$$\Delta E = D\left[b_3^2(1 - \nu\cos^2\vartheta_3) - \sum_{i=1}^{2} b_i^2(1 - \nu\cos^2\vartheta_i)\cos\Theta_i\right], \quad (6.20)$$

where

$$D = \frac{\mu}{4\pi(1-\nu)} \ln \frac{R}{r_0},$$

b_i is the length of the Burgers vector of the ith dislocation and ϑ_i is the angle between the Burgers vector and the dislocation line (this is already determined by the Burgers vector, the plane of glide and Θ_i). Here, naturally, the possible change of the energy of the core has been neglected. From equation (6.20) it can be seen that ΔE is negative only if $\cos \Theta_i$ is sufficiently large, i.e. if the dislocations are nearly parallel. If the individual dislocations do not run up to the boundary surface, Θ_i increases with increasing d and the process comes to a standstill when the equilibrium d value is attained.

Similar considerations apply to the reactions of extended dislocations, but the interaction of two pairs of partial dislocations and also of the stacking faults must be taken into account. Quantitative calculations were carried out for both cases by Hirth [23, 24]. His results are used in the following discussions.

One of the reacting dislocations is always taken as **BA**(d). The other one can be chosen from among twenty-four possibilities. The four typical cases considered below contain ten of these.

(a) **BA**(d) + **AD**(c) → **BD**(c) (Fig. 6.21a). The angle of inclination of the primary Burgers vectors is 120°, the dislocations attract each other, and in the intersecting line of their glide planes a single dislocation is created followed by the decrease of the deformation energy; this can glide further on one of the planes. The reaction takes place only when $\cos \Theta_1 + \cos \Theta_2 \gtrsim 1.15$. Reactions between extended dislocations need not be considered.

(b) **BA**(d) + **DB**(a) → **DA** (Fig. 6.21b). The angle of inclination of the primary Burgers vectors is again 120°, but neither of them is parallel to the intersecting line of the glide planes. The dislocations

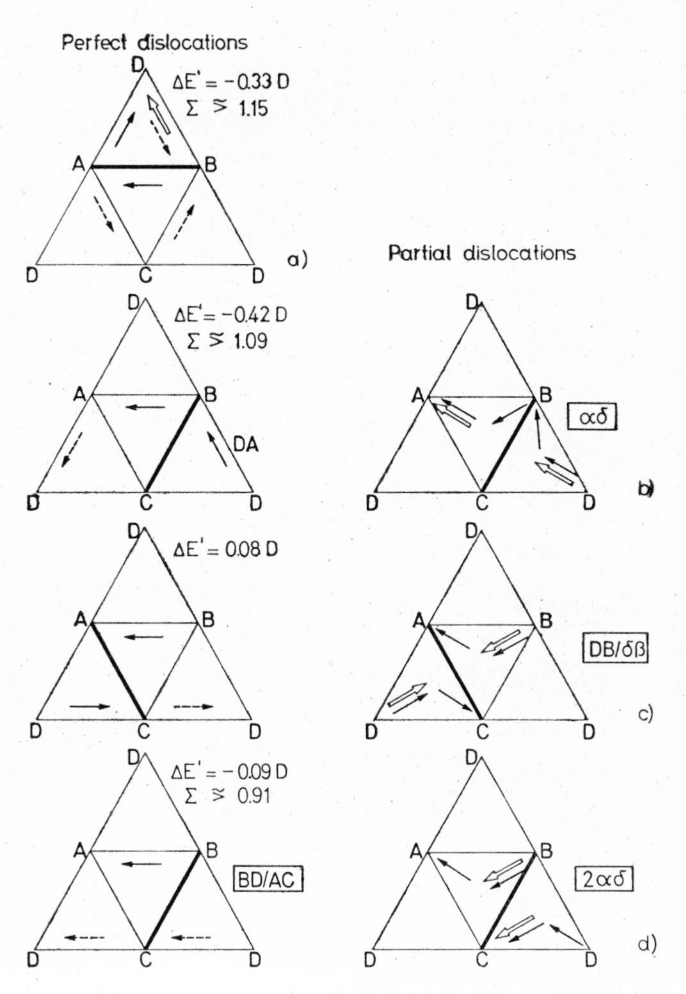

FIG. 6.21. The representation of some reactions in the Thompson tetrahedron. Burgers vector before the reaction: →, and after the reaction: ⇒ (the position of the vector also indicates the glide plane). Burgers vectors of other dislocations reacting similarly with **BA**(d) dislocation: ---→ (these are shown only for perfect dislocations). If the resultant is sessile, its Burgers vector is given within a frame at the right-hand side. The thick line is the line of intersection of the two planes, and is also the direction of the new dislocation. $a^2\ \Delta E'$ is the change of energy if $\Theta_1 = \Theta_2 = 0$ and $\Sigma = \cos\Theta_1 + \cos\Theta_2$

attract one another, and if $\cos \Theta_1 + \cos \Theta_2 \gtrsim 1.09$ a new dislocation is created in the line of intersection; its Burgers vector, however, is not parallel to either of the glide planes containing the line of intersection (Fig. 6.20), and so no further glide is possible. When a reacting dislocation is followed by more dislocations on the glide plane, the resultant impedes their further glide. This configuration is called a *Lomer–Cottrell barrier*. It plays an important role in the phenomena of work-hardening. In the case of extended dislocations the reaction takes place between the two first meeting partial dislocations, for instance in the following way:

$$\mathbf{B}\delta(d) + \delta\mathbf{A}(d) + \mathbf{D}\alpha(a) + \alpha\mathbf{B}(a) \rightarrow \alpha\delta + \delta\mathbf{A}(d) + \mathbf{D}\alpha(a).$$

However, the dislocation $\alpha\delta$ (in which two stacking faults of different planes meet) (Fig. 6.22) cannot glide in either of the planes, and so the two other partial dislocations are also held in position. Thompson calls this a *stair-rod dislocation*.

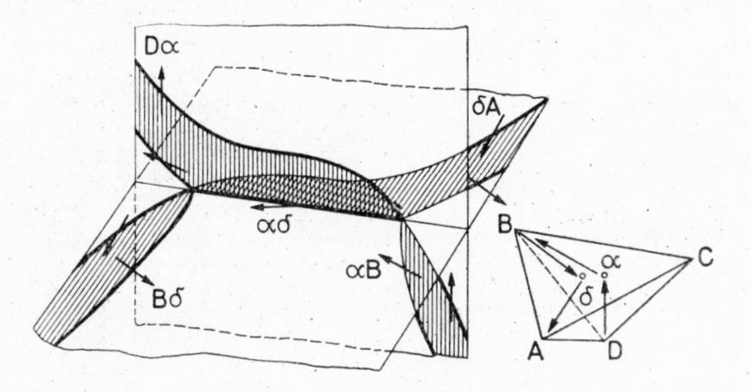

FIG. 6.22. An extended Lomer–Cottrell barrier

Hirth has shown [23, 24] that this formation is not stable, and that a shrinkage into a perfect dislocation would be more favourable energetically. However, a large stress giving rise to a climb acts on the dislocation $\alpha\delta$. If the energy of the stacking fault induced during the climb is $\gamma' < 6\gamma$, then after a climb to a distance of a few atoms an extremely stable, extended Lomer–Cottrell barrier is produced.

(c) $\mathbf{BA}(d) + \mathbf{DC}(b)$ (Fig. 6.21c). The resultant would be a *pure edge dislocation* without a decrease of Σb_i^2, and consequently this reaction would give rise to an energy increase. Hence the two dislocations repel one another, and no reaction takes place between perfect dislocations. For extended dislocations, if they meet as a result of some external stress:

$$\mathbf{B}\delta(d) + \delta\mathbf{A}(d) + \mathbf{D}\beta(b) + \beta\mathbf{C}(b) \rightarrow$$

$$\rightarrow \mathbf{DB}/\delta\beta + \mathbf{B}\delta(d) + \mathbf{D}\beta(b).$$

Again a sessile partial dislocation is created at the intersection of the two stacking faults. Though no climb takes place, a stable Lomer–Cottrell barrier is produced.

(d) $\mathbf{BA}(d) + \mathbf{DC}(a) \rightarrow \mathbf{BA} + \mathbf{DC} \equiv \mathbf{BD}/\mathbf{AC}$ (Fig. 6.21d). A 45°-dislocation is produced, therefore the two dislocations attract each other. If $\cos\Theta_1 + \cos\Theta_2 \gtrsim 0.91$ a Lomer–Cottrell barrier is again formed. In the case of extended dislocations:

$$\mathbf{B}\delta(d) + \delta\mathbf{A}(d) + \mathbf{D}\alpha(a) + \alpha\mathbf{C}(a) \rightarrow 2\alpha\delta + \mathbf{B}\delta(d) + \alpha\mathbf{C}(a).$$

The results are similar to those discussed in (c). As an example let us write out these two last reactions in vector notation:

$$\frac{1}{2}[1\bar{1}0](111) + \frac{1}{2}[110](\bar{1}11) = [110],$$

$$\frac{1}{6}[2\bar{1}\bar{1}](111) + \frac{1}{6}[1\bar{2}1](111) + \frac{1}{6}[121](1\bar{1}1) + \frac{1}{6}[21\bar{1}](1\bar{1}1) =$$

$$= \frac{2}{6}[101] + \frac{1}{6}[2\bar{1}\bar{1}](111) + \frac{1}{6}[21\bar{1}](1\bar{1}1).$$

Whelan [25] examined an austenitic stainless steel with an electron microscope and found a large number of Lomer–Cottrell barriers which he explained as being the result of (b)-type reactions.

Dislocation reactions taking place in the $\{110\}$ planes of bcc crystals have been investigated by Hartley [20] while similar studies for the $\{112\}$ planes were carried out by Teutonico [26].

We must now consider briefly the properties of extended nodes. The reaction of three dislocations intersecting one another generally results in threefold nodes (Fig. 6.22). It has already been mentioned in Chapter 1 that the sum of the Burgers vectors of dislocations running into one node is zero. Hence it also follows that the Burgers vectors of the dislocations associated with one node are coplanar. The structures of the nodes created by the intersection of extended dislocations depend too upon the sequence of the partials. Let us consider the case where the three dislocations lie in the same plane. Such nodes are created if two dislocations of different Burgers vectors move in an identical glide plane. Consider, for instance, a configuration consisting of two sets of parallel edge dislocations **AB**(d) and **CA**(d)

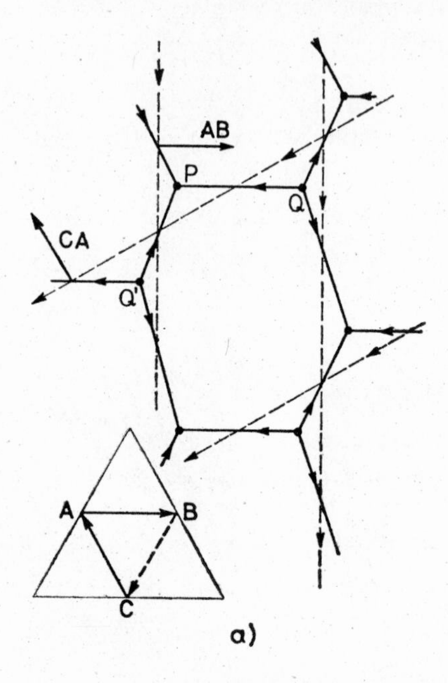

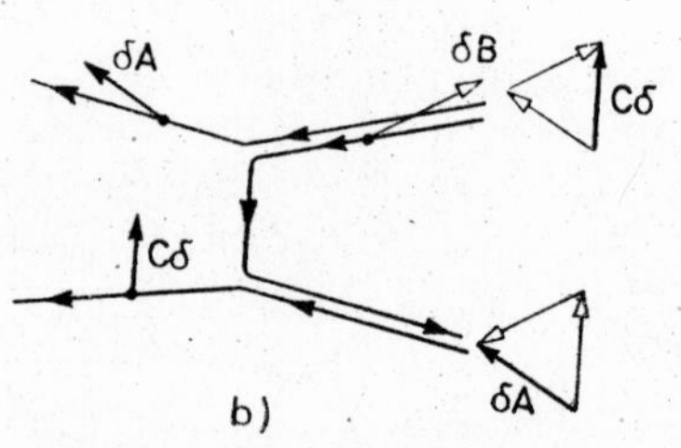

FIG. 6.23. Extended nodes in a two-dimensional network of extended screw dislocations inclined at 60°. The short li es normal to the dislocation line represent the change of the sign of the line element

moving in the same plane. The fourfold nodes produced in the point of intersection spontaneously decompose into threefold nodes in a way similar to the reactions studied earlier. The new dislocation is: **BC**(d) (Fig. 6.23a, segment PQ). In the case of extended dislocations, depending upon the sequence of the partials, in some of the nodes (for instance in the point P) the partials with the same Burgers vector are on the same side of the node, and hence after the reaction the node also spreads out forming an intrinsic stacking fault

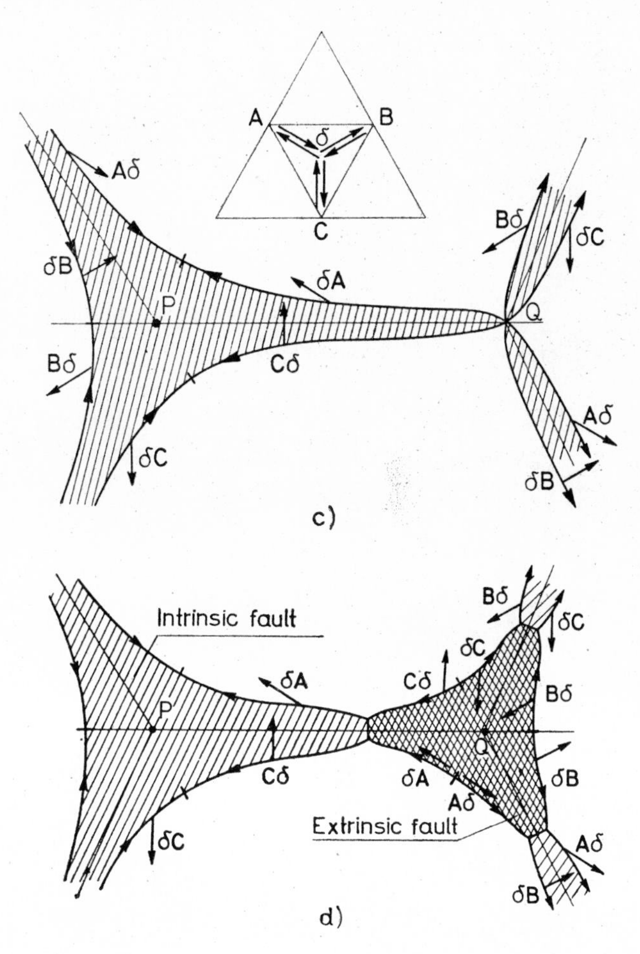

(Fig. 6.23c). In the adjacent nodes the situation is just the reverse. The partials with identical Burgers vectors could meet directly only if their sequence were interchanged. This, however, could be produced only by a further displacement on one of the adjacent planes. For instance, by constructing a new dislocation $\delta\mathbf{B}(d)$ according to Fig. 6.23b between the points PQ, the resultant reproduces the partial dislocations with just the reverse sequence. According to (6.2), however, a displacement on the two adjacent planes results in an extrinsic

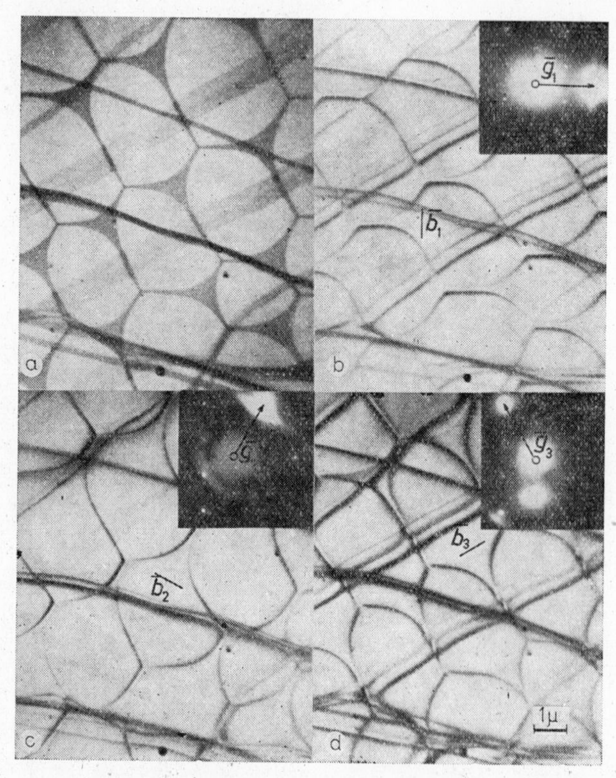

FIG. 6.24. Electron micrograph of a network of extended dislocations in graphite. The contrast of the dislocations depends upon the angle between the Burgers vector and the diffraction vector (**g** in the right upper corners) of the diffraction maximum participating in the image formation. The contrast disappears if **gb** = 0 (Amelinckx [27])

stacking fault around the point Q. Thus, the resulting dislocation network is similar to Fig. 6.23c or d, depending on whether the energies of the extrinsic stacking faults are large, or comparable with those of the intrinsic faults. Figure 6.24 depicts a case which is similar to the previous one (though it must be mentioned that in graphite crystals the structure of the extended dislocations is more complicated than in the close-packed lattices [27]).

In silicon and germanium the extensions of the two kinds of nodes are nearly the same [28].

In a stable dislocation network the two types of nodes alternate. Failing this, the dislocations recombine, at least partly.

6.7. Extended jogs

Not much is known of the exact structure of the jogs produced on extended dislocations. It follows from energy considerations that elementary jogs are constricted (at least if no external stress operates) [29], whereas with large jogs, the jogs too may dissociate. This problem has been investigated so far only for fcc lattices. We consider below only two cases according to Hirsch [30]: the structures of jogs created on screw dislocation **BC**(d) after the intersection of a dislocation **AD**, and on edge dislocation **BC**(d) after the intersection of a dislocation **BD**.

In the first case (screw dislocation), the orientation of the jog is **AD** (Fig. 6.25a), but because of the Burgers vector **BC** it cannot glide on either the b- or the c-plane. With the splitting $\mathbf{BC} \rightarrow \mathbf{B}\gamma + \gamma\mathbf{C}$, however, the Shockley partial $\mathbf{B}\gamma(c)$ can move in the c-plane (Fig. 6.25b) creating an intrinsic fault; moreover, with the possible splitting $\gamma\mathbf{C} \rightarrow \gamma\beta + \beta\mathbf{C}$ of the remaining sessile dislocation, the complex configuration shown in Fig. 6.25b forms. The resulting extended jog is also sessile.

The structure of the jog with direction **BD** formed on an edge dislocation is much simpler. The Burgers vector **BC** can also split in the a-plane ($\mathbf{BC} \rightarrow \mathbf{B}\alpha + \alpha\mathbf{C}$), and thus a glissile extended jog readily forms (Fig. 6.25c).

Assuming that *only intrinsic* faults are produced during the deformation, Hirsch investigated the mobility of extended jogs. He found

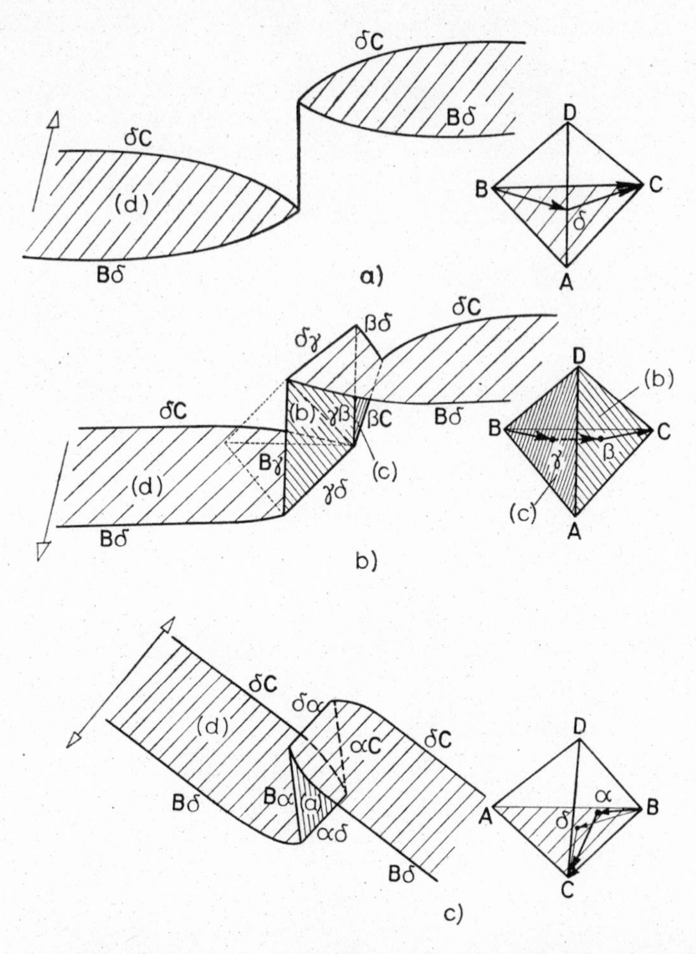

FIG. 6.25. Extended jogs in fcc lattice for the case of acute-angled stacking faults. The arrows indicate the glide direction leading to the configuration considered, and the Thompson tetrahedra show the *splitting off* of the Burgers vectors. (a) Screw dislocation **BC**(*d*) with a *constricted* jog along **AD**. (b) The above dislocation with jog extended on *b*- and *c*-planes. (The dotted lines are the outlines of the Thompson tetrahedron.) (c) Edge dislocation **BC**(*d*) with an extended jog in *a*-plane

that the interstitial jogs (jogs producing interstitial atoms by their non-conservative motion during the deformation) on nearly pure screw dislocations constrict during the motion of the dislocation and move in a conservative way, but the vacancy jogs remain extended and their motion is non-conservative. Nevertheless, as Weertman [31] pointed out, the creation of extrinsic faults cannot be excluded and in this case the situation is just the reverse.

Hirth and Lothe [32] suggested further possible mechanisms, but no observations have been made which permit a choice between the various possibilities.

6.8. Extended dislocations in ordered alloys

The ordering taking place in alloys does not usually change the structure substantially; it merely decreases the symmetry in such a way that the equivalence of sub-lattices (positions) which were equivalent in the disordered state is lost, and – in the ideal case – each

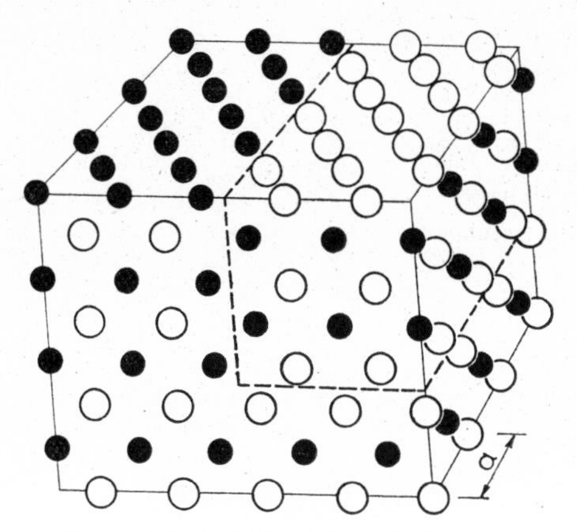

FIG. 6.26. Antiphase boundary in an AB_3-type ordered lattice

sub-lattice contains atoms of only one type. The ordering begins approximately simultaneously in various distant volume elements of the crystal since initially the sub-lattices are equivalent, and if at one place a sub-lattice is filled with say A atoms, at another part of the crystal the same type of sub-lattice is occupied by B atoms. When the increasingly ordered regions meet, a special surface fault, an anti-phase (domain) boundary, develops on the common boundary surface (Fig. 6.26); this can be regarded as a special stacking fault whose relative displacement vector is the (shortest) lattice vector of the disordered lattice.

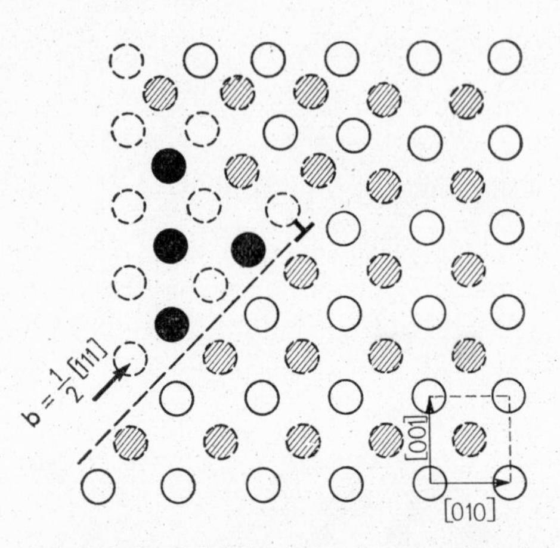

FIG. 6.27. Partial dislocations forming the edge of an antiphase boundary projected onto the (100) plane in CuAuI. The atoms above (below) the plane of the drawing are represented by dotted lines

Since the shortest lattice vectors of the disordered lattice, i.e. the vectors which connect the neighbouring sub-lattices, are no longer lattice vectors in the ordered lattice, the perfect dislocations of the disordered lattice behave as partials in the ordered lattice, and are of necessity connected to antiphase boundaries. The reverse is also true. An antiphase boundary can end within the crystal only in such a partial dislocation (Fig. 6.27), and hence the dislocations always travel in pairs in the glide plane, thus forming a super-lattice dislocation. This dislocation type has been directly observed in several alloys with electron microscope. Naturally both components can dissociate further into partials in the already known way (Fig. 6.28). With the decrease of the long-range order (for instance during a temperature increase), the width of the super-dislocation continually increases and on the attainment of the disordered state, the two parts become independent of each other.

The antiphase boundaries differ from the previously discussed stacking faults only in their much smaller energy, and because they can also be formed without dislocations in the original sense of this concept by constituting a closed surface or a cellular network.

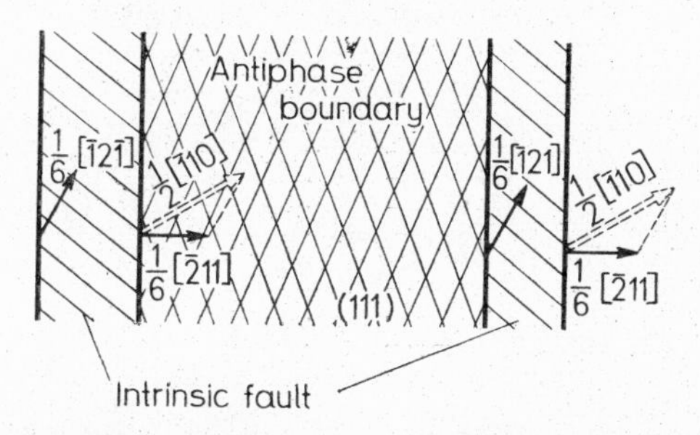

FIG. 6.28. Extended 60° super-lattice dislocation in an ordered Cu_3Au type lattice

The ordering affects the mechanical properties partly because the cross-slip of the super-lattice dislocations is inhibited by the antiphase boundary (though in this case no constriction of the antiphase boundary is necessary for cross-slip), and partly because the intersection of the antiphase boundaries accompanying the ordered state requires an energy input. As regards the details, the reader is referred to the comprehensive work by Stoloff and Davies [34].

6.9. The formation of dislocation loops by vacancy condensation

It has been already mentioned that the high vacancy concentration present in a crystal at an elevated temperature can be frozen-in for a certain time by quenching. If the temperature following the quenching is not very low (in some cases even at room temperature) the vacancies migrate to the dislocations in their vicinity, and become absorbed by non-conservative motion. This process is also promoted by the fact that the stress-field of the dislocations exerts an attraction on the point defects (Section 3.5). The effective range of the attraction, however, is only of the order of 100 Å. The dislocation density is usually sufficiently small for most vacancies to get within this region only after a relatively long diffusion; during this they may join to produce first double and then multiple vacancies. Lattice deformations taking place in the vicinity of impurity atoms also attract the vacancies, and so vacancy clusters develop in the regions more distant from the dislocation.

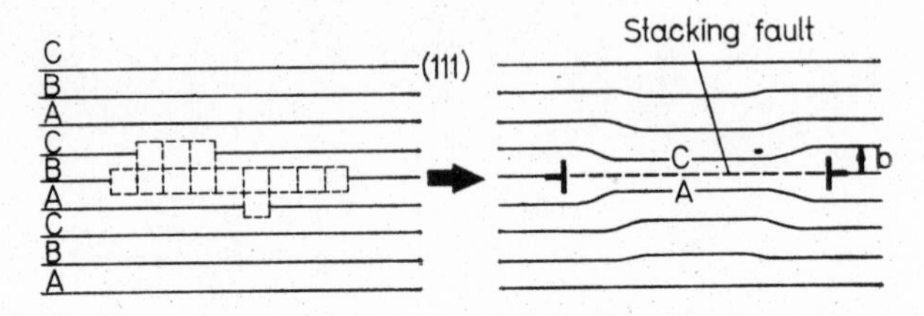

FIG. 6.29. The formation of a dislocation loop in a fcc lattice. On collapsing, the cavity gives rise to a stacking fault which is bordered by a closed, negative Frank partial dislocation

Although the energy of the spherical cavities developing from the vacancy clusters is—at least in isotropic media—always smaller than the energy of the disc of similar volume, in most cases the latter is formed. The anisotropic elastic constants certainly play a decisive role in this, but Frank has also pointed out that the entrance of the newer vacancies has a maximum probability along the edges of the already existing disc-shaped cavities. If the cavity is large enough it collapses; the opposite planes join by creating an extrinsic fault, and a closed, sessile dislocation loop is formed (Fig. 6.29). This process takes place only if the difference between the free surface energy and the energy of the stacking fault is larger than the elastic energy of the dislocation loop. For a loop of radius R formed along the $\{111\}$ plane of the fcc lattice, using relation (2.45) ($b_z^2 = a^2/3$):

$$R^2\pi(2\gamma_0 - \gamma) \geqq 2\pi R \frac{\mu a^2}{12\pi(1-\nu)} \ln \frac{R}{r_0}, \tag{6.21}$$

where γ_0 and γ are the specific energies of the free surface and the stacking fault, respectively. Since $2\gamma_0 \gg \gamma$:

$$\frac{R}{\ln R/r_0} \geqq \frac{\mu a^2}{12\pi(1 - \nu)\gamma_0}. \tag{6.22}$$

According to numerical calculations $2a < R < 6a$. It is also possible to consider cases where the cavity collapses without the formation of a stacking fault (for instance on vacancy condensation along the $\{110\}$ plane of a fcc lattice).

A sessile loop, discussed above, emits vacancies, because the line tension tends to reduce the dislocation length; this effect increases rapidly with decrease of the diameter. On the other hand, depending to some extent on the degree of supersaturation, it can also absorb vacancies from its vicinity. There is a critical supersaturation for every loop diameter; above this the loop grows further, and below it it gradually shrinks until it vanishes [15]. The critical supersaturation increases rapidly if the ring diameter decreases. Hence if a compact material is heat-treated at some lower temperature after quenching, first the smallest rings contract while the large ones continue to grow further. With thin foils the free surface is naturally the main vacancy absorber and with this type of sample the situation is essentially different.

It is an interesting experimental fact that if the energy of the stacking fault is large, electronmicrographs show that the loops consist in most cases of perfect dislocations (without a stacking fault). According to Kuhlmann-Wilsdorf and Wilsdorf [36], this can be explained by the fact that a Shockley partial forms after the collapse of the ring (if the energy of the stacking fault with a surface proportional to R^2 can supply the necessary energy). The resulting displacement is not perpendicular to the layers, and the resulting dislocation is, for instance:

$$\frac{1}{3}[111] + \frac{1}{6}[11\bar{2}] = \frac{1}{2}[110]. \tag{6.23}$$

Thus, a perfect dislocation termed an *R-dislocation* (resultant dislocation, prismatic loop) is obtained; this can glide on a cylindrical surface inclined relative to the plane of the loop. As a rule not every dislocation segment glides uniformly, partly because the segments lie along different planes, and partly because the projection of the shear stress can be different for the various segments. Parts of the larger loops may therefore be starting points of Frank–Read sources.

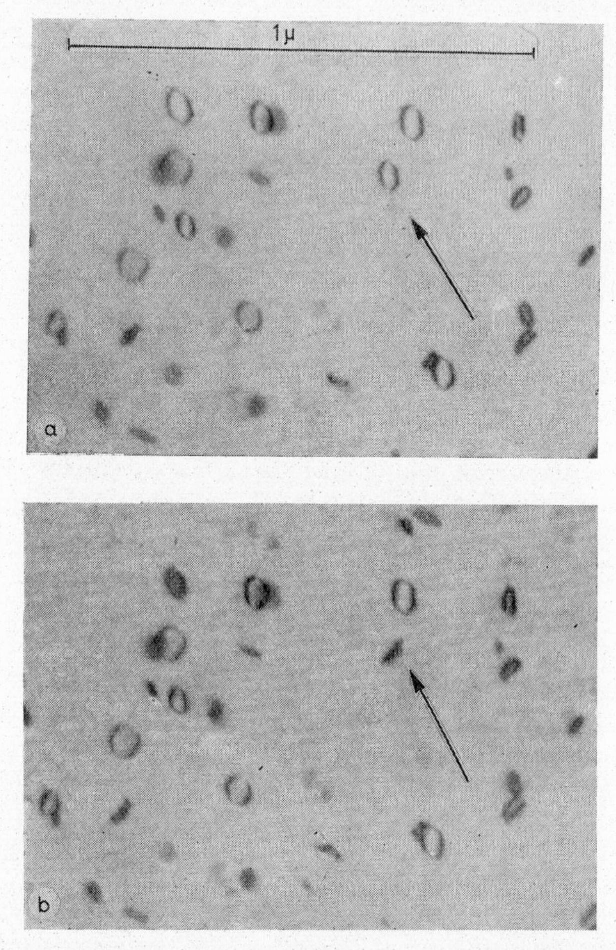

FIG. 6.30. R-dislocations formed by vacancy condensation in Al. The loop marked by arrow has been rotated between the two expositions (Kuhlmann-Wilsdorf and Wilsdorf [36])

It has also been observed that the R-loops frequently change their orientation after their formation (Fig. 6.30), probably because the plane of the ring developed along the {111} planes is inclined relative to the cylindrical glide surface; the dislocation length, however, is smaller in a perpendicular position.

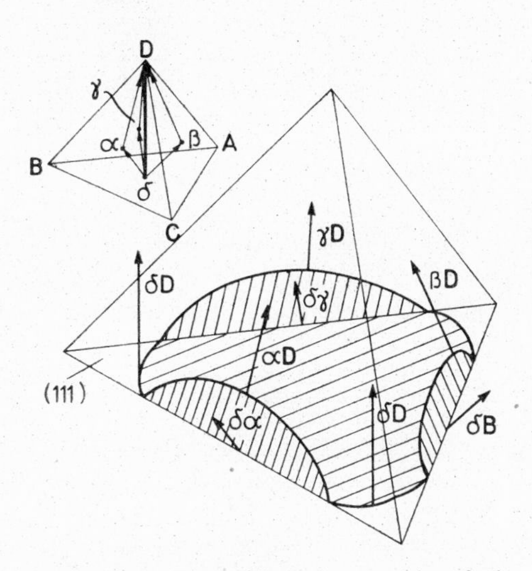

FIG. 6.31. The formation of a stacking fault tetrahedron

Naturally, if the stacking fault energy is small, no R-dislocations can be generated. Sometimes, on the other hand, more complex configurations occur. With gold, for instance, Silcox and Hirsch [37] found a tetrahedral formation whose six edges were six partial dislocations, whereas the four faces formed four stacking faults. This configuration is likewise formed from a dislocation loop. If the energy of an intrinsic stacking fault is small enough, those segments of the dislocation loop which are parallel to the ⟨110⟩ directions split into a Shockley partial and a stair-rod dislocation according to the relation:

$$\frac{1}{3}[111] \rightarrow \frac{1}{6}[011] + \frac{1}{6}[211]\,. \tag{6.24}$$

The resulting partial and stair-rod dislocations produce the tetrahedral formation by creating three intrinsic faults on the other {111} planes as can be seen in Fig. 6.31.

6.10. The energy of stacking faults

It has been shown that the energy of the stacking faults is an important parameter with regard to mechanical properties. Though some efforts have been made to determine this parameter from theoretical considerations, no reliable method is known and unfortunately the experimental data show a considerable scatter. As a rough estimate one may adopt the model according to which the intrinsic fault can be regarded as a single-layer twin crystal whose energy is thus approximately twice the energy of the coherent twin boundary.

TABLE 6.1. *A comparison of stacking fault energies obtained by different methods*

Method	Specific energy of intrinsic stacking faults in erg cm^{-2}							References
	Ag	Au	Cu	Ni	Al	Si	Graphite	
Coherent twin boundary			40	80	200			[3]
Extended dislocations							0.68	[44]
							0.58	[39]
Extended nodes	25		40	150				[40]
	21							[53]
						50		[28]
		30						[55]
Stress-strain curve (τ_3-method)		33	40	95	200			[45]
	43	30	169	410	230			[46]
		10	163	—	238			[49]
				300				[47]
			25		100			[48]
		32						[54]
X-ray line shift	14	67						[56]

This estimate is inaccurate mainly because of the effect of the interaction of non-nearest neighbours. The experimental possibilities are briefly outlined below, and the values obtained for some substances by various methods are compared in Table 6.1 (all the data refer to intrinsic faults).

The electron-microscopic investigation of extended dislocations offers the possibility of direct measurement. If an isolated extended dislocation can be detected microscopically, and if μ, ν and the Burgers vector are known, γ can be determined from the distance of separation by using equation (6.10) or any other similar relation [38–41]. This method is reliable, but it can be applied only for very small γ values ($\lesssim 10$ erg/cm^2) because the electron-microscopic contrast of the dislocations is not sharp enough for the electron-microscopic resolution of a few angstroms to be utilized.

From this point of view the extended dislocation nodes are more favourable. Their extension can be well measured even if the partial dislocations are scarcely separated. The radius of curvature R of the partial dislocations bordering the extended nodes is so large that the line tension T (Section 2.8) just compensates the contracting effect of the stacking fault. The line tension can be determined approximately by calculation, and only the radius of curvature (or some other equivalent parameter) must be measured. Howie and Swann [40] calculated as a first approximation with the line tension of straight dislocations, without taking into account the interactions between the dislocations. Brown [42] revised these calculations accordingly, but his corrected (and substantially increased) values are obviously too large. According to the calculations of Jøssang *et al.* [44], on the other hand, the original values obtained by Howie and Swann are also a little high. Jøssang discusses the experimental errors, and estimates that an accuracy of 10–15% can be obtained by measuring the extended nodes. From the sizes of stacking fault tetrahedra and dipoles the upper limit of the stacking fault energy can be estimated [43].

Another group of methods utilizes the temperature dependence of some parameters of the stress–strain curve to determine the value of γ [45–50]. The energy of the stacking fault affects the yield point, the work-hardening, etc., substantially. The extended dislocation shrinks before the cross-slip or the intersection of other dislocations.

The larger the separation (the smaller the γ value), the larger is the energy input needed. This energy, however, is supplied only partly by the work of the external stress: the rest comes from the thermal energy. The probability of such thermally activated processes is proportional to $\exp(-U/kT)$, where U is the energy of activation (the work done at the expense of the thermal energy) and k is Boltzmann's constant. From the dependences of the respective parameters on the temperature and the rate of deformation, γ can be calculated indirectly. Unfortunately not much can be said of the accuracies of these methods. Dilute alloys cannot be investigated in this way because the alloying directly influences the behaviour of the dislocations [51].

Information can also be obtained from the shifts of the X-ray diffraction maxima, since these are proportional to the stacking fault densities. If the dislocation density is also known, the average width of the extended dislocations can be estimated. The data obtained by this method, however, must be accepted with some caution since the determination of the dislocation density is inaccurate, and in addition the formulae for the line shift refer to stacking faults of large extension [51].

Finally, it should be mentioned that the ratio of the energies of the extrinsic and intrinsic faults is not much larger than unity [52]. Conflicting data are also known; these, however, are based mainly on observations referring to the relative frequency of the two types of faults, and do not take into account the differences due to the mechanisms of their formations.

Chapter 7

Effect of lattice defects on the physical properties of metals

7.1. Mechanical properties of pure metals

7.1.1. Small plastic deformation of single crystals

Crystalline materials react in two ways to mechanical forces. If the force is not too large, the shape of the body changes elastically in direct proportion to the force, and if the force is removed the body resumes its original form. In the case of a large enough force, however, one part of the deformation is not elastic but permanent. According to these properties, those bodies which have perfectly elastic properties up to a certain minimum stress value (the yield stress) are considered to behave ideally mechanically. Above this value, at a given stress, plastic deformation or steady state flow occurs (Fig. 7.1).

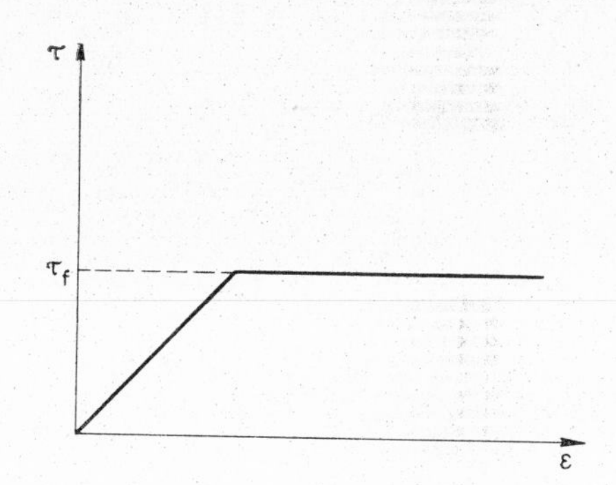

FIG. 7.1. Ideal stress–strain relation

Such ideal substances, however, do not exist. Let us consider under what conditions and to what extent real crystalline materials, and especially metals, approach this ideal state, and what explanation can be given for the deviations, taking into consideration the effect of lattice defects. Only phenomena connected with small plastic deformations are dealt with, when the change of the dislocation density during the deformation can be neglected, and the initial dislocation concentration is not larger than $10^6/cm^2$. (The increase of the yield stress resulting from an extensive deformation, i.e. work-hardening, is treated in Chapter 8.)

At first thought it might be supposed that a perfect crystal, without dislocations, would behave like the ideal body defined above, since it should be elastic up to the theoretical critical shear stress; above this it should abruptly suffer a permanent deformation. From both theoretical considerations and experimental evidence, however, it seems highly probable that no plastic deformation can be induced in perfect crystals.

As soon as the applied stress attained the critical value the permanent glide would not take place simultaneously along a crystallographic plane but, instead, a dislocation would form on the surface of the body. As a result of the extremely large stress, this dislocation would move at a very high rate. At a distance of a few thousand interatomic distances from the surface, however, the force necessary for the movement decreases compared to the initial one by many orders of magnitude [1], and consequently the force on the crystal creates a new dislocation and the whole process starts again. By this means, in a very short time, extremely large slips would occur which would most probably result in fracture. These considerations are supported by experiments on dislocation-free whiskers [2]. The presence of dislocations in the real crystals and the resulting low yield stress, thus, not only facilitate the plastic deformation, but permit it without fracture.

Single crystals which do not contain too large an initial dislocation concentration (10^4–10^6 cm^{-2}) compare well macroscopically with an ideal mechanical body. This is so because—according to the preceding chapters—the plastic deformation (inducing the motion of dislocations) requires a definite stress; below this the deformation must be elastic since without the motion of dislocations, no plastic deformation takes place. The transition into plastic behaviour is thus, at least macroscopically, sharp and the rate of flow becomes stable since, because of the low dislocation concentration, the motion of a single dislocation is not disturbed substantially by the others. Consequently, the rate of motion is determined by the other properties of the crystal. It must be emphasized once more, however, that single crystals behave in this perfect way only macroscopically.

Extremely fine measurements have shown that the transition between the two types of deformation is not abrupt. Below the macroscopic yield stress, at considerably smaller stresses, there already exists a small non-elastic deformation, i.e. glide. Figure 7.2 shows the stress dependence of the deformation along the basal plane of a Zn single crystal [3]. It can be seen that the macroscopic yield stress (τ_f) is approximately 20 p/mm^2, but nevertheless non-elastic deformation can already be observed at about 6 p/mm^2. A considerable part of the non-elastic deformation vanishes immediately after the stress ceases to act, and the residual deformation gradually disappears.

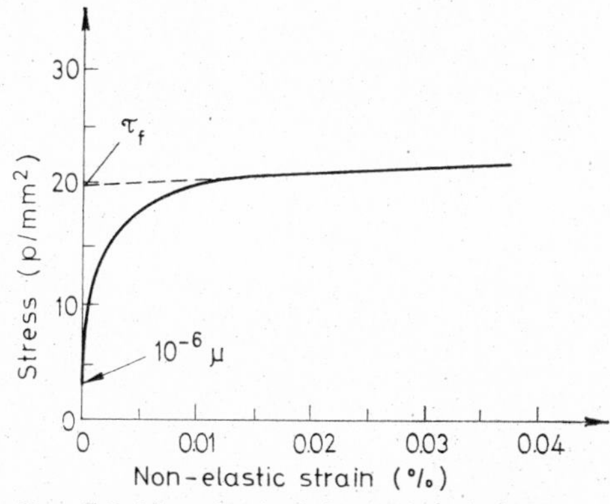

FIG. 7.2. Non-elastic deformation during the glide of a Zn single crystal in the basal plane (according to Robert and Brown [3])

With a stress of 15 p/mm^2 for instance, the non-elastic deformation is 0.002% and its residual part is only about one tenth of this. These phenomena are easy to understand. According to the relation (2.30), if a stress τ acts in the glide plane in the direction of the Burgers vector, then the force exerted per unit length in the glide direction is $F = \tau b$. If a stress acts on the crystal, macroscopic glide begins only when the force acting can already move a considerable number of dislocations. Since the initial dislocation concentration of a well-annealed crystal is of entirely random distribution, the motion of a large number of dislocations occurs only at a relatively large external stress. Nevertheless, there are always dislocations which—because of their favourable positions and orientations – are moved by the force even at a small external stress. In this way, very small slips can develop before macroscopic flow. The ordering of the dislocations left over after heat-treatment, however, presumably results in a minimum total energy. The slips change the favourable dislocation distribution with the effect that a restoring force is generated. If the external stress ceases, this restoring force brings back some of the only slightly slipped dislocations into their original positions.

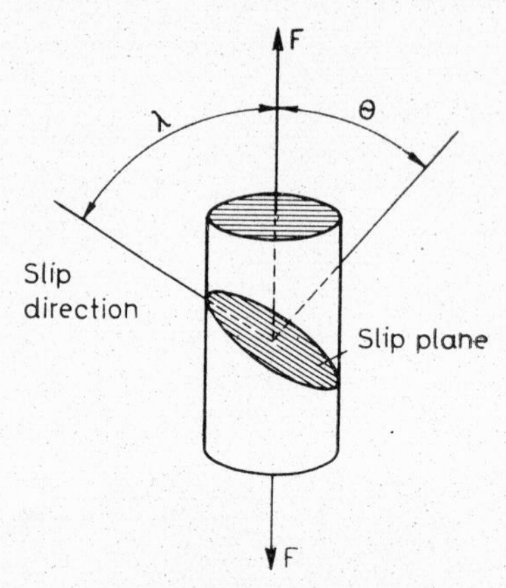

FIG. 7.3. For the determination of the stress acting in a given glide system

Metals can be classified as soft or hard according to their yield stresses. The fcc and hcp metals (these latter only when gliding in their basal plane) belong in the soft group. The hard group contains the bcc and hcp metals (these latter only if the glide does not take place in the basal plane). The macroscopic yield stress is approximately $10^{-5}\ \mu$ for soft metals. These data refer to single crystals only: for polycrystalline metals of both groups, the macroscopic yield stress is about one order of magnitude larger because of grain boundary effects.

In the case of single crystals, the yield stress depends upon the crystallographic directions in which the external forces are exerted. It is a condition of plastic deformation that the stress be large enough in the most favourable glide systems to move the dislocations, or, in other words, that its component in the appropriate direction attain the value of the critical shear stress. Let an axial stress act on a cylindrical crystal and let us denote the angle between the normal of a selected glide plane and the axis of the cylinder by Θ, and the angle between the glide direction and the axis by λ (Fig. 7.3). If a force F is exerted on the cylinder a force $F \cos \lambda$ acts in the glide direction. If the cross-sectional area of the cylinder is denoted by A, the surface along which this force operates is $A/\cos \Theta$. Consequently, in the case of a stress originating from the external force $F/A = \sigma_0$ the shear stress operating in the glide direction in the selected glide plane is

$$\tau = \sigma_0 \cos \Theta \cos \lambda. \tag{7.1}$$

The plastic deformation begins when τ attains the value of the critical resolved shear stress. It can be seen from the above relation that this occurs at an external stress which increases with a decreasing value of the factor $m^{-1} = \cos \Theta \cos \lambda$.

7.1.2. Relation between the macroscopic deformation and the motion of dislocations

Let us now consider the relation between the glide of dislocations and the macroscopic deformation. The effects due to very small stresses result in a single glide; this means that the above-defined component of the external stress attains its critical value only in the glide

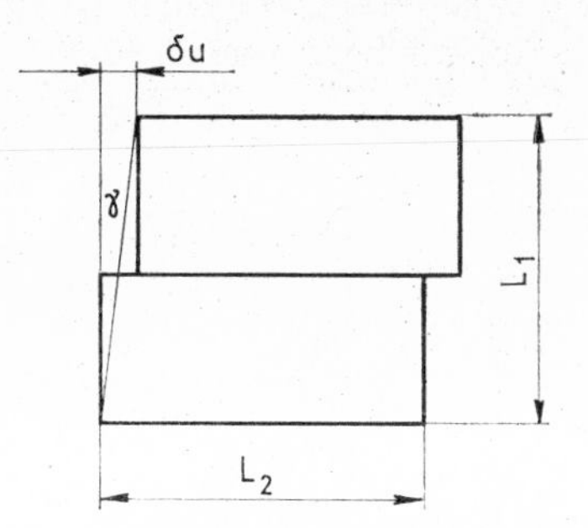

FIG. 7.4. Relation between the motion of a dislocation and the macroscopic shear

plane of the most favourable orientation, and that the dislocations lying in glide planes of other directions remain at rest. It is sufficient in this case to take into account only those dislocations which move in parallel glide planes. If a dislocation completely sweeps a plane of surface A, those parts of the crystal which are separated by this plane become slipped relative to one another by a distance b. On the other hand, if the dislocation moves only along the area δA, then the relative displacement δu of the two crystalline parts is smaller in the ratio $\delta A/A$, and thus $\delta u = (\delta A/A)\,b$. If the size of the crystal perpendicular to the glide plane is L_1, then according to Fig. 7.4 the shear angle is $\delta\gamma = (\delta A/AL_1)\,b$. For simplicity let us assume that there are N' dislocations of length l, and with parallel directions and parallel Burgers vectors, in the crystal, and that these dislocations are displaced by a constant external stress to a distance ds in time dt. The change of the shear deformation for this glide is then

$$d\gamma = N'\frac{\delta A}{AL_1}b = \frac{N'b}{AL_1}l\,ds\,.$$

The expression $N = N'l/AL_1$ gives the length of dislocations per unit volume of the crystal, i.e. the dislocation density. Hence the rate of deformation due to a constant stress is

$$\frac{d\gamma}{dt} = Nb\bar{v}\,, \tag{7.2}$$

where $\bar{v} = ds/dt$ is the average velocity of the dislocations. According to experiment the rate of deformation is constant if a constant stress is applied, and thus in this case the velocity of the dislocation is also constant.

The velocity of the dislocations depends considerably upon the applied stress. Figure 7.5 shows the stress dependence of the dislocation velocity in a Fe–3% Si alloy [4]. It can be seen that the velocity relating to a given stress depends significantly upon the temperature. The motion of dislocations is generally impeded by the intersection of dislocations travelling on their glide planes, and also by their interaction with impurity atoms. An intersection or the break away from an impurity atom occurs by the joint action of the external stress and the thermal activation. The velocity of the dislocations is proportional to the number of elementary processes per unit time, that is to $\exp[-U(\tau)/kT]$, where $U(\tau)$ is the apparent activation energy of an elementary process. This activation energy depends upon the external stress. Figure 7.6 shows the relations between the shear stress and the temperature necessary to keep the velocity constant, and between the macroscopic yield stress and the temperature. It is quite evident from this diagram that there is a clear relation between the yield stress and the stress needed to move the dislocations.

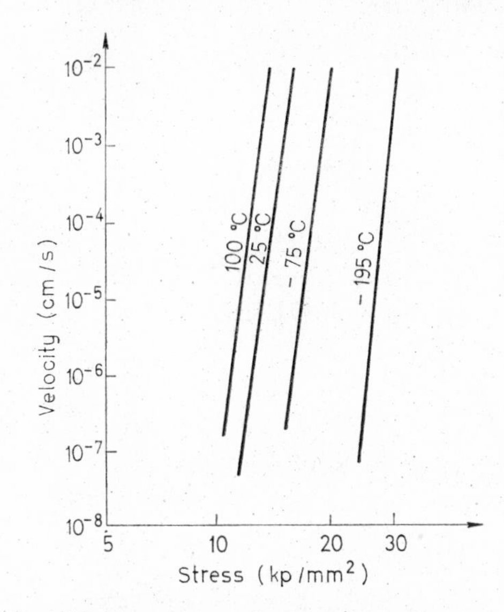

FIG. 7.5. The stress- and temperature-dependence of the velocity of edge dislocations in Fe–3 % Si alloy (according to Stein and Low [4])

The experimental results referring to the stress dependence of the mean dislocation velocity can be described in general by the equation

$$\bar{v} = \left(\frac{\tau}{\tau_0}\right)^n , \tag{7.3}$$

where τ_0 is the stress necessary to attain unit velocity. The constant n is characteristic of the material under investigation and the experimental method. For the alloy Fe–3% Si $n \approx 35$. n generally depends upon the temperature and has widely varying values for different substances [5].

According to Gilman, at low temperature, instead of the thermal activation discussed above, the break free of the dislocations from the obstacles is mainly due to *the tunnel effect* [6]. This occurs at temperatures for which

$$T < T_c \cong 2\tau V/k,$$

where V is the activation volume, which is of the order of magnitude of the atomic volume, and k is the Boltzmann constant. The probability of the break away and together with it, the velocity of the dislocations is

$$\bar{v} = Ae^{-\beta\mu/\tau}, \tag{7.4}$$

where A and β are constants, and μ is the shear modulus. For many materials the stress dependence of the dislocation velocity may be described by this expression [7]. The temperature dependence of the

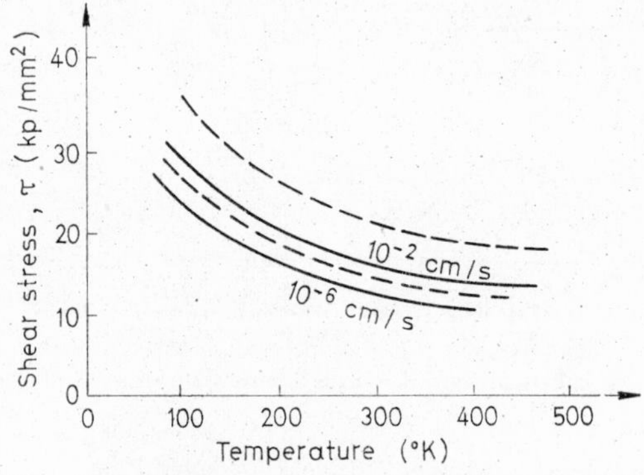

FIG. 7.6. The temperature dependence of the stress belonging to a given dislocation velocity in a Fe–3% Si crystal. The lower of the dotted curves gives the critical resolved shear stress necessary for the extension, while the upper curve gives the stress necessary for 0.3% plastic bending

rate relating to constant stress arises—for instance in the case of interaction with an impurity atom—from the fact that the effective interaction energy decreases because of the thermal vibration of the impurity atom.

7.1.3. Flow stress of pure metals

In pure metals the motions of dislocations may be obstructed by the following effects:

(a) *Frictional (Peierls) forces.* These can probably be neglected in fcc and hcp lattices, but their effects in bcc metals may be considerable at low temperatures.

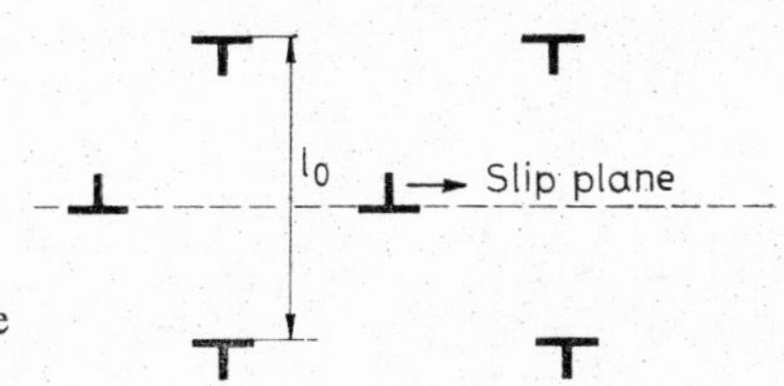

FIG. 7.7. Uniformly distributed edge dislocations

(b) *Line tension.* For the onset of any macroscopic deformation the multiplication of dislocations is necessary. We have already seen that the operation of a source of length l requires to a good approximation the stress τ_F (equation 5.8).

(c) *The stress-fields of other dislocations or dislocation groups.* A dislocation which moves between dislocations with the same or opposite Burgers vector at an average distance from each other of l_0, for example, according to Fig. 7.7 is acted on by an attractive (or repulsive) force of approximately

$$F_a = \frac{\mu b^2}{2\pi l_0}.$$

A stress

$$\tau_a = \frac{\mu b}{2\pi l_0}$$

is needed to overcome this.

Since to a good approximation l and l_0 are equal to the reciprocal of the square root of the dislocation density N, these latter two effects can be expressed by a single formula. Thus the stress necessary to initiate the glide and overcome the elastic interactions (the flow stress) is

$$\tau_E = \tau_F + \tau_a \approx \mu b\sqrt{N}. \tag{7.5}$$

If $N = 10^6$ cm^{-2}, then the flow stress is about $3 \times 10^{-5}\ \mu$ which is in accordance with experiment (see Table 1.1).

(d) *The intersection of other dislocations and the non-conservative motion of the jogs formed by the intersections.* The energy U_0 invested in each elementary step is in this case of the order of magnitude of a few eV. In the meantime the displacement of a dislocation is so small that this energy cannot be provided by the work of the effective flow stress acting on the dislocation. At a temperature $T > 0°$ K, however, a considerable part of U_0 can be provided, with a finite probability, by the thermal energy of the lattice vibrations. (It should be noted that the Peierls potential may also be overcome by a thermally activated process, as can a series of other similar processes, e.g. the contraction of extended dislocations.)

Let us consider the following simple case. Let the energy necessary to displace a dislocation segment of length l' by an average distance of d be U_0. The work of the effective shear stress in this operation is $\tau_T bl'd = \tau_T V$ ($V = l'bd$ is the *activation volume*; $d \approx b$ if the dislocations intersect each other). Thus, the thermal energy to be used (*the activation energy* of the process) is [8]

$$U(\tau) = U_0 - \tau_T V.$$

The rate of deformation is proportional to the probability that this energy is available at the given position, i.e.

$$\dot{\gamma} = \dot{\gamma}_0 \exp\left\{-\frac{U_0 - \tau_T V}{kT}\right\}, \tag{7.6}$$

where $\dot{\gamma}_0$ is a constant which depends upon the nature of the substance, the degree of deformation, and possibly upon the other conditions of the investigation. Let $m = \ln \dot{\gamma}_0/\dot{\gamma}$ (m depends only slightly upon the degree of deformation, and to a good approximation its value is 25 [9]). Thus, from equation (7.6), the stress necessary to move a dislocation is

$$\tau_T = \frac{1}{V}(U_0 - mkT). \tag{7.7}$$

In extremely pure fcc metals, containing initially only a low dislocation density, the shear stress decreases between 0 °K and room temperature by about 15–20% [10]. Calculating with these data, from equation (7.7) one obtains $U_0 \approx 3.5$ eV.

In most cases every effect must be considered in the calculation of the shear stress τ; that is in general:

$$\tau = \tau_E + \tau_T = \tau_E + \frac{U_0}{V}\left(1 - \frac{mkT}{U_0}\right). \tag{7.8}$$

Of course τ_T cannot be negative, but it vanishes if at the given rate of deformation, the thermal energy alone can supply the energy requirement U_0, i.e. if the temperature is

$$T_0(\dot{\gamma}) = \frac{U_0}{mk}.$$

At temperatures higher than this, $\tau = \tau_E$ (Fig. 7.8a). The behaviour in accordance with the above can best be studied in the case of the glide along the basal (001) plane in hexagonal metals. In Fig. 7.8 for example, the straight lines corresponding to the above relation were fitted to experimental results obtained with Cd in 1930. From the dependence of T_0 on the rate of deformation, and from the data of the Figure, $U_0 = 0.7$ eV. With this value, and by assuming that $d = b = 2.97$ Å, one obtains from τ_T that $l' \approx 2\times10^{-4}$ cm, and from τ_E with $\mu = 1900$ kp/mm^2 one has $l \approx 5\times10^{-4}$ cm. These are realistic values.

The latter calculations can be applied in their original form only to those fcc metals whose stacking fault energies are large (e.g. Al). If, because of the small energies of the stacking faults, the partial dislocations become separated from each other by 5–15 atomic distances, the stacking fault between the two partials must first contract before any interaction can take place. The energy necessary for this process is considerably smaller for screw dislocations – because of their smaller separation – than for edge dislocations. For this reason, one cannot tell in advance with complete certainty which dislocation component has the decisive motion. It may be assumed that in many cases neither of them can be neglected, and thus several thermally activated processes take place simultaneously.

Similar experiments with non-metallic materials do not show the above temperature-dependence. From this it may be concluded that with these crystals the motions of dislocations are obstructed mainly by other interactions. Finally, however, everything depends upon the dislocation density, and the interaction between dislocations may be regarded as of secondary importance only in cases of very small dislocation densities.

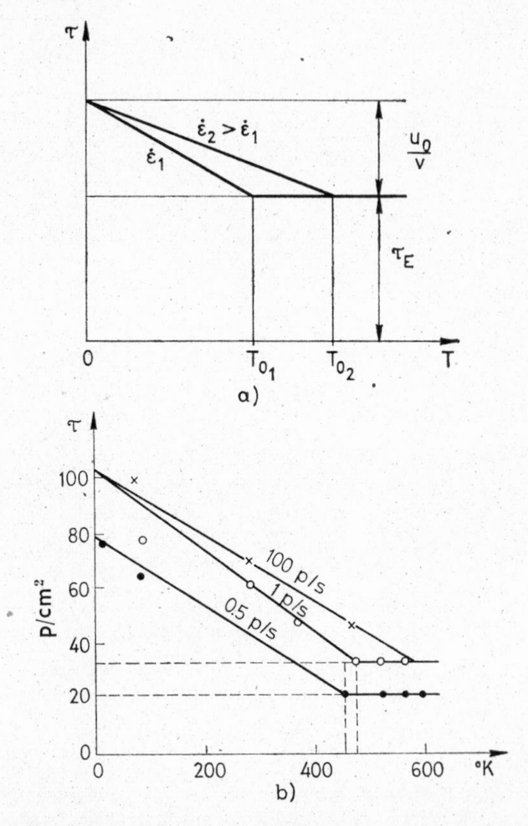

FIG. 7.8. (a) Dependence of the critical stress on the temperature and the deformation rate, and (b) the critical resolved shear stress of Cd single crystals. In the latter, the numbers beside the three types of measurements indicate the rate of change of the load. The measurement series indicated by the lowest empty circles refers to crystals which were grown considerably more slowly than the others; thus these crystals presumably contain fewer dislocations, and consequently the value τ_E is smaller (from data of Schmied and Boas; *Kristallplastizität*, Springer, Berlin, 1935)

It will be seen later that the separation of the flow stress into temperature-dependent and temperature-independent components can also be realized in other cases, even if the processes occurring differ considerably from those which have been considered in this section, since, quite generally, every process can be divided into thermally activated and inactivated processes. A thermally activated process, however, can take place only if the energy required to overcome the obstacle can be supplied during a relatively short time and must be concentrated on a small volume. For a more detailed study, the reader is referred to the literature [12, 13].

7.1.4. *The plastic deformation of polycrystalline metals*

The metals used in practice are never single crystals, but consist of a large number of crystal grains of various size. When a force is applied to a polycrystalline material, the orientations of the glide systems of the various grains compared to the direction of the force differ, and hence the values of the shear stresses in the individual glide planes will also be very different. For this reason, in contrast with single crystals, no clear-cut relation such as equation (7.1) exists between the external stresses and the forces activating the glide. However, the mean value of the orientation factor $m = (\cos \Theta \cos \lambda)^{-1}$ introduced in relation (7.1) can be defined by

$$\bar{m} = \frac{\int N(m)\, m\, dm}{\int N(m)\, dm} \tag{7.9}$$

where $N(m)$ is the number of grains in which the orientation of one or other of the glide planes is determined by the factor m. The integral must be extended to every grain. The factor m has been introduced through the connection of the external tensile stress, σ_0 and the shear stress acting in a given glide plane. The tensile strain and the shear strain in a given glide plane are connected by the same factor. This means that if the shear strain is denoted by γ, and ε is the tensile strain, then:

$$m = \frac{\sigma_0}{\tau} = \frac{d\gamma}{d\varepsilon}. \tag{7.10}$$

By using the factor defined by (7.9), a relation can be found between the single crystal and the polycrystal of the same material. Let us assume that there is no correlation between the orientation of a grain and its yield stress. In this case, the relation between the external stress necessary to induce tensile deformation and the critical resolved shear stress of the single crystal is $\sigma_0 = \overline{m}\tau$. If $N(m)$ does not change during the deformation, then from (7.10) we have

$$\gamma = \overline{m}\varepsilon, \tag{7.11}$$

which connects the tensile and the shear deformations. The resolved critical shear stress, however, is a function of the deformation. The stress–strain relation $\tau = f(\gamma)$ characterizes the work-hardening of the single crystal in question (see Chapter 8). From this, and by using the previous equations the relation between the tensile strain and stress can also be given for polycrystals:

$$\sigma_0 = \overline{m}\tau = \overline{m}f(\gamma) = \overline{m}f(\overline{m}\varepsilon). \tag{7.12}$$

In order to calculate the factor $\overline{m}$, some further conditions must also be taken into account. According to Sachs, provided that every grain deforms along a single glide system independently from its neighbours, in the case of grains of random distribution, $\overline{m} = 2.258$ for fcc metals [14]. This assumption, however, is open to many objections. If every crystal grain could be deformed along one glide system only, cracks and cavities would form along the grain boundaries. In practice, however, the material remains continuous during the plastic deformation. The deformation of the single grains is determined by their neighbours too. Figure 7.9 shows the distribution of glide lines in high-purity polycrystalline aluminium [15]. One can see that several glide systems operate within one grain, and that the deformation is strongly inhomogeneous.

If the volume does not change during the deformation (this condition holds quite well here), then the most general deformation can be accomplished by glides taking place in five various glide systems [16], since the deformation tensor contains six independent components, one of which is fixed by the volume condition. Thus, in the determination of the mean value of the factor m, all five possible glide systems must be considered. Since some of them are in rather unfavourable positions with respect to the external force, their m values are large. If every glide system is taken into account the mean value of m increases. In fact, the experimental study of equation (7.12) led to $\bar{m}$ values between 3.0 and 3.1 [17–19].

FIG. 7.9. Glide lines appearing in high-purity, polycrystalline aluminium after an elongation of a few per cent (Boas and Ogilvie [15])

The plastic deformation of polycrystalline materials has been theoretically investigated more thoroughly by Bishop and Hill [20]. The relation between the stress and deformation has also been determined for deformations more complicated than simple tension, satisfying the continuity conditions for both the material and the stresses. These investigations will not be dealt with further.

In the previous considerations it was assumed that every grain behaves identically and that the deformation is homogeneous and uniform [$N(m) = \text{constant}$]. If this condition is not satisfied, the behaviour of the polycrystals depends also upon the grain size. For instance the relation between the tensile yield stress of polycrystals and the grain size D is given by the following equation (Petch equation [21])

$$\sigma = \sigma_i + k_i D^{-\frac{1}{2}},$$

where σ_i and k_i are temperature-dependent constants.

7.1.5. Creep phenomena

In order to obtain a rapid macroscopic deformation, a stress larger than the yield stress must be applied. In practice, however, plastic deformation also occurs when a load smaller than the yield stress is applied to the crystal for a prolonged time at not very low temperatures. At stresses below the yield stress, macroscopic deformations do not occur immediately but the crystal defects in the substance may move at these stresses if the load is applied for a long time.

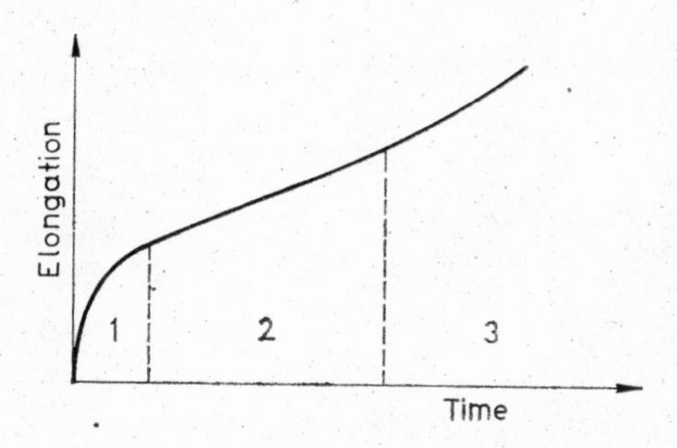

FIG. 7.10. The time dependence of an extension due to a constant stress

This type of deformation is called creep. In these processes the temperature effects are essential in contrast to the plastic deformations occurring at stresses larger than the yield stress where the effect of temperature is very much less.

Figure 7.10 shows a typical creep curve [21]. The curve can be divided into three characteristic stages. The first is that of *primary* or *transient creep*. In this stage the flow rate gradually decreases. This is followed by the stage of *steady state creep*, where the flow rate is approximately constant. The third stage, where the flow rate increases is that of *accelerating creep*.

On plastic deformation the volume of the body remains approximately constant, and therefore in the case of tensile elongation its cross-section decreases with increasing length. Thus, at first sight it seems that the accelerating creep is only a consequence of the increasing stress (at constant external force). This latter effect, however, scarcely contributes to the acceleration, because accelerating creep can also be found with creep induced by compression [22].

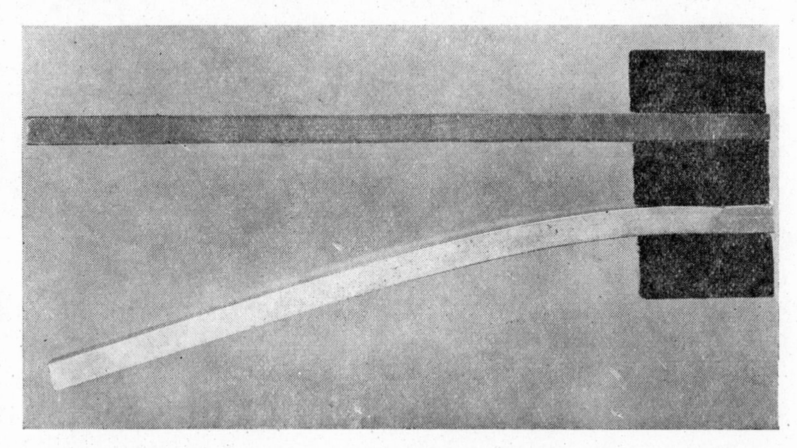

FIG. 7.11. The deformation of cadmium (below) and lead (above) rods during two weeks under the force of gravity (Andrade [23])

It is very important to investigate these phenomena from the technological viewpoint. A unified theory of creep processes does not yet exist. The phenomena are modified by the previous treatment of the samples, the grade of purity, and the nature of the material used. For instance, at room temperature cadmium is more resistant to plastic deformation than lead. Nevertheless, if a lead and a cadmium rod of the same size are both fixed at one end and are subject only to gravitational forces, the cadmium is bent after 14 days, whereas the lead remains practically undeformed [23] (Fig. 7.11). This and many other experimental findings cannot yet be explained satisfactorily. It is certain, however, that the creep phenomena are closely related to thermally activated recovery processes (see Chapter 9). Although many of the experimental results can be attributed to a dislocation mechanism, in some cases they can be explained by grain-boundary motion [13, 24]. In a substance which is free of stresses the dislocation distribution is quite stable. Many dislocations are fixed either by various obstacles or by the Cottrell atmosphere. The motion of these dislocations requires considerable activation energy, which under normal circumstances cannot originate merely from thermal effects. However, an external stress smaller than the yield stress may considerably decrease the value of the effective energy of activation, and in this way the motion of dislocations can be induced by thermal effects. Mostly such processes occur during creep. The slow motion of dislocations becomes directly visible in electron micrographs taken at various time intervals during the flow. Figure 7.12 shows three successive pictures; the arrows indicate characteristic changes, but other changes too can be observed.

Let us briefly review the characteristic properties of the various creep stages. Transient creep can be found at high and low temperatures, and its characteristics depend upon the previous deformation and purity of the sample [13, 25, 26]. (The temperatures here are termed high or low depending on whether they are near or far below the melting point of the sample.) According to experimental observations, the primary creep can be described by the following empirical relation:

$$\frac{d\varepsilon}{dt} = At^{-n}, \qquad 0 < n < 2. \tag{7.13}$$

A depends upon the temperature and the external stress, while the value of n is found in practice to be 1 or $\frac{2}{3}$. The former case is called logarithmic, and the latter Andrade creep.

These phenomena can be interpreted by the following model. If a moving dislocation intersects another one, a jog is produced on it. After this, energy is required to move the jog, but in general this cannot be supplied by the external stress responsible for the creep; consequently, the dislocation is fixed in the point of intersection, and assumes the form shown in Fig. 7.13. The bowing out is such that the line tension is in equilibrium with the shear stress acting in the glide plane. These formations can be well observed in electron micrographs [27]. As a result of this bowing out, the line tension exerts a force to move the jog. This effect, however, though itself not enough

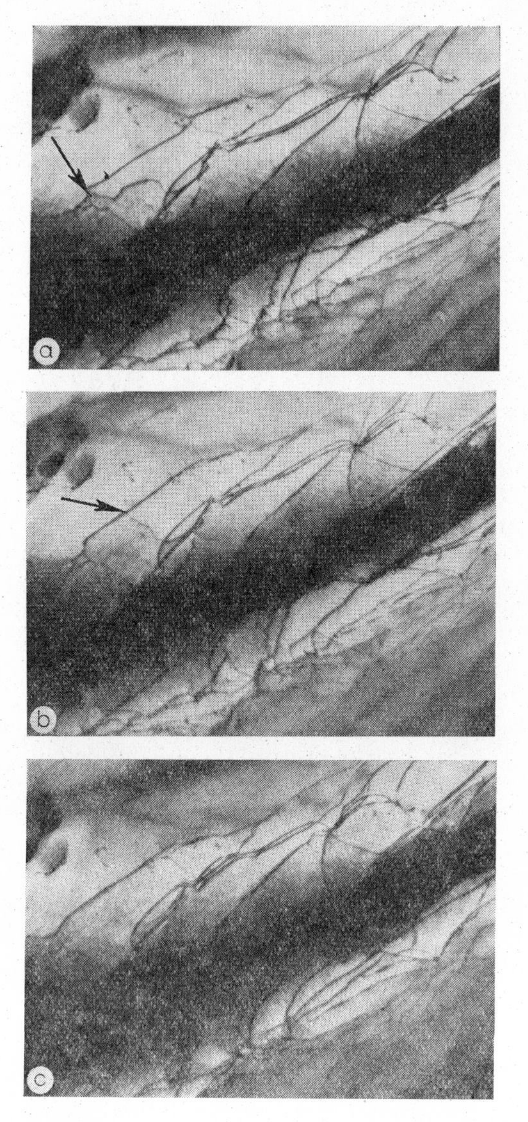

FIG. 7.12. The change of the dislocation structure during creep in 99.999% copper after a 5% extension at 500 °C ($\dot{\varepsilon} = 10^{-5}\text{sec}^{-1}$, 40,000×; micrograph of P. Feltham)

to move the jog, does decrease the activation energy, and thus after a prolonged time there is a finite probability that the jog motion will be thermally activated. Now depending upon the character of the dislocation and the jog this latter moves either in the direction of the

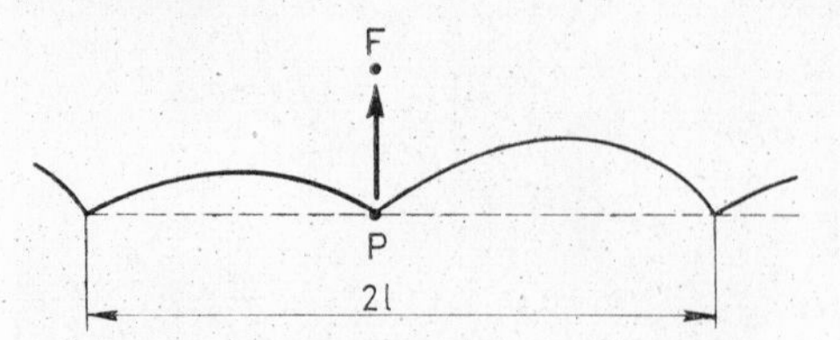

FIG. 7.13. The characteristic form of a dislocation pinned at individual points

point F, and thus the dislocation too can advance, or along the dislocation line, and so the stabilizing effect in the point P ceases. If a stress τ operates in the glide direction in the glide plane, the activation energy of this process which is controlled by the jog motion can be expressed by the equation:

$$U(\tau) = U_j - \tau b^2 l, \tag{7.14}$$

where b is the Burgers vector, l is the average distance of the pinning points, and U_j is the activation energy of the process without the external force. Taking this equation into consideration, the creep rate is determined only by the thermal condition

$$\dot{\varepsilon} = Be^{-U(\tau)/kT}, \tag{7.15}$$

where B generally depends upon the external stress, the number of dislocation segments which can be activated and the atomic vibration frequency. If it is assumed that B does not depend upon the deformation and that τ is proportional to it, equation (7.15) can be integrated, and the result describes the logarithmic creep. For the interpretation of the Andrade creep, according to Mott [28], those local stress changes must be taken into account which are associated with the release of single dislocations from a piled-up conglomeration.

Expression (7.15) is also valid for steady-state creep. This takes place primarily at high temperatures, and is the result of two opposing effects. With increasing deformation the material hardens; at the same time, however, as a result of thermal activation, dislocations are annihilated partly on grain boundaries and partly by meeting dislocation pairs of opposite sign. These latter processes reduce the hardness of the material. The final result is that in this stage, the creep rate is approximately constant. The theory of steady-state creep deals with these ideas in detail [29–31].

Finally it should be mentioned that the development of the third stage of creep is associated with the formation of internal cavities [32].

7.2. Theory of the yield stress of solid solutions

7.2.1. Introduction

The solid solution of any metal with some other substance results in a material which is harder than the pure metal. In other words, impurity atoms or alloying increase the yield stress τ of the metal. This increase depends upon the material and the concentration c of the impurity or alloying atoms, and also upon the way in which these atoms are incorporated into the host material. The yield stress of a solid solution also depends considerably upon the temperature.

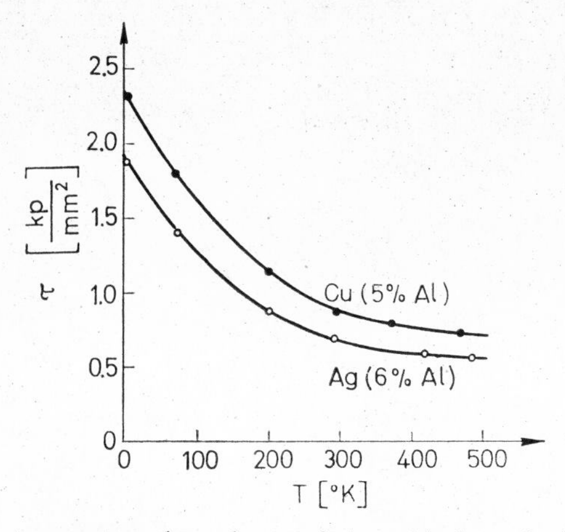

FIG. 7.14. The temperature dependence of the critical resolved shear stress in Cu–5% Al and Ag–6% Al alloys

Figure 7.14 shows as characteristic examples the yield stress versus temperature for the alloys Cu–5% Al and Ag–6% Al [33]. (In the following, only the properties of single crystals are investigated.) It can be seen that at an elevated temperature ($T > 400$ °K) the yield stress is practically constant. The stress relating to this permanent stage is usually called "plateau stress", and is denoted by τ_p. In the temperature range $0 < T < 500$ °K the yield stress is generally composed of two terms. One term has its maximum at $T = 0$ °K and is zero from the beginning of the plateau, whereas the other term is the plateau stress τ_p. Thus,

$$\tau(T) = \tau_0(T) + \tau_p. \tag{7.16}$$

It follows from the temperature dependence of the yield stress that the effect of various alloying materials on the host metal can be compared only in experiments carried out at the same temperature. The experimental results concerning the concentration dependence of the yield stress are not clear-cut. Thus for instance the yield stress measured in the Ag–Al alloy at 4.2 °K is a linear function of the concentration, whereas it increases proportionally to its square root in the Cu–Al alloy [33]. More recently, it was found for several Cu-, Ag- and Au-based alloys that the yield stress (τ_p) is a linear function of $c^{2/3}$ [34].

7.2.2. *Theoretical principles*

Several theoretical attempts have been made in the last thirty years to interpret the yield stress of solid solutions. However, no uniform theory has emerged from these efforts. Although various theories are able to explain a small group of phenomena, they lead to considerable deviations in other cases. Thus, in the following we do not restrict ourselves to any one of these theories, but instead try to review the essential features of the more important ones.

Solid solutions are harder than the pure host metal, because the solute atoms interact with the dislocations and as a result the stress necessary to induce the motion of the dislocation, i.e. the yield stress, increases. The determination of the extent of the increase of the yield stress requires an exact knowledge of the nature of the interaction.

This, however, is not possible with the existing theories. Nevertheless the interaction can be studied (as was seen in Chapter 3) in a manageable form via the linear elasticity on the basis of the continuum theory.

It is certain, however, that the information so obtained cannot be applied to foreign atoms incorporated in the immediate vicinity of the dislocation line; the role of these is of basic importance especially at low temperatures. The solute atoms bonded directly to the dislocation are localized points of stabilization on the dislocation, which can be annihilated by thermal activation. Thus, $\tau_0 (T)$, which depends considerably upon the temperature, must originate from such interactions.

The results obtained from the continuum theory can be applied to solute atoms which are relatively far from the dislocation line. These extend over a large number of atoms, and consequently also over a considerable volume in the vicinity of the dislocations. This results in a non-localized stabilization, and the barrier so developed cannot be overcome by thermal activation of the dislocation. From the results of the continuum theory, it may be expected first of all that they permit the interpretation of τ_p, i.e. that part of the yield stress which is independent of the temperature. The ideas developed show that the experimentally observed form of the yield stress (7.16) is a natural consequence of the interaction between the dislocations and the solute atoms.

7.2.3. *Theories of Mott and Nabarro*

The first attempts to explain the increase of the yield stress of solid solutions were made by Mott and Nabarro. In their first theory they assumed [35] that the average shear deformation brought about by the solute atoms in the matrix is equal to the deformation made by one of these atoms at a distance which is equal to the average distance of the solute atoms. It was further assumed that the stress corresponding to this deformation impedes the motion of the dislocations. According to their results the increase of the yield stress is

$$\tau_p = \frac{8\mu}{3} \eta_V c\,, \tag{7.17}$$

where η_V is the volume size factor defined in Section 3.6.1. If $\eta_V = 0.1$ and $c = 1\%$, then $\tau_p \approx 3\times10^{-3}\,\mu$. (It follows from the averaging that a long-range interaction is considered, i.e. the result characterizes an effect which is independent of temperature.) A change of such an order and a linear concentration dependence have recently been observed in Fe–N single crystals [36] for interstitial N atoms.

Mott and Nabarro later modified this theory [37]. They showed that the stress obtained from their previous considerations is equal to the stress necessary to bend dislocations along which the distance l between the pinning points is equal to the mean distance of the impurity atoms, i.e. $l \approx \mu\, b/\tau_p$. In not very dilute solid solutions, however, as a rule $l \ll \mu\, b/\tau_p$, and therefore the dislocation remains straight and moves without changing its shape. Taking this into account the modified result is

$$\tau_p \cong \mu\eta_V^2\, c^{5/3}\,(\ln c)^2. \tag{7.18}$$

With the previous η_V and c values, $\tau_p = 10^{-4}\mu$. Changes of this order of magnitude can be observed in single crystals of fcc substitutional solid solutions [34], but nevertheless the measured concentration dependence differs considerably from equation (7.18).

Mott made a further modification to the theory [38] by taking into consideration that if $l \ll \mu b/\tau_p$ then the non-zero mean stress acts only on a finite dislocation segment. The result is

$$\tau_p = 1.8\mu\eta_V^{4/3}\, c. \tag{7.19}$$

From this, $\tau = 10^{-3}\,\mu$, which again is comparable with the yield stress observed in Fe–N [36].

7.2.4. *The Friedel theory*

On the basis of the Mott–Nabarro model a theory was worked out by Friedel. In contrast to the previous theory a short-range interaction between the dislocation and the solute atom was assumed. The temperature-independent yield stress increase was expressed by the relation [39]

$$\tau_p = \frac{1}{5}\,\mu\,(\eta_V\, c)^{4/3}. \tag{7.20}$$

Accordingly, $\tau_p = 2\times10^{-5}\ \mu$, which is very small compared with the observed values.

7.2.5. *The Schoeck–Seeger theory*

As a result of the interaction between the dislocation and the solute atoms some ordered distribution of the impurity atoms develops. This impedes the dislocation motion even if the solute atoms are not connected with the dislocations. According to the calculations of Schoeck and Seeger, the effect results, in the case of substitutionally solute atoms, in an increase of the yield stress by c^2 [40]. This can be regarded as temperature-independent only if the change of the ordered structure by diffusion can be neglected with respect to the motion of dislocations. More recent investigations, however, have shown that this latter assumption is not justified [36, 42].

7.2.6. *The Fleischer theory*

Fleischer's theory directly utilizes the interaction between the dislocations and the solute atoms. It was seen in Chapter 3 that the energy of interaction between a straight dislocation and the solute atoms which is due to the size effect and modulus effect can be calculated. During its motion in the glide plane (the x, z-plane) the dislocation leaves behind the impurity atoms. Since the energy of interaction depends upon the distance between the solute atom and the dislocation line, any displacement of the dislocation requires an energy input corresponding to the change of the interaction energy. This means that the force necessary to move the dislocation comes from the gradient of the interaction energy. Assuming that the dislocation moves along the x-axis, from the relations (3.30), (3.32), (3.37), (3.39) and (3.41) with the definition $F_x = -(\partial U/\partial x)$ the following expressions are obtained for the ratio of the forces originating from the modulus and size effects:

$$\frac{F_M}{F_S} = 0.09\,\frac{\eta_\mu}{\eta_V}\,\frac{b}{y}\left\{1 + 2\left(0.3\,\frac{\eta_K}{\eta_\mu} - 1\right)\frac{y^2}{r^2}\right\} \tag{7.21}$$

for the case of an edge dislocation, and

$$\frac{F_M}{F_S} = \frac{1}{8}\frac{\eta_\mu}{\eta_V} \tag{7.22}$$

for a screw dislocation. It can be seen that in this latter case the ratio of the two forces does not depend upon the position of the solute atom. If we consider only the maximum force associated with the solute atoms which are closest to the glide plane and the dislocation ($r = y = b$), the ratio does not depend upon the position in the case of an edge dislocation either, since from (7.21)

$$\frac{F_M}{F_S} = -0.09\frac{\eta_\mu}{\eta_V}\left(1 - 0.6\frac{\eta_K}{\eta_\mu}\right). \tag{7.23}$$

Considering both the size effect and the modulus effect, it can be said that the force experienced by the dislocation depends upon the suitably weighed sum of the quantities η_V, η_μ and η_K. Consequently, the force on a dislocation can in general be written in the following way:

$$F = \beta\mu(\alpha\eta_V + \eta'_\mu), \tag{7.24}$$

where the values of α and β depend upon the character of the dislocation. The sign of α is positive or negative depending upon whether the interaction due to the modulus and size effect is attractive or repulsive. For screw dislocations

$$\eta'_\mu = \eta_\mu \tag{7.25a}$$

and for edge dislocations

$$\eta'_\mu = \eta_\mu\left\{1 + 2\left(0.3\frac{\eta_K}{\eta_\mu} - 1\right)\right\}. \tag{7.25b}$$

The force equation (7.24) is clearly valid, though considerable difficulties arise in selecting the coefficients α and β. Further, the literature value of η'_μ for edge dislocations is not too certain. According to Fleischer's calculations [42] (for the case $\eta_K = 0$):

$$\eta'_\mu = \eta^F_\mu = \frac{\eta_\mu}{1 + \frac{1}{2}\eta_\mu}. \tag{7.26}$$

Saxl, however, has proved that Fleischer's calculation was incorrect [43].

The temperature-independent hardness increase can be accounted for in the following way. It can be assumed that the stress τ_p necessary to move a dislocation is proportional to some power of the quantity $\bar{\varepsilon} = \alpha\eta_V + \eta'_\mu$ in relation (7.24). If the concentration of the solute atoms is c, the mean distance of the solute atoms along the dislocation is proportional to $c^{-1/2}$. τ_p can be taken as inversely proportional to the mean distance of the solute atoms, and hence,

$$\tau_p = \alpha_0 \mu \bar{\varepsilon}^p c^{1/2}. \qquad (7.27)$$

Fleischer found that with $p = 3/2$, $\alpha_0 \approx 1/750$ and in the case of the expression

$$\bar{\varepsilon} = | \eta_\mu^F + 3\eta_V |, \qquad (7.28)$$

which is valid for screw dislocations, there is good agreement with experimental data for a considerable number of materials [44]. This result seems to support the opinion that in many substances the value of the yield stress is determined mainly by the behaviour of the screw dislocations. As a rule, however, it is certain that this conclusion does not hold. In some cases the expressions

$$\bar{\varepsilon} = | \eta_\mu^F | + 16 | \eta_V | \qquad (7.29)$$

and

$$\tau_p = \frac{\mu}{550} \bar{\varepsilon}^{4/3} c^{2/3} \qquad (7.30)$$

have been found to agree fairly well with the experimental results [34]. On the other hand, relation (7.29) indicates the considerable role of the edge dislocations.

To summarize, it can be said that the theory of temperature-independent solution-hardening is still far from complete. Although it is certain that the basic mechanism of the hardening (long-range interaction between dislocation and solute atom) is essentially known, it seems at present that too many chance parameters are introduced in the experimental investigations and in the theoretical models. The further development undoubtedly involves the discovery of these parameters, and the elimination of their disturbing effects from the experimental and the theoretical studies.

7.2.7. *Temperature dependence of the yield stress of solid solutions*

The theory of the temperature dependence of the hardness of solid solutions is another area which has been only partially clarified. If the temperature is decreased from room temperature, the yield stress of the solid solution increases considerably as demonstrated in Fig. 7.14. This effect is still larger with bcc metals where the yield stress of the order of a few kp/mm^2 increases to several tens of kp/mm^2. Many attempts have been made in the last twenty years to explain this strong temperature dependence, but with no satisfactory result. The most probable mechanism of the temperature-dependent process is the following. The solute atoms in the immediate vicinity of the dislocations are bound to them with a well-defined energy U_0. The flow is initiated by the activation of the dislocation motion. This requires the bowing out of certain segments between the fixed points of the dislocation network. If the concentration of the solute atoms is c, then their mean distance along the dislocation is $bc^{-1/2}$ where b is the interatomic distance. The bowing out of a segment of length l requires in this case that the segment moves away from a total of $(l/b)\, c^{1/2} - 1$ impurity atoms. With an external stress τ_0 the measure of the bowing out is determined by the following expression:

$$\Delta L = \Delta l U_d + \left(\frac{l}{b} c^{1/2} - 1 \right) \Delta U_0 , \tag{7.31}$$

where ΔL is the work done by the stress τ_0 during the bowing out, Δl is the change of length of the dislocation segment, and U_d is its energy per unit length. ΔU_0, the change of the interaction energy, is defined by the integral:

$$\Delta U_0 = \int_0^{\Delta x} F_x \, dx, \tag{7.32}$$

where F_x is the interaction force, and Δx is the displacement at which the forces originating from the line tension and F_x are in equilibrium with the forces produced by the external stress. The motion of dislocation requires the input of the total binding energy during the bowing out which takes place with the increase of the length of the segment by Δl. In other words, the activation energy of the dislocation motion is

$$U = U_d \Delta l + \left(\frac{l}{b} c^{1/2} - 1 \right) U_0 . \tag{7.33}$$

If $\Delta L < U$, the thermal energy input necessary to initiate the motion is

$$\Delta U = U - \Delta L = \left(\frac{l}{b} c^{1/2} - 1 \right) (U_0 - \Delta U_0) . \tag{7.34}$$

At a temperature T the rate of flow due to the action of the stress τ_0 is

$$\dot{\gamma} = \dot{\gamma}_0 \, e^{-\Delta U/kT} , \tag{7.35}$$

where $\dot{\gamma}_0$ is a quantity characteristic of the material under investigation. From this:

$$\Delta U = \left(\ln \frac{\dot{\gamma}_0}{\dot{\gamma}} \right) kT = mkT , \tag{7.36}$$

where $m \cong 25$, and, to a good approximation, can be taken as constant. From a comparison of (7.34) and (7.36):

$$\left(\frac{l}{b} c^{1/2} - 1 \right) (U_0 - \Delta U_0) = mkT . \tag{7.37}$$

This result is essentially an implicit relation between the yield stress and the temperature. The quantities l and U_0 contain the value of the yield stress. The activated dislocation length can be regarded as constant for a given material, and thus the explicit equation can be obtained if the relation between ΔU_0 and τ_0 is known. From this condition, however, the basic problem of the theory becomes quite apparent. It has been shown in the previous section that the force F_x is not even unanimously defined in the case of long-range interactions which are relatively well known. This holds much more for forces acting at the core of the dislocation. Thus it can be explained why such a wide range of yield stress versus temperature relations can be found in the literature. So far eight different force equations have been derived. It is very difficult to make a suitable selection since the experimental data are relatively inaccurate, and the numerical results obtained from the various equations do not differ too much from one another. The following linear relations exist between the yield stress and the temperature [46]:

$$\tau_0^{1/2} - T^{2/3}, \qquad \tau_0^{2/3} - T^{2/3},$$

$$\tau_0^{2/3} - T^{1/2}, \qquad \tau_0 - T^{1/2}, \quad \text{etc.}$$

In spite of the difficulties mentioned, one substantial, general conclusion can nevertheless be drawn from relation (7.37). There always exists a temperature T_0 at or above which the thermal effects supply enough energy without any external stress ($\Delta U_0 = 0$) for the dislocation to break away from the solute atoms. This occurs when [neglecting unity compared with $(l/b)\, c^{1/2}$]:

$$T_0 = \frac{lc^{1/2}}{mkb} U_0 . \tag{7.38}$$

If the values $km = 2.2 \times 10^{-3}$ eV/degree, $l = 100\ b$, $c = 0.01$ and $U_0 = 0.1$ eV are substituted into the equation, then $T_0 = 450°$, which is in good agreement with the experimental results.

At the temperature $T = 0$, the increase of the maximum yield stress can be estimated in the following way. The external stress induces the dislocation motion if the force exerted on the dislocation is at least equal to the maximum force originating from the solute atoms, i.e. if

$$\tau_0(0)\, lb \cong \frac{l}{b} c^{1/2} F_0 , \tag{7.39}$$

where $F_0 = -(\partial U/\partial x)_{x=0}$. F_0 is approximately equal to $-\alpha U_0/b$ and $\alpha \approx 1$. With this, from (7.39),

$$\tau_0(0) = \left(\frac{\alpha U_0 c^{1/2}}{\mu b^3}\right) \mu . \tag{7.40}$$

Since $\mu b^3 \approx 1$ eV, with the previous values $\tau_0(0) \approx 10^{-2}\ \mu$.

7.2.8. Yield point phenomena

In this section we briefly review the phenomena of double yield stresses for the case of solid solution. Two yield stresses exist if the stress necessary to initiate the plastic deformation is larger than the stress for any subsequent deformation (Fig. 7.15). In solid solutions this phenomenon is due to the fact that the solute atoms stabilize the dislocations before the deformation. If the dislocation begins to move as a result of some external stress, the stress necessary for further motions is reduced, because after the dislocations leave the solute atoms their stabilizing effect ceases. Similarly the multiplication of dislocations may require a smaller stress once the flow has begun.

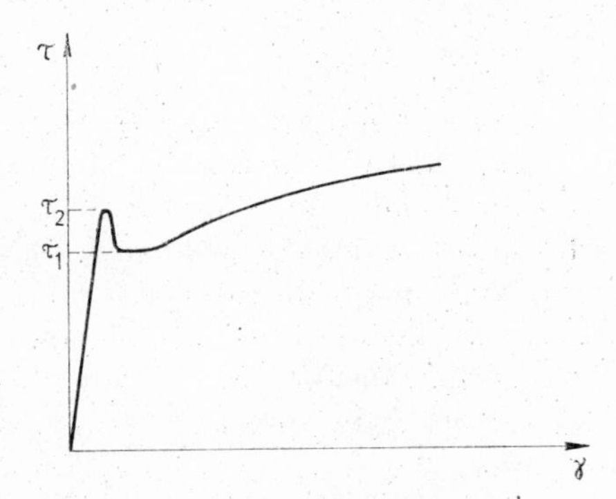

FIG. 7.15. The stress–strain curves of metals containing impurities

Many parameters influence the formation of the double yield stress [47]. For instance, the following quantities are important: temperature, rate of deformation, composition, and in the case of a polycrystalline material the grain size. Since the phenomenon originates in solid solutions from the interaction between the dislocation and the solute atom, it appears more distinctly with interstitially solute atoms in bcc metals, for in this case the interaction is very strong. With fcc substitutional solutions, however, it can be observed only at very high concentrations.

7.3. Theory of precipitation (dispersion)-hardening

7.3.1. Introduction

If the alloying atoms of an alloy are present in the form of finely dispersed particles forming an extra phase, the strength of the material is considerably larger than that of the base metal. This phenomenon is called precipitation- or dispersion-hardening. These two names distinguish between the different origins of the particles. While the former indicates the presence of particles precipitated from a supersaturated solid solution, the latter involves particles formed in some other way (for instance, by internal oxidation). There is no difference between the descriptions of the mechanical effects of these two kinds of particle, and consequently the following investigations are valid for both cases.

The theoretical interpretation of the mechanical properties – yield stress and work-hardening – of the precipitation or dispersion materials is of basic importance in the development of alloys with optimum properties. Accordingly, a wide range of theoretical investigations is to be found in the literature [48–51]. Although the above macroscopic properties are determined by extremely complicated micromechanisms, good approximations can be obtained between the macroscopically measurable quantities and those characteristic of the particles, by the use of certain simplified (but nevertheless apparently realistic) models.

We now review briefly – but the fundamental steps in detail – the theoretical interpretation of the yield stress.

7.3.2. Investigation of the yield stress

The yield stress of a crystalline body is the smallest stress at which the motion of a sufficiently large number of dislocations begins for macroscopic deformation to result. In the case of pure materials this means that dislocation sources (e.g. Frank–Read sources) become operative. In alloys containing particles of a second phase the operation of the Frank–Read sources does not induce any macroscopic change, because the dislocations emitted by the source are arrested among the particles in the vicinity of the source, and a considerably larger stress is needed for their further motion. The yield stress of these materials is thus determined essentially by the properties of the particles. These properties can be summarized as follows:

(a) the nature of the precipitates (dispersion particles) (their rigidity, crystalline structure, elastic constants, etc.);
(b) coherence relations;
(c) size and distribution of the precipitates (dispersion particles).

7.3.2.1. Rigid particles (Orowan mechanism)

The hardening effect is realized by precipitates impeding the motion of dislocations. Macroscopic deformation occurs only if the external stresses are large enough to move the dislocation across the barriers formed by the precipitates. This can be accomplished in one of two ways. If the precipitate acts as a "rigid" obstacle, a dislocation arrested by two precipitates bows out between them if a large enough stress operates, and after the two parts of the bowed out segment surrounding the precipitate join, a dislocation loop is left behind (Fig. 7.16). This process is termed the *Orowan mechanism* [52].

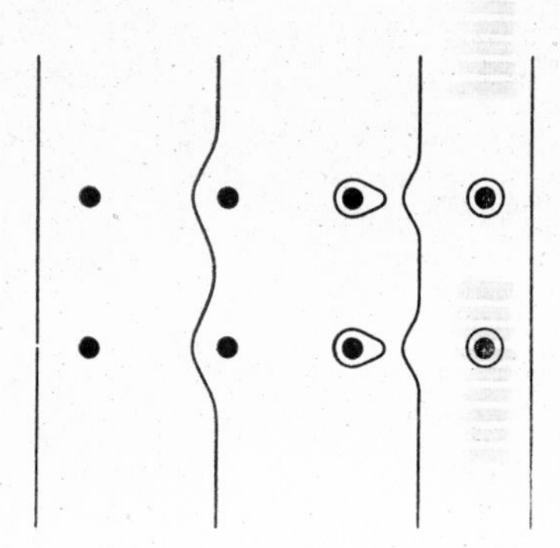

FIG. 7.16. Dislocations passing through "rigid" obstacles

It has been shown in Chapter 5 that for the bowing out between two fixed points a pure screw dislocation segment, a $1/(1 - \nu) \approx 3/2$ times larger stress is required than in the case of a pure edge dislocation. Considering the crystal anisotropy too, this ratio increases for the copper matrix to 1.76. This means that the yield stress is determined in practice by the bowing out of the pure edge dislocations.

Two further corrections must be considered for the determination of the necessary stress. One originates from the fact that the dislocation bows out not between only two anchoring points, but between a series of such points (Fig. 7.17). Because of the attractive interaction of the CC' segments formed around a precipitation, the stress necessary for bowing out is reduced. According to Ashby's estimate [53] this effect must be taken into account with a factor of 0.5–0.7. However, the dislocation starting from the points A and B, which constitute

FIG. 7.17. Bowing out of a dislocation between a series of pinning points

the Frank–Read source, must fully intersect the precipitation in order to ensure the continuous operation of the source. This means that the bowed out segment (not of a pure edge character, e.g. $A'B'$) arriving at the precipitates (as points of anchorage) must also be activated. Consequently, the activation of segments which are not of a pure edge character approximately compensates the stress decrease originating from the above attraction.

Ashby also estimated the stress-decreasing effect which originates from the splitting of dislocations into partials. This is a function of the stacking fault energy. According to Ashby this effect in fcc metals may lead to a decrease by a factor of 2/3. The energy of the stacking fault, however, is only uncertainly known, and further it is not at all certain if the splitting into partials is a similar process in the vicinity of the precipitates as in a pure matrix. For this reason this effect is disregarded.

To summarize, it can be said that if the mean distance of the particles in the alloy is l, the yield stress is practically equal to the stress necessary to bow out a pure edge dislocation between the points of anchorage, that is

$$\tau_f = \tau_{\text{edge}} = \alpha \frac{\mu b}{l}, \tag{7.41}$$

where

$$\alpha = \frac{1}{2\pi}\left(\ln \frac{4l}{b} - 2\right).$$

No great error results in the estimation of τ_f if the quantity $\ln 4l/b$ is regarded as constant. With $l = 10^3 b$, $\alpha \cong 1$.

7.3.2.2. Non-rigid particles

The dislocations arrested before the second phase (precipitation or disperse phase) particles, advance not only by bowing out; they may also pass the second phase by cutting or shearing its particles. The yield stress resulting from this mechanism can be estimated according to Kelly and Nicholson [48] as follows.

Let the stress operating in the glide plane of a dislocation be τ. If the dislocation sweeps an area A, the work done is

$$L = \tau b A \,, \tag{7.42}$$

where b is the Burgers vector of the dislocation. Let us assume that the stress necessary to move a dislocation in the base crystal can be neglected compared to τ. Then the total work (7.42) can be used to shear the particles along the area A. If the energy of intersection is U_p, then the yield stress is

$$= \tau \frac{U_p}{bA} \,. \tag{7.43}$$

Let us consider the process of intersection in more detail. The energy of intersection U_p generally consists of three effects.

1. If the particle forms an ordered structure, then the shear results in the formation of an antiphase boundary. Let the mean width of the precipitation be d. If the dislocation advances during the intersection by a distance b, the surface of the antiphase boundary increases by bd. The energy input to obtain this change is $\gamma_r bd$, where γ_r is the specific energy of the antiphase boundary.

2. During the intersection the particle–matrix boundary surface increases. When the dislocation is displaced by b, the increase of the surface is $(\mathbf{bn})b$, where $\mathbf{n}$ is the unit vector normal to the boundary surface in the point of intersection of the dislocation and the surface. If the specific energy of the boundary surface is γ, the corresponding energy is $(\mathbf{bn})b\gamma$.

3. If the glide planes of the dislocation in the matrix and in the particle are not parallel, then a usually fixed jog is formed as a connection between the two glide planes, and during the further advance of the dislocation, a dipole develops. If the dislocation moves a distance b, the energy increment of this dipole is approximately $(\mu b^3 \cos^2 \alpha)/2$, where α is the angle between the (advancing) dislocation line and the Burgers vector $\mathbf{b}$.

Let the mean distance between the centres of precipitations be l. If a dislocation segment of length l passes the particle and moves a distance b, then the swept area is lb. By using this and the preceding estimations, the yield stress is determined by the following equation:

$$\tau l b^2 = \gamma_r b d + (\mathbf{bn})\, b\gamma + \frac{1}{2}\,\mu b^3 \cos^2 \alpha. \tag{7.44}$$

From this the yield stress is

$$\tau = \frac{\gamma_r \bar{d}}{bl} + \frac{\overline{(\mathbf{bn})}\gamma}{bl} + \frac{\mu b\, \overline{\cos^2 \alpha}}{2l}, \tag{7.45}$$

where the bar indicates the mean value.

Let us consider our result in a little more detail for coherent, spherical precipitations. In this case the third term of (7.45) is zero since the coherency means that the structure of the precipitate fits the matrix exactly and consequently the glide planes coincide. For this reason no dipole is formed in this case. Let the radius of the average circle originating from the intersection of the glide plane and the precipitation be r_k. Then, if the dislocation of length l is to cut completely through the precipitation in this area, equation (7.44) must be satisfied along the diameter ($d_k = 2r_k$) of the precipitation. Thus,

$$\tau l b^2 = d_k b \gamma_r + b^2 \gamma, \tag{7.46}$$

where b^2 is approximately the boundary-surface increment between the precipitation and the matrix.

Instead of the radius r_k it is convenient to introduce the volume fraction f of the precipitations, since this value is known from experiment. If it is assumed that the dislocation travels as a rigid line, then the part of a dislocation segment of unit length which passes the precipitation during a perfectly random motion is equal to the volume fraction of the precipitations. Since the number of precipitations per unit length is $1/l$ and their average width is d_k,

$$f = \frac{d_k}{l}. \tag{7.47}$$

Hence, from (7.46),

$$\tau = \frac{f}{b}\gamma_r + \frac{\gamma}{l}. \tag{7.48}$$

It follows from this result that if the precipitation has an ordered structure and the specific energy of the boundary surface is small, the yield stress does not depend upon the size and mean distance of the precipitates, but only upon the volume fraction. If the precipitation is of a disordered structure, then $\gamma_r = 0$, and the yield stress is inversely proportional to the mean distance of the precipitates.

7.3.2.3. Summary

The mechanism of precipitation- or dispersion-hardening follows from the action of the precipitates in arresting the motion of the dislocations. In the case of rigid particles the dislocations bow out between the precipitates and move further leaving behind a dislocation loop around the precipitates. If, on the other hand, the precipitates are not rigid, the dislocation cuts through them. It should be noted in this connection that equation (7.48) for the yield stress is not used unanimously in the literature. The dependence on the volume fraction varies according to the definition of its relation to the mean diameter d. Three definitions can be given for this relation.

1. The total length of segments touching the precipitates of randomly selected straight dislocations of unit length:

$$f = \frac{d}{l}.$$

2. The surface relating to the precipitates in unit surface of a randomly selected plane:

$$f_s = \frac{d^2\pi}{4l^2}.$$

3. The volume of the second phase per unit volume:

$$f_V = \frac{d^3\pi}{6l^3}.$$

It naturally follows from the preceding considerations that the rigidity of the precipitate is relative. The same second phase can be either rigid or non-rigid if, according to (7.41) and (7.48),

$$\frac{\mu b}{l} \lessgtr \frac{f}{b}\gamma_r + \frac{\gamma}{l}.$$

From (7.48) it can be seen that, for instance, in the case of a coherent precipitation, if

$$r > \frac{\mu b^2}{2\gamma_r}$$

then the precipitation is certain to be rigid, while if

$$r < \frac{b}{2\gamma_r}(\mu b - \gamma)$$

then the cutting mechanism will be dominant.

In connection with the temperature dependence of the precipitation-hardening, it should finally be noted that essentially the same considerations can be applied here as were used in the case of solid solutions. The main problem in this case too is that the interaction between the precipitate and dislocation is only poorly understood.

7.4. The effect of lattice defects on the electrical resistivity of metals

7.4.1. Introduction

It is known from the electron theory of metals that the conductivity of perfect crystals is infinitely large at $T = 0$ °K [54]. If the crystal deviates from its ideal structure, however, i.e. if the atoms are not in geometrically perfect atomic sites, the resistivity of metal is not zero. According to the classical theory, the resistivity due to temperature is a linear function of T. This holds to a good approximation near room temperature.

Since the crystal symmetry is also disturbed by the lattice defects, the resistivity increases if defects are present. The various lattice defects cause different resistivity changes, and therefore the properties of the lattice defects can be studied by measuring the electrical resistivity. Lattice defects can be introduced into a metal in three ways:

1. By irradiating the metal with alpha-particles, neutrons, deuterons, electrons, etc., the lattice defects thus generated are mainly vacancies and interstitial atoms.
2. By quenching from a temperature near the melting point. The defects which become frozen-in by this procedure are practically only vacancies.
3. By plastic deformation. This defect structure depends upon the deformation temperature.

In the next section we deal with the phenomena due to plastic deformation. The other two methods will be discussed in connection with the problems of recovery processes (Chapter 9).

The change of resistivity caused by lattice defects is usually determined by assuming that the thermal resistivity $\rho_i(T)$ of the defect-free crystal and the resistivity $\Delta\rho$ due to the lattice defects are additive (Matthiessen's rule), that is,

$$\rho(T) = \rho_i(T) + \Delta\rho, \tag{7.49}$$

where the term due to the effect of lattice defects does not depend upon the temperature.

7.4.2. The extraresistivity of the lattice defects

In a study of the resistivity change due to plastic deformation, an exact knowledge of the degree of purity of the sample, its previous treatment and the temperature of measurement is extremely important. Reliable measurements can be carried out mainly at low temperatures, when the term $\rho_i(T)$ of equation (7.49) is small, and consequently the relative resistivity change is considerably larger. Figure 7.18 shows the resistivity change of Au, Ag and Cu as a function of the shear strain during simultaneous torsion and extension. The measurements were made at 78 °K [55]. It can be seen that as a result of the deformation, the resistivity changes considerably. The resistivity measurement permits the determination of the number of defects produced, and the study of the mechanism of their generation. If the resistivity changes

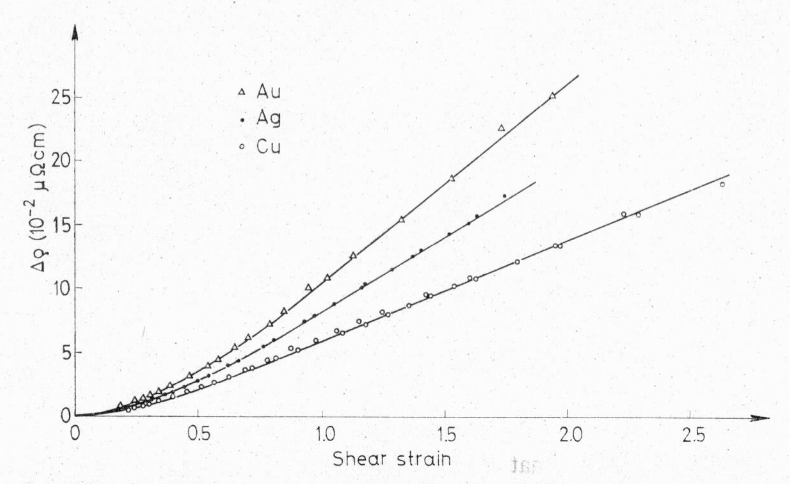

FIG. 7.18. The change of the electrical resistivity of polycrystalline Ag, Au and Cu during simultaneous extension and torsion

for the individual lattice defects are known from theoretical calculations, then the defect concentration can be derived from the measurements, and conversely the theoretically obtained resistivity values can be checked if the defect concentration is known. According to theoretical and experimental data, the resistivity increase due to vacancies for fcc metals [56, 57] is

$$\rho_{vac} = 1.5 \, \mu\Omega \text{ cm}/\% \text{ vacancy}.$$

The effect of the interstitial atoms on the resistivity is more difficult to estimate than that of the vacancies. In the latter case, it is enough to take into account the scattering effect of the holes, while the scattering due to the deformation of the surroundings of the vacancies can be neglected. For interstitial atoms, however, this part of the scattering is larger. Thus the resistivity increase due to the interstitial atoms [56, 57] is

$$\rho_{\text{int}} = 5\ \mu\Omega\ \text{cm}/\%\ \text{interst.}$$

The successful theoretical interpretation of the contribution of the dislocations of the resistivity has not yet been achieved. Either results which were much smaller than the experimental values [59], or results with divergent resistivities were obtained [60].

The experimental investigation of the resistivity of dislocations has shown that Matthiessen's rule does not hold in this case [61]. Thus, with copper for instance, the resistivity per unit length of the dislocation at 4.2 °K is 1.3×10^{-19} ohm cm^3, while at 77 °K it is 1.6×10^{-19} ohm cm^3 [62].

The experimental results indicate that the resistivity change due to deformation is to a good approximation proportional to the plastic work [55, 57]. In the following section a model is discussed on the basis of which this result can be interpreted.

7.5. Increase of defect concentration during plastic deformation

If a dislocation moves in its glide plane, of necessity it intersects every dislocation which crosses this plane. Each intersection results in a jog which can move in general only if a point defect is also generated. In the following, a relation is determined between the macroscopically measurable parameters and the concentration of defects generated during the deformation.

Let us deal first with the generation of point defects. Let us assume that if the yield stress is increased by $d\tau_r$, then dn_f new Frank–Read sources begin to operate in unit volume, while during the increase of the shear strain by $d\gamma$ each new source emits n_0 loops which on average are displaced along a surface A. The strain increment can then be given by

$$d\gamma = n_0 dn_f bA. \tag{7.50}$$

Let the mean distance of the dislocations intersecting the glide plane be l_d. The number of dislocations intersected by the loops is then

$$N_0 = \frac{n_0 dn_f A}{l_d^2} = \frac{d\gamma}{bl_d^2}.$$

The number of point defects generated – dn_p – is proportional to the number of dislocations intersected by the expanding loops N_0, and to the average number n_1 of point defects generated by the jog produced by one intersection; that is,

$$dn_p = \alpha_1 n_1 N_0 = \alpha_1 \frac{n_1}{bl_d^2} d\gamma\,, \tag{7.51}$$

where α_1 is constant.

If the jog follows the dislocation during its motion only by climb, the number of point defects generated is proportional to the number of the atomic steps. Consequently, n_1 is proportional to l_s/b where l_s is the mean free path ($l_s^2 \sim A$). Let us assume that the mean free path is proportional to the mean distance of the dislocations intersected. Then from relation (7.51)

$$dn_p = \alpha_2 \frac{d\gamma}{b^2 l_s}\,, \tag{7.52}$$

where α is constant. If it is further assumed that the stress required to move the dislocations is inversely proportional to the mean free path (i.e. according to the previous assumption, to the mean distance of the dislocations intersected) so that

$$\tau_r = \alpha_3 \frac{\mu b}{l_s}\,, \tag{7.53}$$

then (7.52) can be rewritten in the following form:

$$dn_p = \frac{A_p}{\mu b^3} \tau_r d\gamma\,, \tag{7.54}$$

where A_p is a constant containing the previously introduced proportionality factors. The point defect concentration is obtained by dividing dn_p by the number of atoms per unit volume ($1/b^3$) and integrating:

$$c_p = \frac{A_p}{\mu} \int \tau_r d\gamma\,. \tag{7.55}$$

Our result shows that – as long as the assumptions hold – the concentration of point defects generated during the plastic deformation is proportional to the plastic work. This relation was first determined by Saada by another method [63]. From a special model he obtained for the numerical value

$$0.3 < A_p < 0.9. \tag{7.56}$$

Similar considerations can be carried out for dislocations too. The dislocation density increment is $dN \sim n_0 dn_f l_s$. By applying relations (7.50) and (7.53),

$$dN = \frac{A_d}{\mu b^2} \tau_r d\gamma , \tag{7.57}$$

where A_d is constant. From this the dislocation density is

$$N = \frac{A_d}{\mu b^2} \int \tau_r d\gamma . \tag{7.58}$$

Thus the dislocation density too is proportional to the plastic work. This means that at sufficiently low temperatures, the number of point defects is proportional to the number of dislocations.

The value of the constant A_d can be estimated in the following way. According to (2.25) the energy per unit length of a dislocation to a good approximation is μb^2. From theoretical and experimental studies, the energy stored by the dislocations is about 5% of the total plastic work [64]. Hence,

$$A_d = \frac{\mu b^2 N}{\int \tau_r d\gamma} \cong 0.05. \tag{7.59}$$

From a comparison of expressions (7.54) and (7.57), the ratio of the point defect and dislocation increments can be estimated from the relation

$$\frac{dn_p}{dN} = \frac{A_p}{A_d b}, \qquad \text{that is} \qquad \frac{4}{b} < \frac{dn_p}{dN} < \frac{18}{b} . \tag{7.60}$$

Chapter 8

Work-hardening

8.1. General features

The explanation of work-hardening represents one of the most important contributions to the applications of dislocation theory. Unfortunately there is so far no unified explanation that correctly describes al-phenomena quantitatively. One reason for this is that each phenomenon is the result of many mechanisms, which possibly act indepenl dently of each other. With macroscopic measurements, only the sum of these effects is observed, the separation of the various elementary processes is rather uncertain and the same macroscopic behaviour can frequently be accounted for by various microscopic models. Of course, today widespread methods already exist by which the single dislocations, their motions and interactions can be observed, but with these methods, only thin layers, or layers close to the surface, can be examined and nothing proves that the phenomena take place in the same way in the bulk material.

Various essentially different theories have recently been developed to explain work-hardening; each of these contains several empirical, or very inaccurately determinable, parameters. The basic models too appear to be oversimplified. Our present knowledge does not permit the selection of one of these theories as correct, and the discarding of the rest as inaccurate. For this reason, attempts will be made below only to summarize, mostly qualitatively, the more important ideas. For details the reader is referred to the literature (e.g. [1]).

First let us review the general experimental findings. The majority of these investigations have been carried out on fcc materials. Accordingly, in this chapter only these structures are dealt with, although references are made to the behaviour of bcc and hexagonal metals.

8.2. The stress–strain curve of fcc crystals

The stress-dependent glide occurring in the plastic deformation of fcc single crystals can be described by a curve as in Fig. 8.1. Three characteristic stages can be distinguished. The parameters for the various stages depend in very different ways upon the temperature, the direction of the stress and the impurity content. These characteristic changes are summarized, mainly on the basis of experimental results, in Seeger [2], see Table 8.1.

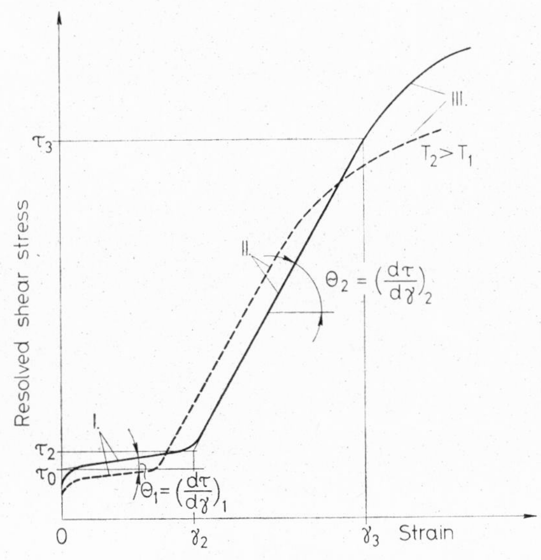

FIG. 8.1. Typical stress–strain curve of fcc metals

Stage I begins after the elastic deformation at the critical stress τ_0. This is the *easy glide*, or the stage of single glide. According to investigations of the surface with an optical metallographic microscope or an electron microscope, this stage is characterized by long (a few 100–1000 μm), straight, uniformly placed slip lines lying relatively close to each other (10^2–10^3 Å) (Fig. 8.2).

The slip lines which can be observed electron-microscopically consist of fine steps due to edge dislocations intersecting the surface.

TABLE 8.1. *The changes of the parameters of the stress–strain curve**

Parameter considered	The tendency of change		
	With increasing temperature	With the change of stress orientation**	With substitution impurities
τ_0	—	slightly +	strongly +
τ_2	—	slightly +	+
$\tau_2 - \tau_0$	0	slightly +	+
τ_3	strongly —	slightly +	+
Θ_1	0	strongly +	strongly +
Θ_2	0	+	0
Θ_3	—	0	?
γ_2	—	strongly —	strongly +
γ_3	—	—	slightly —

* According to H. G. van Bueren, *Imperfections in Crystals*, North-Holland Publ., Amsterdam, 1961.

** Measured from the middle of the unit triangle of the stereographic projection.

Their length in practice determines the diameters of the dislocation loops perpendicular to the Burgers vector, i.e. the distance of the screw components from each other. Screw dislocations emerging on the surface do not produce steps. The lines which can be observed with the optical microscope have composite structures: they consist of a multitude of elementary slip lines. The observations indicate that in this case only a single active glide system operates, and if a group of dislocations is once emitted by a source, the group moves to a relatively large distance. The first stage lasts – depending upon the direction of the strain – from a shear of a few tenths of a percent to a few per cent.

The work-hardening of hexagonal metals is frequently similar in its whole behaviour to that of fcc samples in stage I. At low temperatures (below room temperature) it is independent of the temperature and of the orientation. At higher temperatures the work-hardening depends on the temperature. This phenomenon can be explained by the mutual annihilation by climb of dislocations of opposite sign. Stages II and III do not occur, because in hexagonal metals glide can take place mainly only in the basal plane. For this reason the work-hardening of hexagonal metals and stage I of the work-hardening of fcc metals can theoretically be investigated together. If, on the other hand, glide systems outside the basal plane also become active, this naturally no longer holds, and with hexagonal samples too, stages II and III appear.

FIG. 8.2. Glide lines on the surface of a copper single crystal in stage I (0–9.4% glide, 10,000×) (Mader and Seeger [3])

The most characteristic feature of the much steeper stage II, the linear stage, is that the rate of hardening Θ_2 is practically independent of the temperature and the impurities, and even its change in the case of different materials appears only through the change of the shear modulus μ. (In the centre of the stereographic triangle $\Theta_2/\mu \approx$ $\approx 1/300$; this may increase near the corners to approximately 1/150.) This is surprising inasmuch as according to experiment, the developing dislocation structure depends considerably upon, for example, the specific energy of the intrinsic stacking faults. Thus, it must be concluded that the work-hardening is not too sensitive to the details of the dislocation structure.

The length L of the slip lines in stage II is only of the order of 10 μm, and it decreases with the increase of the shear deformation γ

according to the equation

$$L = \frac{\Lambda}{\gamma - \gamma_0}, \tag{8.1}$$

where the value of Λ for copper and nickel is approximately 4×10^{-4} cm, and $\gamma_0 \approx \gamma_2$ (Fig. 8.1). The distance between the new slip lines simultaneously grows, while their distribution becomes more and more inhomogeneous. (In the course of the investigation the surface is polished after every intermediate strain and in this way those slip lines which are developed during the new deformation are always observed [2, 3].)

According to transmission electron-microscopic investigations, the dislocation structure is strongly inhomogeneous. In some places "tangled", relatively dense dislocation paths are formed, whereas

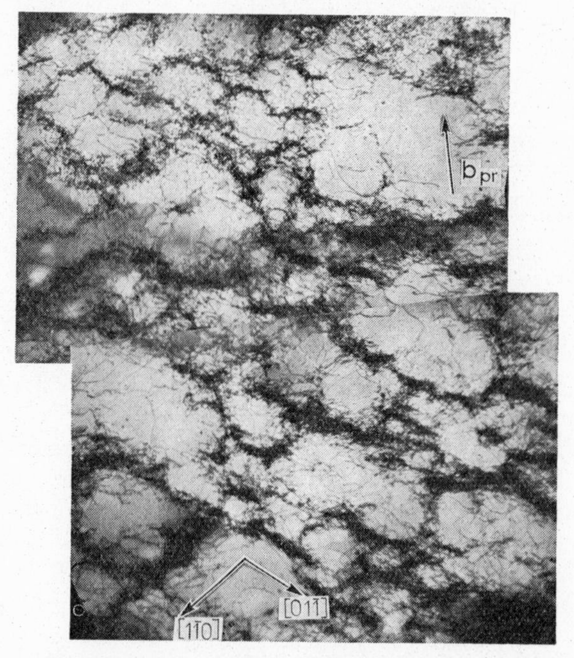

FIG. 8.3. The dislocation structure of copper in stage II of the stress–strain curve. (Transmission electron-micrograph, Essmann [4])

elsewhere, relatively clean, dislocation-free domains develop. The regions of large dislocation density are usually extended along the slip lines or constitute walls perpendicular to them (cell formation, Fig. 8.3) [4, 5]. With an increasing deformation, the cell structure becomes continuous and more and more dense but according to Kuhlmann-Wilsdorf [1] the total volume of the walls consisting of reasons of high dislocation density does not change after the formation of the cell structure. At the beginning of stage II, the number of dislocations in the secondary glide system is relatively small but their density later surpasses that of the primary dislocations. In spite of this, the secondary glide remains small since the displacement of the secondary dislocations is insignificant.

In connection with the formation of the average dislocation density, it has been already mentioned in the previous chapter that its increase is proportional to the increment of the plastic work [see equation (7.58)]; consequently it appears natural that – at least in the linear stage – a relation similar to equation (7.5) exists between the stress and the dislocation density. According to investigations carried out by Bailey and Hirsch [6] on silver and copper, the shear stress τ necessary for the further deformation can be given within very broad limits by the equation

$$\tau = \alpha_t \frac{\mu b}{l}, \tag{8.2}$$

where $\alpha_t \approx 0.2$, and l is the average distance between dislocations in the "tangled" volumes. (A similar formula is frequently quoted in the literature, but instead of $1/l$, the square root of the average dislocation density is used, and the proportionality factor $\alpha \lesssim 1/2$. The data proving this relation are tabulated, for example, by Otte and Hren [7].)

By measuring the magnetic susceptibility of deformed Ni, Kronmüller and Seeger concluded that in stage II the wavelength of the fluctuation of the internal stresses is considerably larger than the average distance between the dislocations [8, 9]. This can most simply be explained by the presence of piled-up dislocation groups, though it may also be caused by any other inhomogeneous dislocation distribution. At the end of stage II, together with continually shortening slip lines, deformation bands are increasingly frequently developed.

Along these the slip lines take the characteristic form shown in Fig. 8.4. The number of these bands increases as the temperature decreases, but even more so as the rate of the deformation increases. At the same time, however, the band structure becomes increasingly irregular. The formation of the deformation bands is attributed to the effect of dislocation groups of opposite Burgers vectors piled up against each other. The bands appear only after the start of the secondary glide system, and consequently they do not occur in hexagonal metals.

In the course of further deformation, beginning with the point (γ_3, τ_3), the rate of hardening continually decreases, and the change

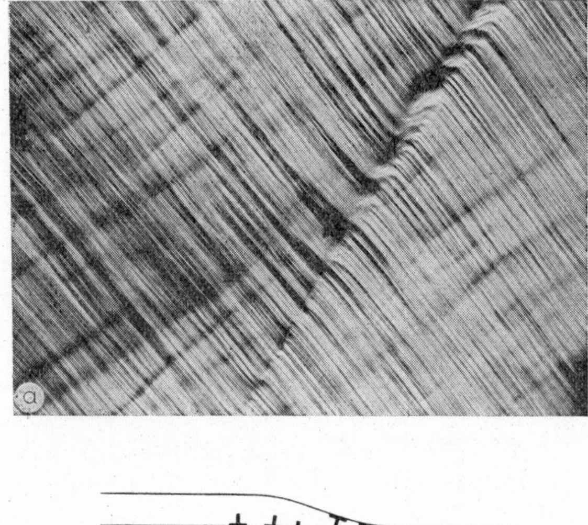

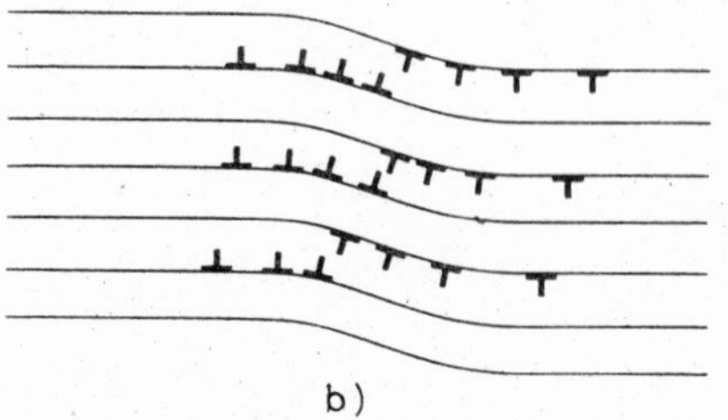

FIG. 8.4. (a) Deformation band on the surface of a copper single crystal in stage III (220×). The central strong band was formed in stage II (Mader and Seeger [3]). (b) Dislocation model of the formation of the deformation band

of the stress can be described by the equation

$$\tau = \Theta_3 (\gamma - \gamma')^{1/2}, \tag{8.3}$$

where $\gamma' =$ constant (stage III, parabolic hardening). It is significant that the stress limit τ_3 decreases exponentially with increasing temperature, and Θ_3 shows a similar tendency (Bell [10] has shown that because of the smooth transition from the linear to the parabolic stage, $\Theta_3^2 = 2\Theta_2\tau_3$). In this stage the glide lines cluster into short bands a few micrometres in length (Fig. 8.5). The bands are situated as if some of them were a slightly displaced continuation of the others. At

FIG. 8.5. Glide bands on the surface of a copper single crystal in stage III (5300×). The traces of cross-slip can be observed at several places (Mader and Seeger [3])

some places, the short bands or the band system connecting the bands and lying too along a possible glide direction, can be observed well. Accordingly, the dislocations which start to move in one band, pass over by cross-slip after a certain time to another parallel lattice plane, and continue their glide there. The track of the cross-slip is probably not always clearly visible. This can be explained by considering that it is relatively rare that dislocations gliding originally in one plane continue their glide in a single common plane after the cross-slip.

It can also be assumed that only a fraction of the dislocations slip to another plane and generate a new source.

All this holds independently of the magnitude of the energy of the stacking faults—at least when the temperature is not too high. The starting point of stage III, however, depends mainly on the stacking fault energy. With Al, in which the stacking fault energy is relatively large, stage II in practice is already absent near room temperature since τ_3 (the starting stress of the stage) is so small.

In Section 7.1.3 it has already been shown that the flow stress can be divided into a temperature-dependent part τ_T (relating to the thermally activated processes) and a stress τ_E which is independent of the temperature. This is true along the whole deformation curve. In stage I, the part dependent on the temperature (and together with it also the rate of deformation) makes a contribution still comparable with that of the temperature-independent part, but at the beginning of stage II the ratio becomes small, and from here on it no longer depends upon the deformation *(Cottrell–Stokes law)*. Accordingly, it was earlier thought that both parts originate directly from the same dislocation network. This, however, has no foundation whatsoever.

The behaviour of bcc metals differs from that of fcc metals inasmuch as, at low temperatures, the temperature-dependent contribution τ_T is very large (this is indicated by the appearance of long, straight screw dislocations) since, in these crystals, the Peierls forces are larger and the mobility of the screw dislocation is small. At low temperatures, the three-stage deformation curve is not observed. It appears only close to or above room temperature, when the thermal activation is already able to compensate for the large mobility difference between the edge and screw dislocations. The initial flow stress in this case, however, is considerably larger, stage II is less defined and the hardening is somewhat smaller ($\Theta_2/\mu \approx 1/540$). The resulting dislocation structure is similar to that in fcc metals, with the difference only that the cross-slip of the screw dislocations can take place much more easily (it frequently occurs even in stage I). The various experimental results have been reviewed by Hirsch [11] and Šesták and Seeger [12].

8.3. The "forest" theory of work-hardening

In this and the next sections some of the more important theories of work-hardening are summarized. The majority of the models used are very imperfect, and tell us very little about how the assumed dislocation structure develops. Several semi-quantitative theories have been derived, sometimes for the same model, these differing in the semi-empirical parameters introduced.

As the most simple model, let us assume that the new dislocations generated during the deformation, move within a dislocation structure already present in the material (the "dislocation forest") with a mean free path l_s. The average distance of the dislocations in the dislocation "forest" is $l = 1/\sqrt{N}$ (where N is the dislocation density). It has been pointed out in Section 7.1.3, and was also mentioned among the experimental results, that the stress necessary to move the dislocations in this case is inversely proportional to the average distance of the dislocations:

$$\tau = \alpha\mu b\sqrt{N} \tag{8.4}$$

and if it is also assumed that the mean free path is proportional to l, then according to equation (7.57) the change of the plastic work is proportional to the change of the dislocation density (the new dislocations enrich the "forest" after their motion):

$$\tau d\gamma = \alpha_0 \mu b^2 dN. \tag{8.5}$$

The above proportionality has been proved experimentally by Kovács for stages II and III [13]. From the measurement of the stored energy of deformed fcc metals, $\alpha_0 \approx 20$. With the repeated use of equation (8.4)

$$2\tau d\tau = \alpha^2 \mu^2 b^2 dN.$$

By comparing this with equation (8.5), the hardening is immediately obtained:

$$\Theta_2 = \frac{d\tau}{d\gamma} = \frac{\alpha^2}{2\alpha_0}\mu. \tag{8.6}$$

If the average value $\alpha = 0.35$ is inserted, then $\Theta_2/\mu = 1/325$. Thus the above model accounts for the linear stage, but it does not explain many details (e.g. what determines l_s). Similarly, the formation of the inhomogeneous dislocation structure is not accounted for. The model cannot be applied to stage I, and difficulties also arise in an explanation of stage III. Presumably equation (8.4) is not valid in that stage. Kovács has shown [13] that in the parabolic stage – if equation (8.5) is valid – the following relations hold:

$$\tau = \alpha' \mu b^{2/3} N^{1/3} \tag{8.7}$$

and

$$\Theta_3^2 = \frac{4}{3} \Theta_2 \tau_3 . \tag{8.8}$$

This latter formula differs somewhat from Bell's equation given above.

8.4. The "jog" theory of work-hardening

This theory deals only with stage II. According to the basic idea of this theory, the motion of dislocations is obstructed by sessile jogs either already existing in the source or developed during the motion of the dislocations [14, 15]. If the dislocations are closer to each other than the sources, the resultant of the stress-field of the parallel dislocations of identical Burgers vectors is comparable in a large part of the volume with the flow stress. Immediately after the glide begins, a part of the stress becomes neutralized by secondary slip. Naturally, in this way further jogs are formed; these slow the motion of the primary dislocations, with the result that finally the tangled dislocation network is formed and no further glide takes place. In this case, however, it must be assumed that the secondary glide is comparable with that of the primary system.

By investigating the properties of the large jogs and extrapolating the results to jog lengths of a few atomic distances, Hirsch has shown that the interstitial jogs generally move in a conservative way along the dislocations [16]. The motion is thus a thermally activated process which at low temperatures must be maintained in part by a temperature-dependent stress. The activation energy of the process is smaller than that necessary to generate interstitial atoms. The vacancy-jogs, on the other hand, move along the dislocations more easily in a non-conservative way (with the creation of point defects) than in a conservative way. Of course, at large enough temperatures the conservative motion begins here also, and this again decreases the flow stress. This point defect formation, however, is not a thermally activated process at low temperatures. In principle, the thermal motion may help to make the jog jump further, but if the vacancy so

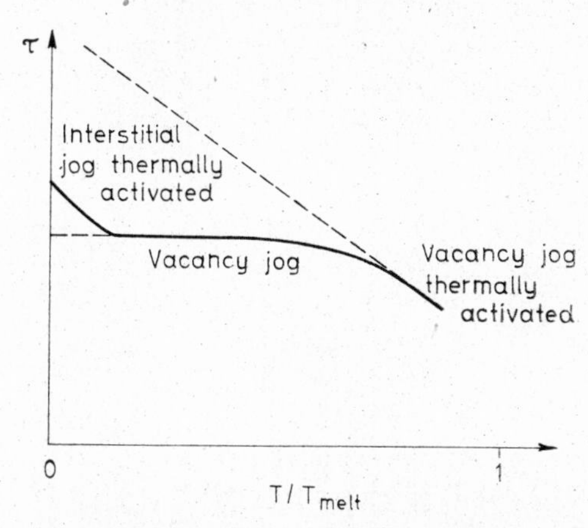

FIG. 8.6. Temperature dependence of the flow stress of cold-worked copper according to Hirsch and Warrington [17]

developed does not diffuse away at once, the reversed process pulls the jog back again. This means that the thermal activation plays a role only if it provides (with a high probability) the total energy of self-diffusion (i.e. the sum of the activation energies of the formation and the motion of a vacancy). According to calculations by Hirsch [16] for the case of screw dislocations, the temperature necessary for this process for Al is *ca.* 200 °K and for gold is *ca.* 550 °K. (The disappearance of the second stage of the hardening curve takes place with the same metals at 250 °K and 700 °K, respectively.) Consequently, on cold-working, the vacancy jogs produce a temperature-independent hardening (Fig. 8.6).

Let us now estimate the degree of hardening. Let the jog density along the dislocations be m, of these let the density of vacancy jogs be $m_v = fm$, and let the energy of formation of a single vacancy be

$E_F = \beta\mu b^3$ (for copper $E_F \approx 1$ eV, and $\mu b^3 \approx 5$ eV, and consequently $\beta \approx 0.2$). The activation volume is b^2/m_v (see Section 7.1.3), and so the stress necessary for the non-conservative motion of the jogs is

$$\tau = \beta\mu b m_v. \tag{8.9}$$

Let us assume that the primary glide $d\gamma$ is followed by the secondary glide $gd\gamma$ ($g < 1$), which is proportional to the primary one, and that is equal to the displacement between the two ends of a primary dislocation of unit length perpendicular to its glide plane which is then bdm, whence

$$dm_v = fdm = fgd\gamma/b.$$

With this equation, from (8.9):

$$\Theta_2 = \frac{d\tau}{d\gamma} = \mu\beta fg\,. \tag{8.10}$$

The values of f and g must still be estimated. It was first pointed out by Cottrell [18] that the dislocations moving in a secondary system ("moving forest") produce principally interstitial jogs on the primary dislocations and only a small fraction of the jogs produce vacancies [19]. If it is assumed that $f = 0.05$, then in accordance with experience the value of g must be approximately 1/3 to give a proper degree of hardening. The secondary glide, however, is strongly orientation-dependent, and in the centre of the stereographic unit triangle the extent of the secondary glide is only about one-tenth of the previous value. In addition, the model does not take into account that some of the jogs may recombine during the deformation. Consequently, although equation (8.1) and the Cottrell–Stokes law among others can be interpreted by the "jog" theory, this theory does not account satisfactorily for the whole process of work-hardening.

8.5. The "meshlength" theory of hardening

This theory was developed by Kuhlmann-Wilsdorf [1, 20] by assuming that the flow stress is determined in the long run by the stress necessary to bow out dislocation segments of a given dislocation network and to initiate the multiplication.

8.5.1. *Stage I of the stress–strain curve*

The crystal initially contains few dislocations. Thus, the motion of dislocations initiated by an external stress is practically unobstructed by the other dislocations, and only the obstacles due to the Peierls stress, the point defects, etc., must be surmounted. Hence, the flow stress is relatively small and no considerable hardening takes place until the whole volume is gradually filled with dislocations to such an extent that they mutually obstruct each other's motion. A weak hardening results which is determined by the following effects:

(a) Long-range fluctuations of the applied stress due to the inexactness of the experimental conditions. For this reason the glide starts at the position where the stress is the largest, this volume element becomes hardened, and gradually larger and larger volumes attain the flow stress. (This is still not a real hardening!)

(b) During the intersection of the already existing dislocations, a continually increasing number of jogs is formed in proportion to the rate of deformation, but jogs may also be formed because of the indeterminate character of the dislocation axis.

(c) The glide may also be obstructed as a result of the interaction between the dislocations and the point defects generated by the motion of the jogs.

No quantitative formulation of the above concepts is known.

8.5.2. *Stages II and III of the stress–strain curve*

When the dislocation density of the deformed lattice becomes large enough, a spontaneous rearrangement of the dislocations into small-angle grain boundaries takes place either by cross-slip or by conservative climb. The driving force is the elastic interaction between the dislocations; this is inversely proportional to their distance, and consequently it becomes effective only after the attainment of a critical dislocation density which depends upon the temperature, the stacking fault energy, the direction of the external stress, etc. Because of the local stress changes during the rearrangement, the originally relatively straight dislocations become curved at many places (see

Section 2.5.7) and new nodes are formed. In this way there develops a "cellular structure" consisting of volume elements which are relatively free of dislocations, and the surrounding "tangled" dislocation networks; in this structure the long-range stresses are practically screened. [It should be mentioned in this connection that according to Seeger *et al.* the motion of the dislocations is impeded by the long-range internal stresses (see Section 8.6).] Within the cells the dislocations can move relatively freely, and consequently the stress needed for further hardening is determined by the length of the dislocation segments suitable for the initiation of new sources, while the free path of the moving dislocations is, on the other hand, a function of the cell dimensions.

During further deformation, new dislocation groups are formed within the single cells, and the interactions of these create new cell walls. Kuhlmann-Wilsdorf also pointed out that in the course of this, the earlier formed walls become increasingly massed together as a result of the increasing external stress. Thus, the total volume containing the dislocations – disregarding a transitional stage – remains unchanged in spite of the increasing average dislocation density; the average cell dimensions merely decrease.

No essential structural change takes place during the transition into stage III either. Only if the cell dimensions become small enough is there a very small probability that the new dislocations will obstruct each other's motions within the cell. For this reason the formation of new cell walls and the further shortening of the free-path length slow down considerably and possibly even cease, and only the dislocation density of the walls increases further.

In order to give the theory a quantitative form, first the relationship between the dislocation density and the flow stress (i.e. the length of the dislocation segments forming the basis of the sources) must be established. As the next step the relation between the dislocation density and the degree of deformation is investigated.

Let the average dislocation density be N, and the average length of the dislocation segments in the high-density volume fraction be $\bar{l}$, and let the ratio of their volume to the total volume be f. (It has been found experimentally that $f \approx \frac{1}{5}$ and is constant.) In this case

$$N = fm/\bar{l}^2, \tag{8.11}$$

where m is a geometrical factor which is a function of the dislocation structure. For a regular cubic network $m = 3$, while in irregular three-dimensional structures its value is larger.

Kuhlmann-Wilsdorf assumed that $m \approx 5$ ($mf \approx 1$) [1, 20]. (It should be noted in this connection that as far as possible the original notations of the various authors have been retained, and consequently various quantities may be denoted by the same letters in the various models; however, we have tried to avoid the reverse procedure.)

The stress necessary to operate the sources is naturally not determined by $\bar{l}$, but with great probability by the largest value l_m. The nodes formed on the intersection of two dislocation segments, divide each segment into two or three parts (depending upon whether fourfold or threefold nodes are formed), and thus in the extreme case, the length of one part may be three times longer than the average. It may therefore be reasonably assumed that

$$l_m = n\bar{l}, \qquad n \approx 3. \tag{8.12}$$

According to equations (5.6) and (5.7), the stress necessary to initiate the operation of a source of length l_m ($r_0 = b$) is approximately

$$\tau = \frac{\mu b}{2\pi l_m} \ln \frac{l_m}{b}. \tag{8.13}$$

By applying the previous two relations:

$$\tau = \frac{\mu b\sqrt{N}}{2\pi n\sqrt{mf}} \ln \frac{n\sqrt{mf}}{b\sqrt{N}} \approx \frac{\mu b\sqrt{N}}{6\pi} \ln \frac{3}{b\sqrt{N}}. \tag{8.14}$$

Comparing this with equation (8.4)

$$\alpha = \frac{1}{6\pi} \ln \frac{3}{b\sqrt{N}}. \tag{8.15}$$

The coefficient α, however, depends upon N to only a small extent and the values so obtained are in good agreement with data otherwise obtained (e.g. if $b = 2.5$ Å and $N = 4 \times 10^8$ cm^{-2}, $\alpha = 0.43$, while if $N = 1.6 \times 10^{11}$ cm^{-2}, $\alpha = 0.28$).

Let us assume now that the cells have roughly circular cross-sections (their diameters being the length L of the slip lines). If a cell is traversed by one dislocation per unit volume, the strain increment is

$$d\gamma = L^2\pi \frac{b}{4}. \tag{8.16}$$

The corresponding change of the dislocation density is

$$dN = \beta L\pi. \tag{8.17}$$

The value of the proportionality factor β is affected by two conditions. Firstly, the primary dislocations are in part annihilated (this is estimated to be 50% of the total length) and, in addition, new dislocations are created in the secondary glide systems which do not contribute considerably to the glide but, nevertheless, the density of the secondary ("forest") dislocations is generally larger than that of the primary ones. Consequently, $1/2 \leqq \beta \leqq 3/2$.

It is assumed that L and $\bar{l}$ decrease proportionally to each other as the deformation increases

$$L = g\bar{l}. \tag{8.18}$$

The proportionality factor g cannot be estimated theoretically; it is known only that it must be a fairly large number (we shall see later that with fcc lattices $g \approx 200$ appears to be realistic). Using equation (8.11) and substituting $\sqrt{mf/N}$ for $\bar{l}$

$$L = g\sqrt{\frac{mf}{N}}, \tag{8.19}$$

and from equations (8.16) and (8.17)

$$d\gamma = \frac{Lb}{4\beta}\,dN = \frac{bg\sqrt{mf}}{4\beta}\frac{dN}{N}. \tag{8.20}$$

After integration:

$$\gamma - \gamma_0 = \frac{bg\sqrt{fm}}{2\beta}\sqrt{N} \approx \frac{1}{2}bg\sqrt{N}, \qquad \gamma_0 = \text{const.} \tag{8.21}$$

Let us now express the dislocation density from this equation. Substituting its value into (8.14), the stress necessary to initiate the deformation is:

$$\tau = \frac{\mu b (\gamma - \gamma_0)}{\pi n m f g} \ln \frac{n m f g}{2\beta (\gamma - \gamma_0)} \approx \Theta_2 (\gamma - \gamma_0). \qquad (8.22)$$

In fact the relation corresponding to linear hardening has been obtained. The dependence of the coefficient Θ_2 on γ is very small, and would not occur at all if the logarithmic factor were neglected in the stress expression. There are, however, experimental data [21] which show a small change of this nature in Θ_2, and this has been considered as an important proof of the Seeger theory (see Section 8.6.2).

Unfortunately, the value of Θ_2 cannot be established without knowing the value of g, but in the case of $g = 200$ and $\beta = 1$, $\Theta_2/\mu = 1/300$.

The relation describing stage III is obtained if the cell dimensions are regarded as practically constant (L_c). The strain increment is now

$$d\gamma = bL_c \frac{dN}{4\beta}, \qquad (8.23)$$

and after integration

$$\gamma - \gamma' = \frac{bL_c}{4\beta} N, \qquad \gamma' = \text{const.} \qquad (8.24)$$

If this is combined with equation (8.14), which determines the necessary shear stress, the following equation is obtained:

$$\tau = \frac{\mu \sqrt{\beta b}}{\pi n \sqrt{m f L_c}} \ln \left(\frac{n^2 m f L_c}{4\beta b (\gamma - \gamma')} \right)^{1/2} (\gamma - \gamma_0)^{1/2} \approx \qquad (8.25)$$

$$\approx \frac{1}{6\pi} \left(\frac{b}{L_c} \right)^{1/2} \ln \frac{g L_c}{4\pi (\gamma - \gamma')} (\gamma - \gamma')^{1/2},$$

i.e. the work-hardening follows a parabolic law. Unfortunately, this formula too contains a parameter (L_c) which cannot be calculated theoretically.

Finally, let us investigate how the theory expresses the change of length of the slip lines. On the basis of equations (8.19) and (8.21)

$$\gamma - \gamma_0 = \frac{bg^2fm}{2\beta L} \approx \frac{bg^2}{2L}. \tag{8.26}$$

With equation (8.1) this gives

$$\Lambda = \frac{bg^2fm}{2\beta} \approx \frac{bg^2}{2}. \tag{8.27}$$

If g is taken as 200 and $b = 2.5$ Å, one obtains $\Lambda = 5 \times 10^{-4}$ cm in good agreement with experience.

Passing over a detailed discussion of the theory, it is mentioned here merely that it accounts satisfactorily for the work-hardening of bcc metals [22] and also explains the *Cottrell–Stokes law* [1]. (The temperature-independent part of the stress τ_E, discussed previously, and τ_T due to the other effects occur additively.) In spite of this, it is questionable to explain work-hardening exclusively in terms of the stress necessary to initiate the sources. The presence of long-range stresses can scarcely be denied, and there are also further details which do not correspond with the experimental evidence, e.g. the magnitude of the stored energy [23], and the observation that a very small increase of the external stress may create new slip lines during the deformation, but that the glide no longer increases along these slip lines if the stress is further increased.

8.6. The "long-range stress" theory of hardening

According to Seeger *et al.* [2, 24, 25] the hardening is primarily caused by the piling-up on obstacles of dislocations emitted from the sources, and as a result the long-range stress-field due to the piled-up dislocation groups obstructs the further operation of the source.

8.6.1. Stage I of the stress–strain curve

From the experimental results it can be concluded that in stage I the deformation, i.e. the motion of dislocations, takes place in accordance with the following conditions:

(a) The displacement of the dislocations emitted by the sources is approximately constant, and large relative to the average distance of the sources from each other. Some of the dislocations sooner or later reach the free surface.
(b) Only a few obstacles are present in the crystal. These were formed during the crystal growth, and to a first approximation their number may be regarded as constant.
(c) Within each glide plane the distance between adjacent dislocations is larger than the average distance of the sources, and consequently there is scarcely any question of piling-up.
(d) The number of active sources is practically constant. (This condition, of course, does not apply to the short, initial part of the deformation when the number of active sources increases continuously with time.)

According to Table 8.1, Θ_1 and $\tau_2 - \tau_1$ are almost independent of temperature, and so it seems that τ_T does not change in stage I; consequently, it is sufficient to deal only with the temperature-independent component [τ_T is included in τ_0 of equation (8.29)].

When a dislocation is emitted by a source, it travels unobstructed to a distance L_0. Here it encounters a dislocation of another glide plane at a distance y_0 from it and thus the stress acting on it can no longer move the two dislocations with respect to each other (Fig. 8.7). As was seen in Chapter 5, the stress-field of the dislocation so arrested decreases the effective stress acting on the source, and so a stress larger by this amount is necessary to emit the next loop. For the sake of simplicity, let us investigate the frequently observed case when the edge dislocation travels much faster than the screw, and so the

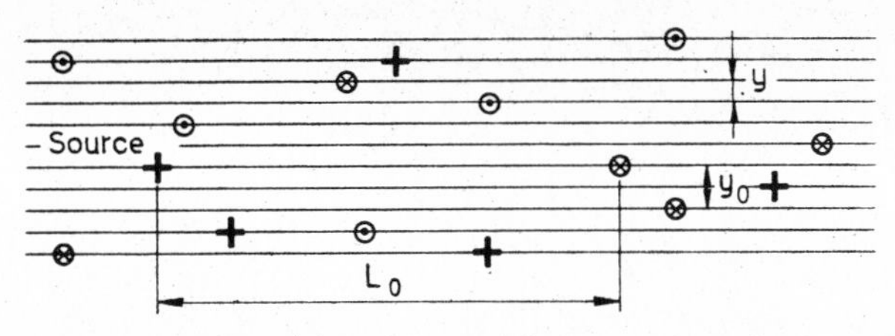

FIG. 8.7. Dislocation distribution in stage I of the stress–strain curve according to Seeger. (Positive and negative screw dislocations perpendicular to the plane of the drawing)

internal stress-field is determined primarily by the latter. Then, if n is the number of dislocations emitted by one source:

$$\frac{d\tau}{dn} = \frac{\mu b}{2\pi L_0}. \tag{8.28}$$

The hardening must be expressed not with L_0, but with parameters which can be measured. Thus, let us find the smallest distance y_0 between the two glide planes so that the external stress be just unable to move the two parallel dislocations with respect to each other. If τ_0 is the critical resolved shear stress (necessary to overcome other obstacles), then in the limiting case the condition that two parallel dislocations travelling on glide planes a distance y_0 apart will go on moving is

$$\tau - \tau_0 = \frac{\mu b}{4\pi y_0}. \tag{8.29}$$

(This equation is not valid if $\tau \approx \tau_0$.) A relation has thus been obtained for y_0 which together with the density of the sources and dislocations determines also the value of L_0. By investigating the probability of the occurrence of the above condition, Seeger *et al.* have shown [24] that

$$y_0 L_0 = \frac{\sqrt{y L_s^3}}{2n}, \tag{8.30}$$

where y is the average distance of the glide lines and L_s is the length of the glide lines (i.e. the distance of moving off of the screw components of the dislocation loops from each other). From the above relations one finally obtains the following equations:

$$\tau = \tau_0 + \frac{n\mu b}{2\pi}(y L_s^3)^{-1/4}, \tag{8.31}$$

and

$$\frac{d\tau}{dn} = \frac{\mu b}{2\pi}(y L_s^3)^{-1/4}.$$

The question arises whether the number of the dislocations depends on the deformation. If the number of sources per unit volume is N', the glide due to one dislocation per source is

$$\frac{d\gamma}{dn} = \frac{9}{16} bL_eL_sN', \tag{8.32}$$

where the factor 9/16 arises from the fact that the average separation of opposite parts of the piled-up dislocation loops of the source is three-quarters of the total slip length (see Section 5.3); L_e is the length of the slip line of the edge dislocation (as has been mentioned, this line does not result in steps of the surface). By using the relation for the number of sources (the reciprocal value of the volume per source)

$$N' = (L_sL_ey)^{-1}.$$

From (8.32)

$$\frac{d\gamma}{dn} = \frac{9}{16}\frac{b}{y},$$

and thus finally,

$$\Theta_1 = \left(\frac{d\tau}{d\gamma}\right)_1 = \frac{8\mu}{9\pi}\left(\frac{y}{L_s}\right)^{3/4}. \tag{8.33}$$

This relation is in good agreement with experiment. For copper, experimentally, $\Theta_1 = 0.70$ kp/mm^2, while with the measured values $\gamma = 0.15$, $y = 330$ Å and $L_s = 620$ μm, from equation (8.33), $\Theta_1 = 0.77$ kp/mm^2.

The agreement with experiment is perhaps better than could be expected from the approximations used. In (8.28) the degree of work-hardening is somewhat overestimated, because as the external stress exceeds the value corresponding to equation (8.29) the arrested dislocation moves further, and this again results in a decrease of the internal stress obstructing the operation of the source. After the emission of the following loop, its reacting stress contributes to the somewhat smaller internal stress. Simultaneously, however, in (8.32) $d\gamma$ too was overestimated. As has been mentioned already, no piling-up takes place in this case, and so the factor 9/16 is large. (If the loop distribution were completely uniform one should write 1/9 instead.)

The question of why the hardening of fcc metals depends on the orientation of the external stress must be discussed. This orientation dependence is only important with crystal diameters which exceed a few tenths of a millimetre (and thus L_s). The phenomenon can be explained very simply. Let us visualize the glide systems again with the Thompson tetrahedron (Fig. 6.19). If the orientation of the external stress is $\langle 100 \rangle$, $\langle 110 \rangle$ or $\langle 111 \rangle$, its projection in the direction of the glide is of the same magnitude for at least two glide systems. For this reason the glide may start simultaneously in several systems, which results in the formation of Lomer–Cottrell obstacles (Section 6.6) (or in other obstructing processes). Consequently, with this orientation of the external stresses, the condition of the constant number of obstacles is not valid. The formation of new obstacles is accompanied by an increased hardening. The quantitative considerations referring to stage I thus apply only to orientation intervals which are not too wide and can be arranged between the above directions, and in which no multiple glide takes place.

In the case of pure basal glide, hexagonal metals at low temperature always harden in accordance with stage I. At higher temperatures, Θ decreases considerably because of thermal recovery processes. With Zn crystals this occurs between -40 °C and -30 °C.

It should be mentioned here that Hazzledine [26] explains stage I by the formation and interaction of dislocation multipoles.

8.6.2. Stage II of the stress–strain curve

During the transition into stage II the previously long-distance glide decreases considerably in every direction, together with the simultaneous activation of the secondary glide systems. This is borne out by investigation of the glide lines. A study of the temperature dependence of the flow stress shows that the temperature-dependent part hardly changes with increasing deformation. The abrupt increase of the hardening cannot simply be explained by a sudden increase by secondary glide of the density of the "dislocation forest" to be intersected [25]. The hardening becomes manifest by a considerable increase of the temperature-independent τ_E; this appears to be "catalysed" by the secondary glide through the continual formation of new obstacles.

Seeger *et al.* [25] identified these obstacles with Lomer-Cottrell obstacles formed by the reactions of dislocations travelling in various glide systems. In spite of the fact that it yields values for the degree of hardening which are in good agreement with experiment, this concept can be criticized from several points of view. Nevertheless, it must be dealt with, if only in outline, because the results of the macroscopic experiments can be explained in this way, and – taken rigorously – the actual calculations do not utilize the presence of the Lomer–Cottrel obstacles, but merely consider that τ_E results from long-range internal stress fluctuations.

For the sake of simplicity, let us consider a "two-dimensional" example. The loops which have travelled a suitable distance from the sources are replaced by straight dislocation pairs, and the number of sources is characterized by the density N' (defined on a surface perpendicular to the dislocations). If n dislocations are generated per source and their average distance of separation is L, then the following differential relationship is obtained between the glide and the increase of the number of sources:

$$d\gamma = LnbdN'. \tag{8.34}$$

Substituting L from (8.1) and integrating

$$N' = \frac{\gamma - \gamma_0}{2nb\Lambda}. \tag{8.35}$$

Here the dislocations cannot be regarded as isolated units moving independently of each other; their motions are limited, which results in piling-up. Within one glide plane the distance between the neighbours is smaller than the distance between the glide planes (glide lines at the surface). Consequently, as an approximation, every piled-up dislocation group can be replaced by one single dislocation with the Burgers vector nb. The average distance of such groups of the same sign is $N'^{-1/2}$, and thus from (8.4),

$$\tau = \alpha\mu nb\sqrt{N'}.$$

Using this expression

$$\Theta_2 = \left(\frac{d\tau}{d\gamma}\right)_2 = \alpha\mu\sqrt{\frac{nb}{2\Lambda}}. \tag{8.36}$$

With a somewhat more rigorous mathematical treatment Seeger *et al.* [25] obtained the value

$$\Theta_2 = \alpha\mu\sqrt{\frac{nb}{3\Lambda}}. \tag{8.37}$$

According to experimental results with copper single crystals (in the case of an orientation corresponding to minimal hardening), in the overwhelming majority of the glide lines of the screw dislocations, the step height, independently of the magnitude of deformation, is approximately 50 Å ($n = 20$), and $\Lambda = 4\times10^{-4}$ cm [1]. Thus with the value $\mu = 4\times10^3$ kp/mm², from (8.37) $\Theta_2 = 13$ kp/mm², which again is in surprisingly good agreement with the measured value of 13.5 kp/mm².

In this way we have still only established a relation which holds between the various experimentally determined quantities. It is expected from the theory, however, that it will predict, at least approximately, the degree of work-hardening, since according to Table 8.1 it is almost constant, and scarcely changes from one material to another.

In Section 5.3 we dealt with the distribution of the piled-up dislocations, and by integrating (5.23) in the case of obstacles at a distance $\pm R = L/2$ from the source, the number of emitted dislocation loops is

$$n = \frac{\tau R}{\mu b} = \frac{\tau L}{2\mu b} = \frac{\tau\Lambda}{2\mu b(\gamma - \gamma_0)}. \tag{8.38}$$

If now $\tau/(\gamma - \gamma_0)$ is simply replaced by Θ_2, and the value $\Lambda = 4\times10^{-4}$ cm as measured for screw dislocations is used,

$$n = \frac{\Lambda\Theta_2}{2\mu b} = 27, \tag{8.39}$$

which again is in good agreement with experiment. It should be noted that the same combination nb/Λ occurs here as in (8.37). If it is eliminated from the two equations

$$\Theta_2 = \frac{\mu}{24\pi^2} \approx \frac{\mu}{240}. \tag{8.40}$$

In reality this is an upper limit, since the stress-field of the surrounding dislocations partly supplements the external stress and the effective stress is $\beta\tau$ where $\beta < 1$. The factor β depends only on the arrangement of the dislocations, which changes together with the crystal orientation to only a small extent, and is not influenced by the substitutional impurities. The row concerning Θ_2 in Table 8.1 can thus be understood.

Seeger's concept, as outlined, interprets correctly the surface slip lines and every macroscopically measurable quantity (e.g. in the case of Co and Co–Ni single crystals the change of the magnetic susceptibility during the deformation is correctly obtained from the theory [7]). The situation is less favourable with data obtained from transmission electron microscopy. Unfortunately the behaviour of the thin foils, especially when exposed to an intense electronic bombardment, differs considerably from that of the bulk material. Thus, with copper foils which are thinner than a few micrometres, for example, stage II cannot be observed [27]. These phenomena are also influenced by surface oxide layers.

The dislocation structure formed in the bulk material cannot be studied during their motion, but suitable specimens can be made from deformed single crystals. This is, of course, a somewhat cumbersome method, and investigations of this nature have therefore yielded far from sufficient experimental data. Nevertheless the pictures so far have unanimously shown the following:

(a) Disregarding materials in which the stacking fault energy is verysmall ($\gamma \leqq 5$ erg cm^{-2}), piled-up dislocation groups have mainly been observed on grain boundaries, and otherwise the already mentioned "cellular structures" develop.

(b) The number of Lomer–Cottrell obstacles is much smaller than assumed by the Seeger theory, and even these obstacles are, practically without exception, much too short to obstruct the total length of the dislocations emitted by an active source. Sources blocked on every side do not occur at all.

(c) It is difficult to understand why a piled-up dislocation group – whose members repel each other with a considerable force – does not start to move backwards after the termination of the external force. Seeger assumes that the motion of the loop is also obstructed from within by Lomer–Cottrell obstacles, but from what has been said previously this possibility does not seem to be feasible.

Hirsch and Mitchell [28] pointed out that the theory needs further refinement, and that only half of the observed work-hardening can be explained by the fluctuations of the long-range stresses. The replacement of the piled-up dislocation group by one single dislocation

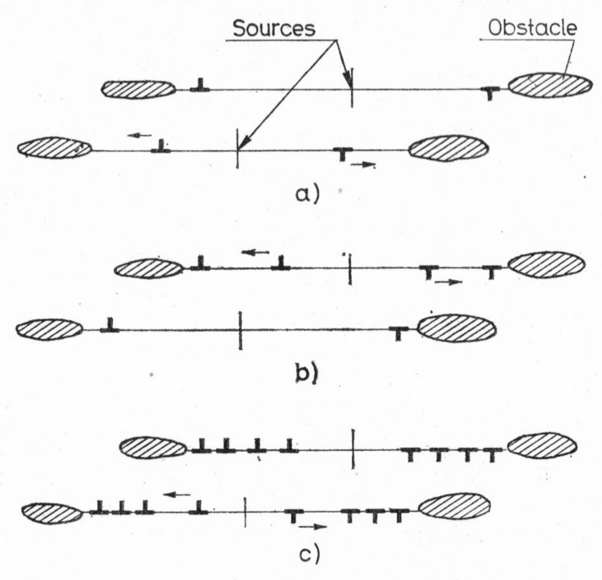

FIG. 8.8. Cooperative multiplication of dislocations after Hirsch and Mitchell [28]

with Burgers vector nb also appears to be an oversimplification, for the resultant stress is taken at distances comparable with the extension of the piled-up group. In addition the theory should also take into consideration that the strength of the obstacles depends upon the distance from the given glide plane too. The most important fact realized, however, is that the slip lines develop by the co-operative motion of the dislocation groups by and large in the following way. Let us assume that as a result of an external stress a dislocation loop is emitted by the source S_1, and that this loop becomes obstructed in the vicinity of the source S_2 (Fig. 8.8a). The stress-field of this dislocation promotes the activation of source S_2, but as its dislocations come near to the first source, their stress-field in turn promotes the operation of this source, and so on. In this way the glide occurs abruptly at certain places as a result of very small stress increments. For the details the reader is referred to the original paper.

8.6.3. Stage III of the stress–strain curve

According to Seeger's concept, stage III begins when the screw components of the piled-up dislocation group continue their motion by cross-slip in another glide system due to the large stress concentration, or they slip over to a parallel glide plane where no obstacles impede their motion. The segment which slips over possibly acts as new source. The extended dislocations, even if they are extended to only a small extent, for this purpose must first contract in a short segment whose length depends upon the stress. The contraction needs a further energy input. This of course is partly promoted by the repulsion between the adjacent dislocations lying close beside each other. The start of the cross slip is consequently a typically thermally activated process, the energy of activation of which can be calculated with the Peierls model of extended dislocations [29]. On this basis it is quite evident that the stress necessary to initiate stage III decreases rapidly with increasing temperature. The increase of the stacking fault energy similarly leads to the decrease of τ_3, as a result of the decrease in the separation of the partial dislocations.

However, apart from what was said in relation to stage II, there is a further serious argument against the probability of the above mechanism. The basis of the whole process is the cross-slip of the piled-up screw dislocations. From Fig. 6.21, however, it can be seen that an obstacle parallel with the Burgers vector cannot be created by the dislocation reaction between the primary and secondary glide systems. Such obstacles can be generated only by reactions between the secondary systems (and even then only in a few cases), and are therefore not very frequent. Consequently, even if many piled-up dislocations exist they are practically never screw dislocations.

The cross-slip is quite clearly explained by Hirsch [16, 30] for example. In his view the cross-slip in fcc metals starts from extended jogs; some of the large jogs have structures such that part of the extended dislocation lies in the plane of the cross-slip if the dislocation forming the jog disintegrates into several partial dislocations, and in the vicinity of the jog the separation of the partials is smaller. It is thus sufficient to start the cross-slip if the extended dislocation segment contracts in one point (Fig. 8.9). The extended dislocation segment so developed in the plane of the cross-slip, steadily increases as a result of a stress acting in the **CB** direction at the point of contraction, and simultaneously becomes more and more bowed out.

This process is thermally activated in the same way as the Seeger cross-slip, and consequently it results in a similar $\tau_3\,(T)$ relation. However, the activation energy in this case is considerably smaller,

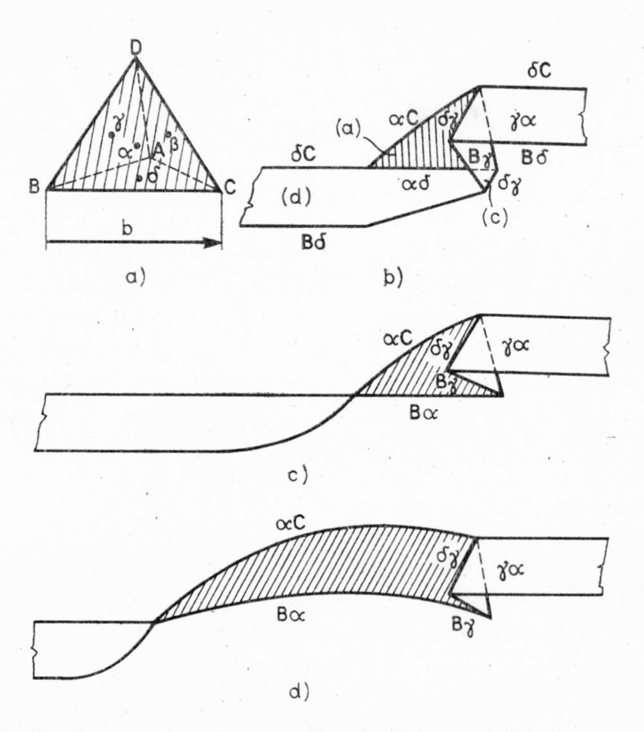

FIG. 8.9. The onset of cross-slip according to Hirsch [16]. The shaded part lies in the plane of the cross-slip, the letters indicate the Burgers vector of the partial dislocations in accordance with the Thompson tetrahedron

and therefore no concentration of the stress due to the piled-up dislocations is necessary to initiate the process.

It is not difficult either to interpret work-softening on the above basis. It has frequently been observed that whenever the deformation is interrupted, and then later resumed at some higher temperature, the stress flow on the resumption of the deformation decreases after a transitional maximum and then slowly increases again (Fig. 8.10).

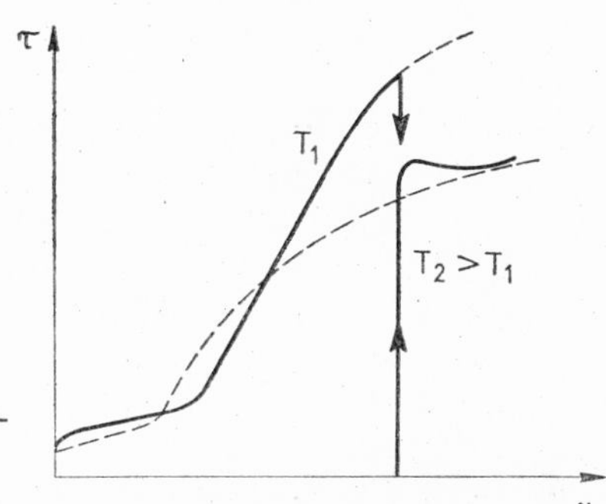

FIG. 8.10. Deformation-softening during the temperature change

This apparent yield point effect is caused by the rapid dissolution of the equilibrium dislocation structure, already formed at a lower temperature, by the thermally activated cross-slip at the repeated high-temperature deformation. For this reason, the stress asymptotically approaches the lower value which corresponds to the higher temperature.

The sudden decrease of the deformation rate also has an apparent softening effect, since the temperature-dependent part of the flow stress becomes smaller with the decrease of the rate.

8.7. Stability of the deformed state

In this section we deal quite briefly with the question of why the dislocations do not slip back to their sources after the deforming stress has ceased to operate. It should be mentioned in advance that the possibility of a small-scale back-slip is not entirely excluded, but its macroscopic effect may be neglected.

The stabilization is due to the following causes:

(a) Frictional forces (Peierls stress), etc., quite independent of the direction of motion, obstruct the backwards motion.

(b) Obstacles formed during the deformation – especially in the case of extended dislocations – also remain after the stress has ceased to act.

(c) The dislocations are so situated in relation to each other that their stress-fields mutually compensate each other at large distances. Hence, the dislocations must be lifted over a potential barrier in the case of a motion in the reverse direction. For this, however, approximately the same stress is needed as for the original motion.

(d) Jogs formed during the motion cannot disappear, but the interstitial and vacancy jogs change their roles, and the back-slip becomes obstructed by the previously interstitial jogs.

If a stress of opposite direction begins to operate after the termination of the stress, the deformation in the opposite direction begins considerably sooner than with the previous flow stress (Fig. 8.11,

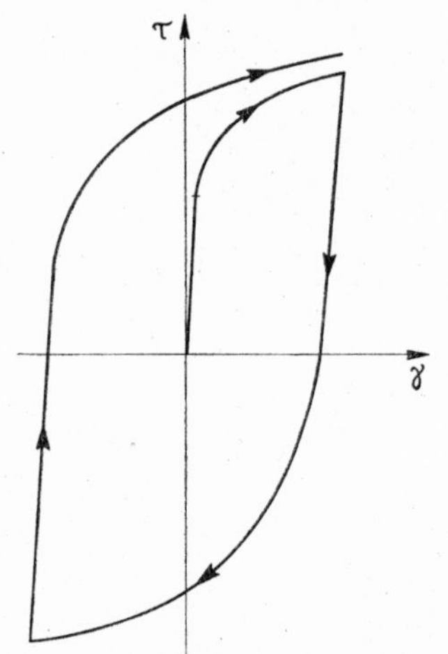

FIG. 8.11. Stress–strain relation for strains of alternating direction in the case of a small deformation (Bauschinger effect)

Bauschinger effect), since, because of the internal stresses, some of the dislocations already move backwards at very small external stresses. The Bauschinger effect is the strongest in stage I of the hardening. This is quite understandable, since the stabilizing effect of the jogs and obstacles increases rapidly with the onset of stage II.

In general, if a stress is applied to the material after its work-hardening so that the deformation takes place in a glide system differing from the previous one, the hardening of the first activated glide system affects the other systems. A considerably larger stress is necessary to activate the new glide system than was needed to start the first one. This *latent hardening* sometimes exceeds the hardening of the primary glide system [2].

Chapter 9

Effect of heat-treatment on the defect structure of metals

9.1. Introduction

Because of lattice defects, plastically deformed metals contain internal stresses. The free energy of a body is larger in the deformed state than at the same temperature before deformation, and so this state is thermodynamically unstable. Nevertheless, at not too high a temperature, the deformed state is still stable because the thermodynamic driving forces cannot overcome the effects impeding the motion of the lattice defects and thus produce a lower free energy. With increasing temperature, however, certain defects attain the thermal energy necessary to move them, and consequently the defects gradually become rearranged or disappear. This results in a change of the physical properties of the metal.

When a metal is plastically deformed at very low temperatures a defect structure is created which is stable at the temperature of deformation, and which generally contains every possible defect type. If the temperature is gradually increased, the process requiring the least energy will become activated at some temperature and defects of certain types then either disappear or are rearranged. If the temperature is further increased, changes of the defect structure requiring higher activation energy are initiated and so on. With such heat-treatment the deformed structure can either be changed or practically dissolved. If the change of measurable physical properties is studied during the heat-treatment, conclusions concerning the nature of the individual processes and their activation energies can be drawn. In the following sections, the experimental methods and results in this field are briefly reviewed.

9.2. Processes during heat-treatment

The fundamental process occurring during heat-treatment is the diffusion of vacancies, or in general, of point defects. It has been shown in Chapter 3 that the activation energy of the motion of point defects is relatively small. Let us consider a body whose point defect content at very low temperatures is larger than the thermal equilibrium concentration. If the body is heated, the point defects acquire the thermal energy necessary to move at a still-low temperature. This is the reason why the surplus point defects introduced into the crystal, for instance, by plastic deformation, are annihilated by diffusion at various sinks (grain boundaries, dislocations) at relatively low temperatures. Since the decrease of the point defects produces a measurable decrease in the electrical resistivity, this process can profitably be followed by resistivity measurements. The change of the point defect concentration causes little change in the mechanical properties of the metal.

The vacancy diffusion becomes important, in relation to mechanical properties, when the number of vacancies necessary to maintain thermal equilibrium in the body is relatively large. These activate further processes so that, because of the thermal equilibrium, the number of vacancies remains constant at a given temperature.

Such a process activated by vacancies is as follows. Vacancies diffuse to jogs where they may become absorbed by initiating a non-conservative jog motion. By this means, the dislocation containing the jog is moved by internal stresses since the fixing effect at the previous jog site ceases. The process, called polygonization, leads to a rearrangement of the dislocations in a dislocation structure of lower energy. During the rearrangement, the lattice orientation in the deformed crystal changes from site to site because of the randomly arranged dislocation structure. Finally, in the course of polygonization, the dislocations become arranged according to certain rules in individual regions of the crystal. As a result, many subgrains (polygons) form which have slightly differing orientation.

A simple example of polygonization is depicted in Fig. 9.1. It has been seen in Section 2.5.4 that between parallel edge dislocations lying in parallel glide planes there is an interaction which tends to arrange the dislocations in a plane perpendicular to the glide plane. Consequently, if the dislocation motion can be thermally activated, then the set of edge dislocations shown in Fig. 9.1a becomes arranged in "dislocation walls" as in Fig. 9.1b. Each dislocation wall separates two grains whose orientations differ only slightly. If the distance between the glide planes is d, then the orientation difference ϑ is given according to Fig. 9.1b by the equation

$$\frac{b}{d} = 2 \sin \frac{\vartheta}{2}. \tag{9.1}$$

Further heating following polygonization induces a new process called re-crystallization. During polygonization, the defect structure was rearranged so that the number of dislocations changed only slightly, but at the same time their arrangement became energetically

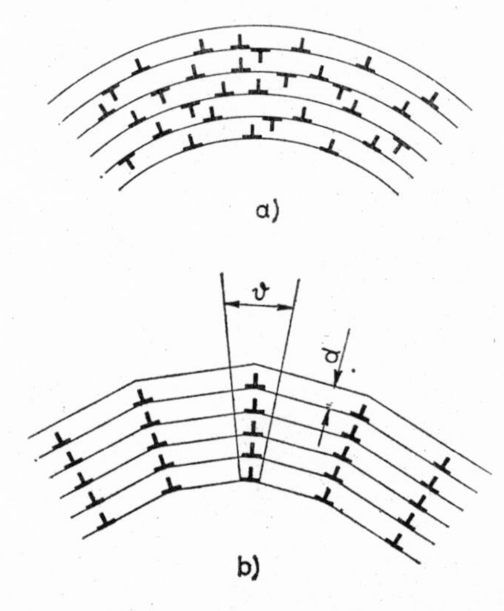

FIG. 9.1. The arrangement of dislocations in a bent rod (a) before and (b) after polygonization. The formation of sub-grains, as can be seen in (b), results in a lower energy state

more favourable. The re-crystallization annihilates a considerable number of dislocations, and during this process, new grains are formed which grow at the expense of the old ones. After a few hours of heat-treatment at a suitable temperature, the material is practically free from internal stresses. All those thermally activated processes connected with the rearrangement of the defect structure before the re-crystallization, are called crystal recovery processes.

The heat-treatment can, in general, be carried out in two ways: either with varying or constant temperature. In the first case the temperature is continuously increased and the above processes are gradually activated. Instead of continuous heating, a stepwise heat-treatment can be applied. In this case, the sample is kept for some time (for example 15 minutes) at increasingly higher temperatures (isochronal annealing). This method is particularly advantageous if parameters are to be studied which cannot be measured *in situ* during the heat-treatment. Thus, for instance, in a study of the electrical resistivity the change of resistivity due to heat-treatment at various temperatures can be measured only at low temperature. With the other type of annealing, the time-dependence of thermally induced phenomena is studied at suitable temperatures (isothermal annealing). In this way, information is obtained about the course of the process with time (the kinetics).

Let us now return to the recovery due to the diffusion of vacancies. Consider the time dependence of the internal stresses at a constant temperature where the vacancy concentration due to thermal equilibrium is large enough. To describe the process quantitatively, a suitable model is applied. The motion of thermally activated vacancies results in a non-conservative motion of jogs, and as a result the dislocations can slip into positions of more favourable energy. The activation energy of such a slip obviously depends upon the internal stress at the site of the moving jog, while the rate of the process is a function of the temperature. [This is analogous to transient creep with the difference only that the quantity τ of (7.15) is now the resultant of the internal stresses.] The change of the internal stresses is called stress relaxation; its rate can be expressed by the equation

$$\frac{d\tau}{dt} = -Ce^{-(U_0 - \tau b^2 l)/kT}, \tag{9.2}$$

or after integration,

$$\tau = \tau_0 - \frac{kT}{b^2 l} \ln\left(1 + \frac{t}{t_0}\right), \tag{9.3}$$

where τ_0 is the stress at time $t = 0$, and t_0 is a constant of integration whose value can be determined from the equation

$$\tau_0 = -\frac{kT}{b^2 l} \ln \frac{Cb^2 l t_0}{kT}. \tag{9.4}$$

The logarithmic time dependence (9.3) has been experimentally verified for a large number of materials [1]. The energy characterizing the mechanism of the change depends upon the previous deformation and also upon the temperature. This process requires the presence of a large number of jogs and vacancies. A strongly deformed material contains a large number of jogs and, in such a case, the rate of the process at a given temperature is in practice determined by the energies of formation and activation of motion of the vacancies, that is by the self-diffusion energy of the material in question. In the case of creep, however, because of the considerably larger stress, other processes may also become activated at lower temperatures. In these cases U_0 may be smaller than the energy of self-diffusion [2].

It should be mentioned that re-crystallization takes place in polycrystalline metals by the motion of grain boundaries. The driving force of this process originates from the fact that the local dislocation density and hence the deformation may be very large in some parts of the grains. With the displacement of the grain boundary, the extensive inhomogeneity of the stresses decreases. This process can develop only if the increase of the surface energy is smaller than the decrease of the energy stored in the grain [3].

9.3. Change of the physical properties during heat-treatment

The rearrangement and annihilation of the lattice defects result in the change of many physical properties. The measurement of these changes yields valuable information on the properties of defects. Let us consider, for instance, the energy stored in the stress-field of the lattice defects in a crystal. A very simple answer can now be given to the question that arose in the first section, of why the stored energy is released at different stages during heat-treatment. In a metal plastically deformed at low temperatures, every type of defect is generated. If this metal is subjected to a gradual heat-treatment the defects of various types become activated at different temperatures, and their annihilation decreases the internal energy of the crystal.

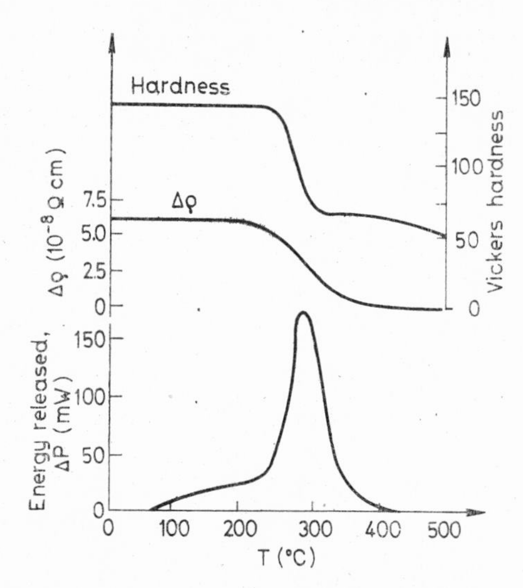

FIG. 9.2. The change of the physical properties during annealing in copper deformed by torsion (according to Clarebrough *et al.* [4])

The measurement of the stored energy is extremely important. Many experiments have been carried out in this field in the last sixty–eighty years, but satisfactory results have been obtained only in the last fifteen years. Figure 9.2 shows the stored energy released during the continuous heating of copper deformed by torsion, and also, the simultaneously measured changes of electrical resistivity and hardness [4]. The bottom curve shows the energy released per unit time. It can be seen that the curve has a sharp maximum.

X-ray investigations have shown that the re-crystallization begins at the same time as the rapid energy release. Simultaneously, the electrical resistivity and hardness considerably decrease. This indicates that during re-crystallization the number of dislocations, and thus the residual internal stresses, are reduced to a considerable extent.

The measurement of the stored energy permits the identification of the onset of re-crystallization by the well-defined temperature of maximal energy release. Naturally this temperature, i.e. the development of the re-crystallization process, depends upon the magnitude of the previous deformation. With very large internal stresses, the thermodynamic instability of the internal stresses occurs at lower temperature (see Fig. 9.3 and Table 9.1).

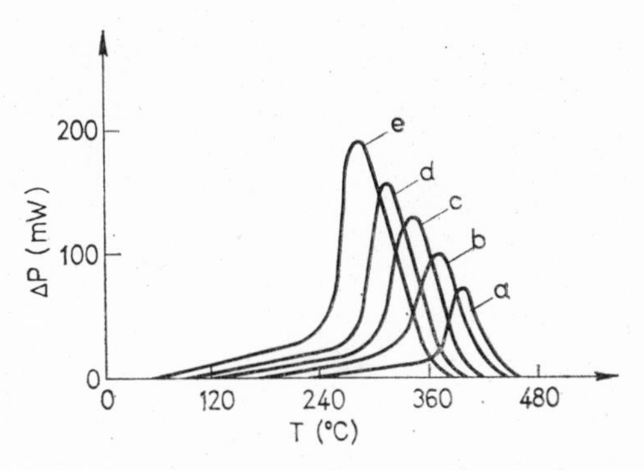

FIG. 9.3. The change of the re-crystallization temperature of copper as a function of the deformation (Clarebrough *et al.* [4]) (see also Table 9.1)

TABLE 9.1. *The dependence of the stored energy and re-crystallization temperature on the degree of torsional deformation*

Symbols in Fig. 9.3	Degree of deformation, nd/l*	Stored energy in cal/g	Temperature of re-crystallization, °K
a	0.47	0.100	400
b	0.94	0.145	370
c	1.41	0.198	345
d	2.15	0.263	315
e	3.03	0.331	290

* d = diameter of the cylindrical sample, l = its length, n = number of twists.

Similarly, the re-crystallization is also influenced considerably by the purity of the material. In an impure material the Cottrell atmosphere pins down the dislocations, and so the re-crystallization begins only at a higher temperature (Fig. 9.4). The results show clearly that in copper deformed at room temperature, in practice, only the dislocations are stable. Because of their low activation energy of motion,

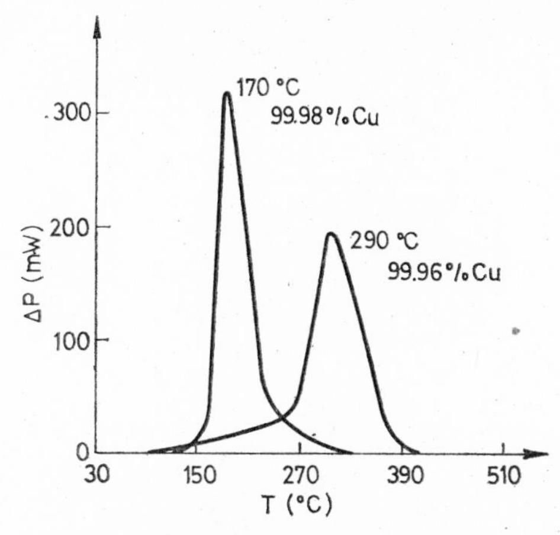

FIG. 9.4. The effect of impurities on the re-crystallization temperature. Both materials were deformed till fracture (according to Clarebrough *et al.* [4])

the vacancies and other point defects possess considerable mobility, even below room temperature, and consequently become annihilated at once during the deformation.

Nickel is very suitable for investigating the energy stored by point defects, since in this metal, the vacancy mobility becomes high only above 200 °C. Figure 9.5 shows the change of the physical properties of Ni with temperature. It can be seen that the first maximum is accompanied by a considerable decrease of the electrical resistivity while the hardness remains unchanged. It follows therefore that this process can only involve the annihilation of vacancies. This is supported by density measurements. The vacancies increase the volume of the body, i.e. decrease its density. The annihilation of vacancies thus must result in a density increase (Fig. 9.6).

The release of the stored energy can also be investigated as a function of time at constant temperature (isothermal annealing). Such measurements were carried out on polycrystalline silver by Bailey and Hirsch [5] (Fig. 9.7). The stored energy was released in two stages. Simultaneous X-ray and electron-microscopic investigations showed

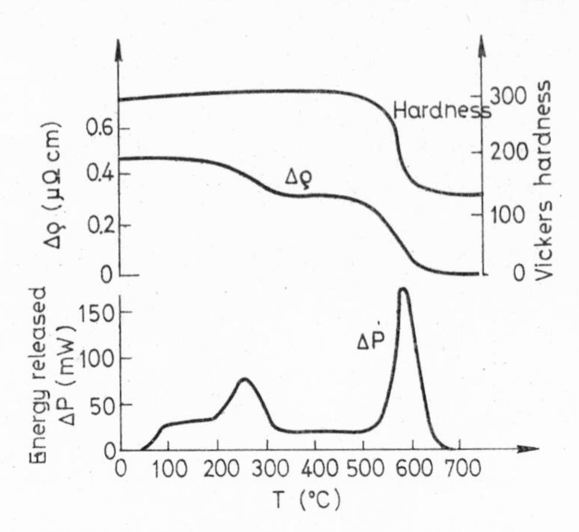

FIG. 9.5. The change of the physical properties during annealing of nickel deformed by torsion (according to Clarebrough *et al.* [4])

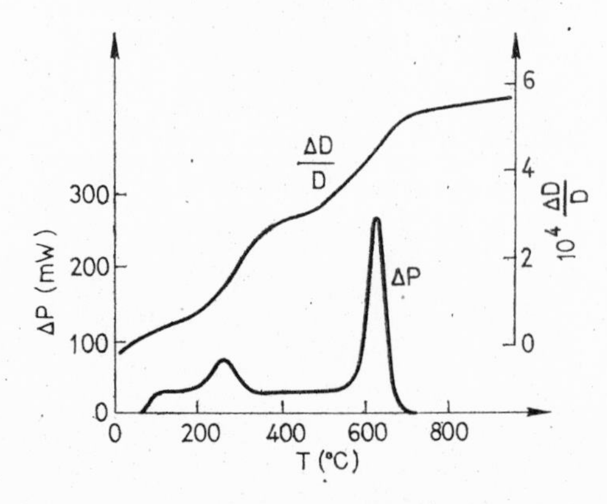

FIG. 9.6. The change of the density of deformed nickel during annealing (Clarebrough *et al.* [4])

that the first (AB) section involves only recovery, but in the C stage re-crystallization also begins. The energy released during the recovery is about half the total energy stored.

The theoretical study of the stored energy presents an extremely difficult problem, since the exact dislocation distribution in the crystal should be known. The calculated results depend upon the nature of the model used. According to the results of Seeger and Kronmüller [6], approximately 7% of the mechanical work done during the deformation of the crystal remains stored in the stress-field of the dislocations. This result is in fairly good agreement with the measurements of Bailey and Hirsch.

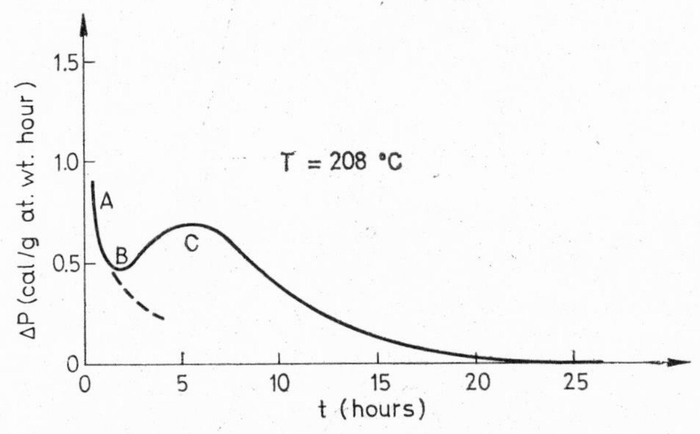

FIG. 9.7. The release of the stored energy from silver during isothermal annealing (according to Bailey and Hirsch [5])

The experiments discussed so far were carried out at relatively high temperatures where processes involving the dislocations primarily occur. However, the internal energy of a crystal deformed at low temperature can be changed in a measurable way by many other processes, for instance the interactions of defects of various types can be induced. These can be measured only by heat-treatment at low temperature because their activation energy is low.

Van den Beukel investigated the mechanism of the release of stored energy at low temperature by an ingenious method [7]. The measurements for copper and nickel are shown in Fig. 9.8. According to these results, the release of the stored energy at low temperature is induced by many different processes together. The energy release accompanying these processes is very small. It can be concluded from this that the processes involve the rearrangement (by interaction with one another and with other defects) and annihilation of the point defects. The nature of these processes is still not satisfactorily known, and many more experimental data are probably needed for a deeper understanding of the mechanisms.

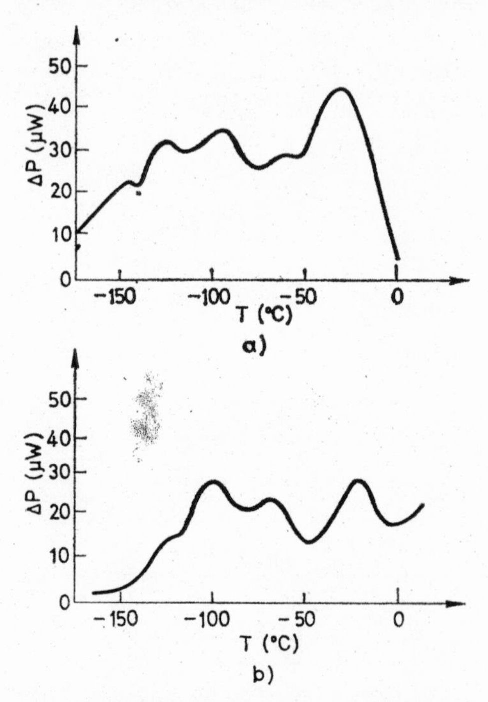

FIG. 9.8. The release of the stored energy from (a) copper and (b) nickel deformed at low temperature (78 °K) (according to van den Beukel [7])

In the last decade intensive studies have been carried out in this direction. The change of the defect structure and the properties of the individual defects have been investigated with extremely sensitive methods. The detailed analysis of the results obtained would exceed the scope of this book. In the following we briefly review the possible processes which take place at low temperatures and for further details the reader is referred to the literature [8].

The changes occurring during heat-treatment are usually classified by five temperature regions or stages, on the assumption that the processes taking place in these stages have relatively well-defined activation energies. These stages are briefly summarized for fcc metals in Fig. 9.9.

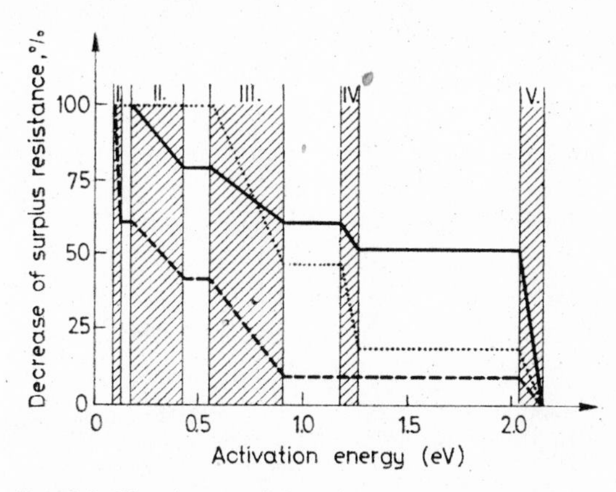

FIG. 9.9. The change of the electrical resistivity due to the annealing of fcc metals subjected to various effects (continuous line, plastic deformation; dotted line, quenching; broken line, irradiation) (according to van Bueren, H. G.: *Imperfections in Crystals*, North Holland Publishing Co., Amsterdam, 1961)

Stage I: the temperature range from −230 °C to *ca.* −150 °C. This stage has been found only in irradiated substances. The activation energy of this process is about 0.1 eV. In this interval, the extra resistivity of the irradiated samples decreases by about 40% or 90% (the latter refers to electron irradiation). This process can probably be explained by the annihilation of interstitials and larger point defect groups. More exact investigations indicate a fine structure, i.e. complex processes within the stage [8].

Stage II: the temperature interval from −150 °C to −80 °C. The activation energy in this stage is approximately 0.2–0.5 eV. The resistivity decrease is *ca.* 20% in plastically deformed metals [9], and in irradiated substances either 20% [10] or, in the case of electron irradiation, 10%. The process can probably be accounted for by a vacancy-interstitial recombination.

Stage III: this stage extends from −80 °C to *ca.* +50 °C. It can be found in plastically deformed materials as well as in quenched and irradiated substances. The extra resistivity decreases by *ca.* 20%, 30% or 60% [10, 11]. The activation energy is approximately 0.7 eV. This process is probably connected with the annihilation of vacancies, though no unified picture of the mechanism has developed.

Stage IV: this extends from 50 °C to 250 °C in plastically deformed and quenched metals. The activation energy is approximately 1.2 eV, and the decrease of the extra resistivity is *ca.* 10% or 30%. The value of the activation energy indicates that the processes taking place in this stage involve the migration of vacancies.

Stage V: processes which take place above 250 °C. The residual resistivity of the plastically deformed and quenched metals completely disappears during the changes in this stage [12], and furthermore the original mechanical properties (yield stress, etc.) are restored. The activation energy is about 2 eV, or even more. This process is clearly connected with the annihilation and rearrangement of dislocations.

Appendix A

Concepts and notations of crystal geometry

A.1. Lattice criterion, unit cell, crystal systems

A common fundamental property of every crystalline material is that its atoms are arranged in a definite systematic order, i.e. they form a three-dimensional lattice. More exactly, three vectors, **a**, **b** and **c**, can always be found which do not lie in the same plane and for which it stands that if the lattice is displaced by a vector

$$\mathbf{r} = n_1\mathbf{a} + n_2\mathbf{b} + n_3\mathbf{c} \qquad (n_1, n_2, n_3 = 0, \pm 1, \pm 2 \ldots),$$

then it corresponds exactly with its previous position. Any translation which brings the lattice into coincidence with its original position can be produced by the three numbers n_1, n_2, n_3 according to the above equation. The vectors **a**, **b**, **c** satisfying this lattice criterion are termed primitive translation vectors. Three such vectors always span a parallelepiped: the primitive unit cell. The lattice may be regarded as the sum of many unit cells situated side by side without any gaps.

In a given lattice, the primitive translation vectors (i.e. the unit cell) can be defined, of course, in an infinite number of ways. Theoretically, these are all equivalent. The lattice is usually described by the shortest of the equivalent vectors, whose mutual angles are closest to 90° (Fig. A.1).

The unit cell is described by the edge lengths a, b and c, and the angles α, β and γ. α denotes the angle between the vectors **b** and **c**, and so on.

The atoms forming the crystal lattice may satisfy various symmetry operations besides the lattice criterion. Depending upon the main symmetry elements of the lattice, crystals can be classified into seven crystal systems which are listed in Table A.1.

It frequently occurs that as a result of the lattice symmetry, mutually perpendicular preferred directions exist in the lattice, yet no rectangular primitive unit cell can be found in the crystal. The requirement that the unit cell should possess the main symmetry properties of the lattice is nevertheless justified. For this reason the concept of the centred cell has been introduced, this cell containing lattice points on its faces

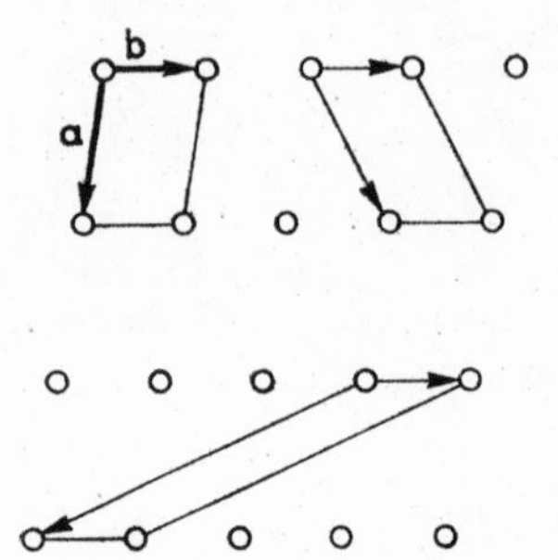

FIG. A.1. Various primitive unit cells in a lattice

or in its inside in addition to those at the corners (Fig. A.2). In this way the specifications of the third column of Table A.1 are realized. It can be shown that only three types of centre are required (Fig. A.3). The base-centred cell contains lattice points in the centres of two opposite faces. This type is denoted by C (or A or B) if the centred face intersects the **c** (or **a** or **b**) axis. In the body-centred cell one lattice point is in the geometrical centre; and in the face-centred cell there is a lattice point in the centre of every face.

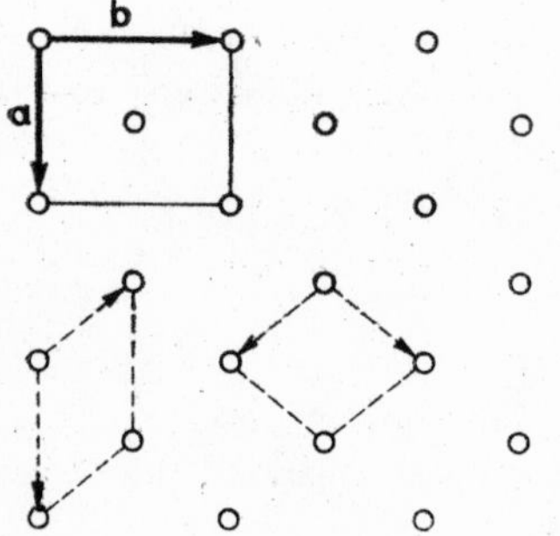

FIG. A.2. Centred cell in a two-dimensional lattice

TABLE A.1. *Crystal systems*

System	Highest symmetry operation	Unit cell dimensions	Centred cells
Triclinic	Center of symmetry	$a \neq b \neq c$ $\alpha \neq \beta \neq \gamma$	
Monoclinic	Twofold axis in one direction	$a \neq b \neq c$ $\alpha = \beta = 90^\circ \neq \gamma$ or $\alpha = \gamma = 90^\circ \neq \beta$	A (or B, C)
Orthorhombic	Twofold axes in three perpendicular directions	$a \neq b \neq c$ $\alpha = \beta = \gamma = 90^\circ$	C (or A, B) I F
Trigonal	Threefold axis in one direction	$a = b = c$ $\alpha = \beta = \gamma \neq 90^\circ$	
Hexagonal	Sixfold axis in one direction	$a = b \neq c$ $\alpha = \beta = 90^\circ \neq$ $\neq \gamma = 120^\circ$	
Tetragonal	Fourfold axis in one direction	$a = b \neq c$ $\alpha = \beta = \gamma = 90^\circ$	I
Cubic	Threefold axes along the cube diagonals	$a = b = c$ $\alpha = \beta = \gamma = 90^\circ$	I F

It is easy to see that not every centred cell is needed in every system. Thus, for instance, it is quite clear that in a triclinic system only a primitive (P) cell is used, since no type of centring would have any

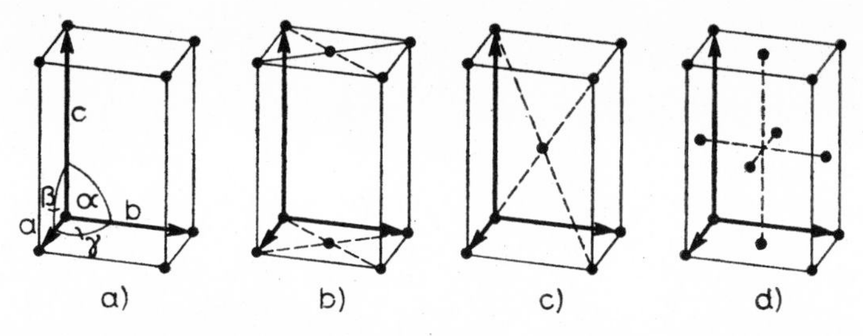

FIG. A.3. Centred cells in a three-dimensional lattice: (a) primitive (P), (b) base-centred (C), (c) body-centred (I), (d) face-centred (F)

sense in this system. The various cell types which are necessary in the individual crystal systems are given in the fourth column of Table A.1. A total of fourteen types of unit cell exist. In the literature they are referred to as the fourteen Bravais lattices *(Bravais cells)*.

A.2. Lattice planes, Miller indices

A parallel set of planes can be laid in various ways across the lattice points of any lattice so that

1. every plane passes through lattice points;
2. each lattice point rests on a plane of the set;
3. the distance between adjacent planes of a given set is constant, and the individual planes can be brought to coincidence by simple translation.

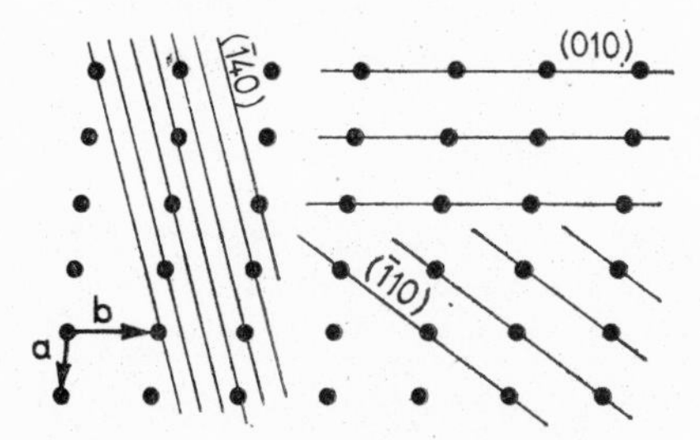

FIG. A.4. Lattice planes (straight lines) in a two-dimensional lattice

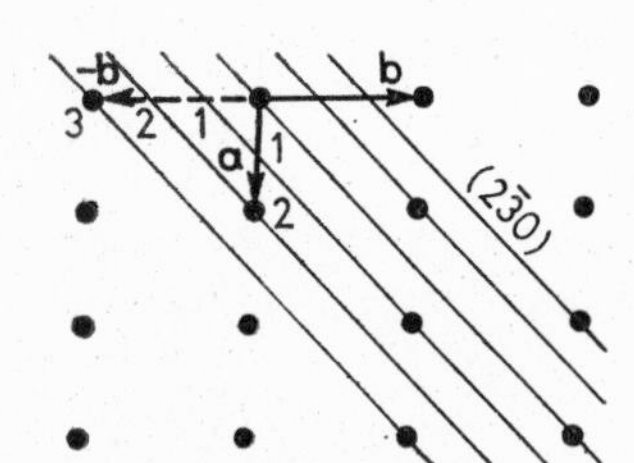

FIG. A.5. Derivation of the Miller indices

The planes meeting these conditions are lattice planes. It can be seen from Fig. A.4 that adjacent lattice planes are further apart, the larger is the density of the lattice points. If the coordinate system (which may be oblique-angled) is placed along the edges of the unit cell, it can be seen that the intercepts which the planes make with the axes, i.e. their distance in the **a**, **b** and **c** directions, are a/h, b/k and c/l, or their multiples, where h, k and l are integers. These integers are used to define the individual sets of planes, and are called Miller indices. The indices can be obtained in the following way. Starting from the origin of the coordinate system, set off in the direction of a lattice plane along the a-, b- and then the c-axis, and count the number of planes passed until the next lattice point is reached (Fig. A.5). The three numbers obtained are characteristic of the set of lattice planes and are written without a comma or bracket (e.g. 121) if they refer to an X-ray (electron, neutron) diffraction maximum produced by a set of planes. If, however, a plane or a set of planes is to be characterized, the three are put in parentheses, e.g. (521). Negative numbers are indicated by short, horizontal lines above the number. If numbers of two or more digits occur, in order to avoid any misunderstanding a comma is used to separate them, e.g. $(11, \bar{4}, 2)$. If the set of planes is parallel to one of the axes the corresponding index is naturally 0.

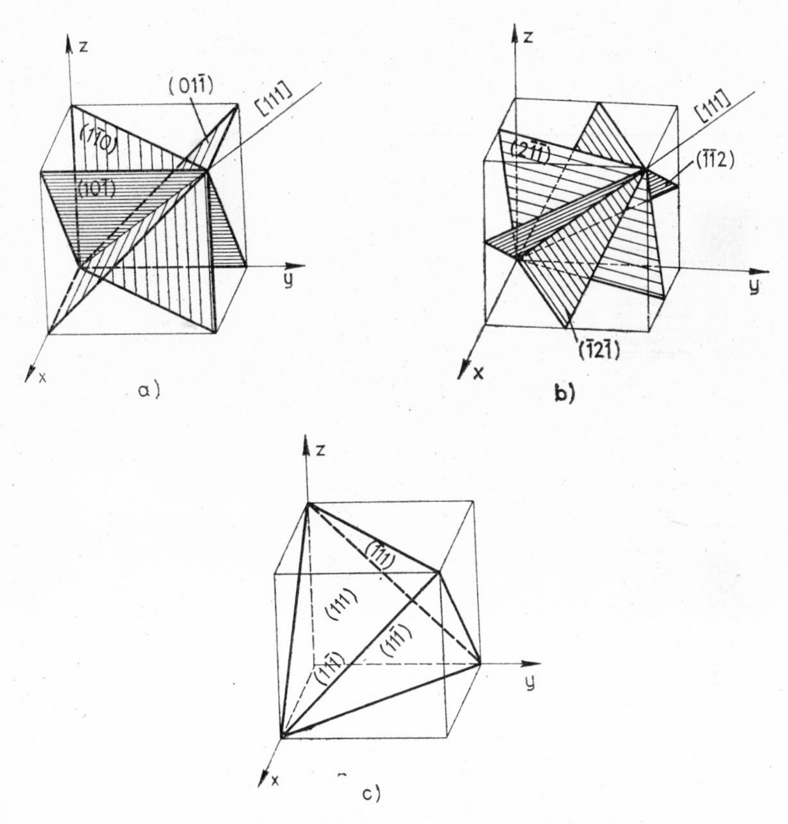

FIG. A.6. The important planes of a cubic lattice

A few examples follow. Figure A.6 depicts some lattice planes of the cubic lattice. Besides the planes (110), (011), $(0\bar{1}1)$ shown in the figure, the planes $(1\bar{1}0,)$ (101), $(10\bar{1})$ also exist. These six planes differ from each other only in their orientation; they are otherwise perfectly equivalent, and the symmetry operations of the cubic system transform them into each other. The sum total of such planes (faces) are the *crystallographic forms* which are denoted by a brace. Thus $\{110\}$ denotes either the total of the six planes listed, or else anyone of them. Figures A.6b and c show the $\{211\}$ and $\{111\}$ planes, respectively.

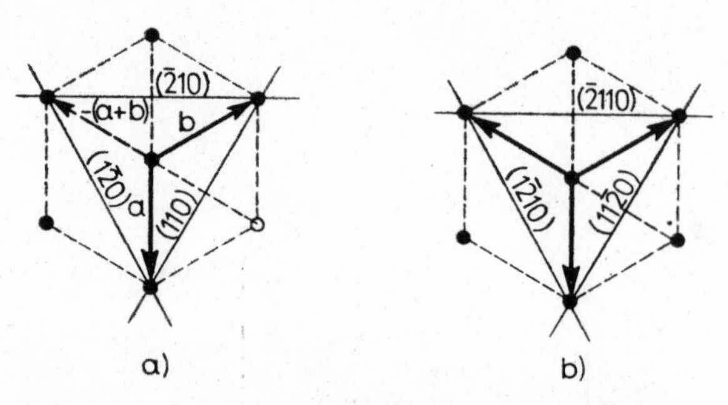

FIG. A.7. Triple and quadruple indices in the hexagonal system

The indices of the hexagonal system deserve special mention. Here, in the basal plane normal to the sixfold (or threefold) axis, there are three equivalent directions at angles of 120° to each other. Two of these axes are arbitrarily specified to define the unit cell. For this reason the (110), (120), $(\bar{2}10)$ indices which are of the same form do not reflect this similarity (Fig. A.7). This difficulty can be simply eliminated if the direction of the vector $-(\mathbf{a}+\mathbf{b})$ is also regarded as an axis and the Miller indices are accordingly completed with fourth index $i = -(h+k)$ which is written in the third place: $(hkil)$. If this index is not written out, its place is usually indicated by a point. The indices of the three planes mentioned are thus:

$$(11\bar{2}0),\ (1\bar{2}10),\ (\bar{2}110) \quad \text{or} \quad (11.0),\ (1\bar{2}.0),\ (\bar{2}1.0).$$

A.3. Lattice vectors, zone

A vector in the lattice is always given by its relative components in the directions of the *a*-, *b*- and *c*-axes. These components are put in square brackets. The edge vectors of the unit cell are thus [100], [010], [001]. As with the lattice planes, here too we speak of forms, the notation being angle brackets ⟨ ⟩. Thus, for instance, in the cubic system ⟨110⟩ indicates the face diagonal vectors of the cubic unit cell.

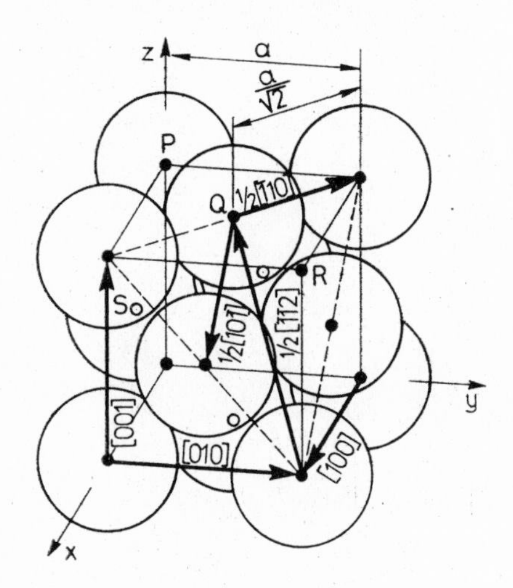

FIG. A.8. Vectors in the face-centred (close-packed) cubic lattice

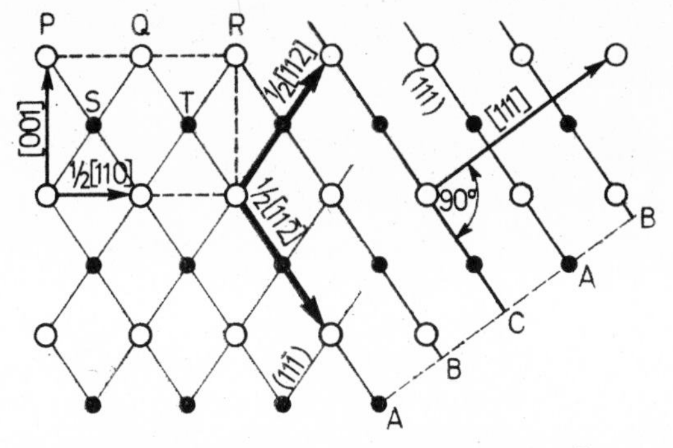

FIG. A.9. Face-centred cubic lattice projected onto the $(1\bar{1}0)$ plane. The lattice points lying in the plane of the drawing are denoted by empty circles, and the atoms lying at a distance of $a/2\sqrt{2}$ above and below this plane by full circles

If a group of lattice planes has a common straight line (which is parallel to every plane), the planes are said to belong to a common zone and the straight line is the zone axis.

The zone is characterized by the axial lattice vector $[u\ v\ w]$. A plane $(h\ k\ l)$ belongs to the zone $[u\ v\ w]$ only if

$$hu + kv + lw = 0.$$

This is Weiss' zone law from which, however, it does not follow that the vector $[h\ k\ l]$ is perpendicular to $[u\ v\ w]$ or to the $(h\ k\ l)$ plane, since in the general case one is dealing with an oblique-angled coordinate system and working with relative coordinates. For cubic system, naturally, these latter conditions are also valid.

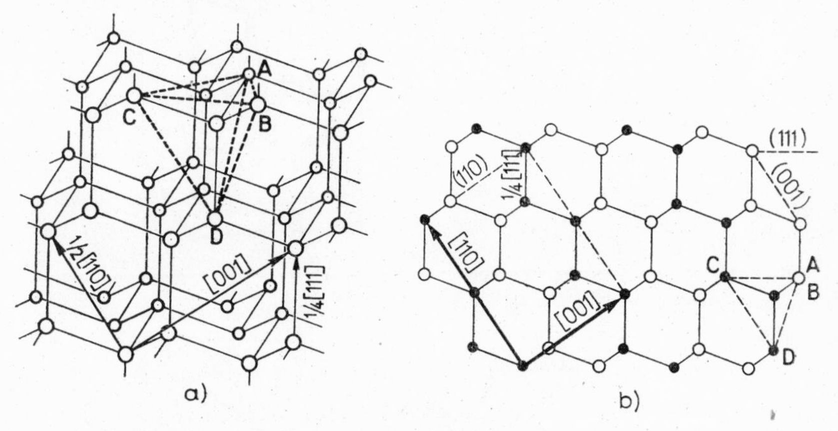

FIG. A.10. The perspective view (a) of the diamond lattice and (b) its projection on the $(1\bar{1}0)$ plane. In the latter the full circles are in the plane of the figure, and the empty circles denote atoms at a distance $a/2\sqrt{2}$ from this plane. The dotted line in the right lower corner represents a tetrahedral structure

Figures A.8 and A.9 depict the face-centred cell corresponding to the cubic close-packed lattice, and the vectors occurring most frequently in a treatment of dislocations. Finally a projection of the diamond lattice is given in Fig. A.10.

Appendix B

Principles of the theory of elasticity

B.1. The dilatation tensor

In the phenomenological description of the elastic deformation of solid bodies, continuous material distribution is assumed and also the density, dm/dV, is regarded as a continuous function of the position throughout the total volume of the body.

As a result of some force, the solid body is deformed and consequently its shape and volume change. (In the following the word deformation always refers to elastic deformation.) During the deformation every point of the body generally moves from its original position. Mathematically the deformation can be characterized by defining the displacement of every point of the body as a function of the position.

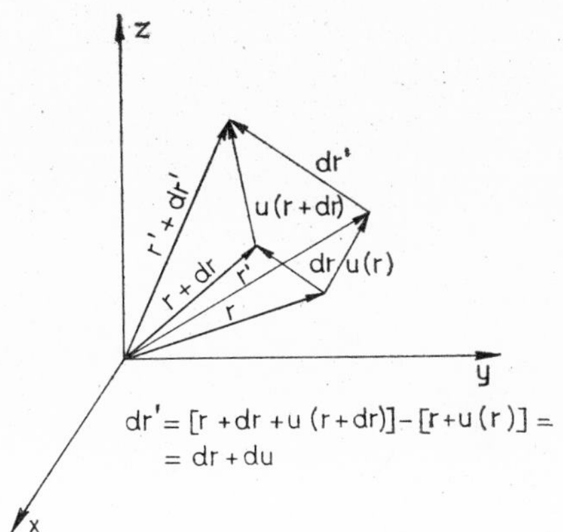

FIG. B.1. Displacements in a deformed body

A point of the undeformed body defined by the vector $\mathbf{r}\,(x, y, z)$ is transformed after deformation into the point $\mathbf{r}'\,(x, y, z)$. The displacement during the deformation is

$$\mathbf{u} = \mathbf{r}' - \mathbf{r} = \mathbf{u}\,(x, y, z) = \mathbf{u}\,(\mathbf{r}). \tag{B.1}$$

u is the vector of displacement. The deformation is fully determined if **u** is known as a function of the position. As a result of the deformation the distances between given points of the body are changed. Let the vector connecting two infinitesimally separated points be $d\mathbf{r}$ before the deformation. After the deformation these points are connected by the vector $d\mathbf{r} + d\mathbf{u}$ (Fig. B.1). Let us consider the change of the distance between these two points. If ds and ds' are the distances between the points before and after the deformation, respectively, then

$$(ds')^2 = (d\mathbf{r} + d\mathbf{u})^2 = ds^2 + 2d\mathbf{u}\, d\mathbf{r} + d\mathbf{u}^2. \tag{B.2}$$

If only infinitesimal deformations are considered, the third term of (B.2) can be neglected, since it is of the second order. According to (B.1)

$$d\mathbf{u} = \frac{d\mathbf{u}}{d\mathbf{r}}\, d\mathbf{r}$$

and hence,

$$ds'^2 = ds^2 + 2\left(\frac{d\mathbf{u}}{d\mathbf{r}}\, d\mathbf{r}\right) d\mathbf{r}\,. \tag{B.3}$$

The derivative of a vector–vector function is a tensor, i.e. a quantity which on multiplying by a vector is again a vector [1]. The derivative $d\mathbf{u}/d\mathbf{r}$ in the second term of (B.3) is also a tensor. Let us determine the matrix forming of this tensor. In the case of a rectangular coordinate system the differential of the function $\mathbf{u} = \mathbf{u}\,(x, y, z)$ is

$$d\mathbf{u} = \frac{\partial \mathbf{u}}{\partial x} dx + \frac{\partial \mathbf{u}}{\partial y} dy + \frac{\partial \mathbf{u}}{\partial z} dz\,.$$

Since

$$\mathbf{u}(x, y, z) = u_x(x, y, z)\,\mathbf{e}_x + u_y(x, y, z)\,\mathbf{e}_y + u_z(x, y, z)\,\mathbf{e}_z\,,$$

(where $\mathbf{e}_x$, $\mathbf{e}_y$, $\mathbf{e}_z$ are unit vectors in the directions of the x-, y-, z-axes), then

$$\frac{\partial \mathbf{u}}{\partial x} = \frac{\partial u_x}{\partial x}\mathbf{e}_x + \frac{\partial u_y}{\partial x}\mathbf{e}_y + \frac{\partial u_z}{\partial x}\mathbf{e}_z\,,$$

$$\frac{\partial \mathbf{u}}{\partial y} = \frac{\partial u_x}{\partial y}\mathbf{e}_x + \frac{\partial u_y}{\partial y}\mathbf{e}_y + \frac{\partial u_z}{\partial y}\mathbf{e}_z\,,$$

$$\frac{\partial \mathbf{u}}{\partial z} = \frac{\partial u_x}{\partial z}\mathbf{e}_x + \frac{\partial u_y}{\partial z}\mathbf{e}_y + \frac{\partial u_z}{\partial z}\mathbf{e}_z .$$

Hence

$$d\mathbf{u} = \left(\frac{\partial u_x}{\partial x}dx + \frac{\partial u_x}{\partial y}dy + \frac{\partial u_x}{\partial z}dz\right)\mathbf{e}_x + $$

$$+ \left(\frac{\partial u_y}{\partial x}dx + \frac{\partial u_y}{\partial y}dy + \frac{\partial u_y}{\partial z}dz\right)\mathbf{e}_y + $$

$$+ \left(\frac{\partial u_z}{\partial x}dx + \frac{\partial u_z}{\partial y}dy + \frac{\partial u_z}{\partial z}dz\right)\mathbf{e}_z .$$

Finally, $d\mathbf{r} = dx\mathbf{e}_x + dy\mathbf{e}_y + dz\mathbf{e}_z$ and thus

$$d\mathbf{u}\ d\mathbf{r} = \frac{\partial u_x}{\partial x}dx^2 + \frac{\partial u_x}{\partial y}dxdy + \frac{\partial u_x}{\partial z}dxdz + $$

$$+ \frac{\partial u_y}{\partial x}dxdy + \frac{\partial u_y}{\partial y}dy^2 + \frac{\partial u_y}{\partial z}dydz + $$

$$+ \frac{\partial u_z}{\partial x}dxdz + \frac{\partial u_z}{\partial y}dydz + \frac{\partial u_z}{\partial z}dz^2 .$$

On rearranging:

$$d\mathbf{u}\ d\mathbf{r} = \left[\frac{\partial u_x}{\partial x}dx + \frac{1}{2}\left(\frac{\partial u_x}{\partial y} + \frac{\partial u_y}{\partial x}\right)dy + \right.$$

$$\left. + \frac{1}{2}\left(\frac{\partial u_x}{\partial z} + \frac{\partial u_z}{\partial x}\right)dz\right]dx + \left[\frac{1}{2}\left(\frac{\partial u_x}{\partial y} + \frac{\partial u_y}{\partial x}\right)dx + \right.$$

$$\left. + \frac{\partial u_y}{\partial y}dy + \frac{1}{2}\left(\frac{\partial u_y}{\partial z} + \frac{\partial u_z}{\partial y}\right)dz\right]dy + $$

$$+ \left[\frac{1}{2}\left(\frac{\partial u_x}{\partial z} + \frac{\partial u_z}{\partial x}\right)dx + \frac{1}{2}\left(\frac{\partial u_y}{\partial z} + \frac{\partial u_z}{\partial y}\right)dy + \right.$$

$$\left. + \frac{\partial u_z}{\partial z}dz\right]dz .$$

Let us introduce the following notations:

$$\varepsilon_{xx} = \frac{\partial u_x}{\partial x},$$

$$\varepsilon_{xy} = \frac{1}{2}\left(\frac{\partial u_x}{\partial y} + \frac{\partial u_y}{\partial x}\right),$$

$$\varepsilon_{xz} = \frac{1}{2}\left(\frac{\partial u_x}{\partial z} + \frac{\partial u_z}{\partial x}\right), \qquad \text{etc.} \tag{B.4}$$

From these

$$d\mathbf{u}\ d\mathbf{r} = \left(\frac{d\mathbf{u}}{d\mathbf{r}}\, d\mathbf{r}\right) d\mathbf{r} = (\varepsilon_{xx}\, dx + \varepsilon_{xy}\, dy + \varepsilon_{xz}\, dz)\, dx +$$

$$+ (\varepsilon_{yx}\, dx + \varepsilon_{yy}\, dy + \varepsilon_{yz}\, dz)\, dy + (\varepsilon_{zx}\, dx + \varepsilon_{zy}\, dy + \varepsilon_{zz}\, dz)\, dz =$$

$$= \left[\begin{pmatrix} \varepsilon_{xx} & \varepsilon_{xy} & \varepsilon_{xz} \\ \varepsilon_{yx} & \varepsilon_{yy} & \varepsilon_{yz} \\ \varepsilon_{zx} & \varepsilon_{zy} & \varepsilon_{zz} \end{pmatrix} \begin{pmatrix} dx \\ dy \\ dz \end{pmatrix}\right] \begin{pmatrix} dx \\ dy \\ dz \end{pmatrix}. \tag{B.5}$$

It can be seen from this expression that the matrix of the tensor $d\mathbf{u}/d\mathbf{r}$ is

$$\boldsymbol{\epsilon} = \begin{pmatrix} \varepsilon_{xx} & e_{xy} & \varepsilon_{xz} \\ \varepsilon_{yx} & \varepsilon_{yy} & \varepsilon_{yz} \\ \varepsilon_{zx} & \varepsilon_{zy} & \varepsilon_{zz} \end{pmatrix} = [\varepsilon_{ik}], \quad (i, k = x, y, z). \tag{B.6}$$

$\boldsymbol{\epsilon}$ is the dilatation tensor. With its help the change of the distance between the two points can be determined. For this reason $\boldsymbol{\epsilon}$ is a quantity characteristic of the extent of deformation. It follows from the definition of the quantities ε_{ik} that the tensor $\boldsymbol{\epsilon}$ is symmetrical, and $\varepsilon_{ik} = \varepsilon_{ki}$. In order to find out the meaning of the components of the dilatation tensor, let us consider two points at a distance dx from each other on the x-axis before the deformation. The distance between these two points after the deformation according to (B.2) and (B.3) [now $d\mathbf{r} = (dx, 0, 0)$] is

$$(dx')^2 = dx^2\,(1 + 2\varepsilon_{xx}),$$

i.e.

$$\frac{dx'}{dx} = \sqrt{1 + 2\varepsilon_{xx}} \cong 1 + \varepsilon_{xx}, \tag{B.7}$$

from which

$$\varepsilon_{xx} \cong \frac{dx' - dx}{dx}.$$

Hence ε_{xx} gives the relative elongation of the line segment in the x-direction. The other elements of the main diagonal are of similar type. It can be further proved [2] that the mixed index components

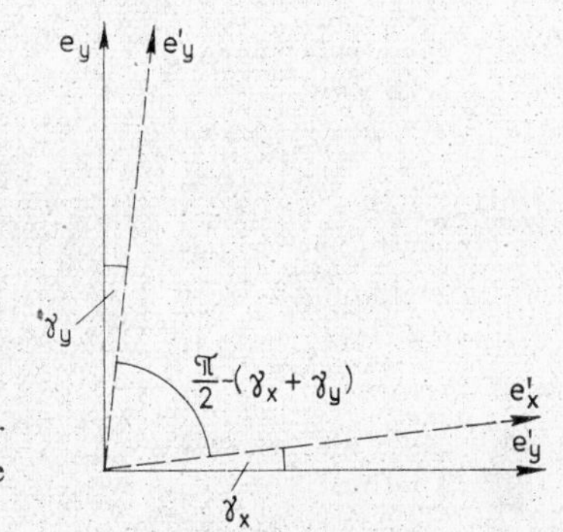

FIG. B.2. Relation between the shear deformations and the turning of the coordinate axes

of the dilatation tensor are equal to half the change of the angle between the corresponding coordinate axes. Thus, for example (Fig. B.2),

$$\varepsilon_{xy} = \frac{1}{2}(\gamma_x + \gamma_y). \tag{B.8}$$

On the strength of their meaning, the elements of the main diagonal are called strains (dilatations), while the mixed index components are shears.

A deformation is also accompanied by a volume change. According to (B.7) the value $\Delta v = dx\, dy\, dz$ becomes

$$\Delta V' = dx\,(1 + \varepsilon_{xx})\, dy\,(1 + \varepsilon_{yy})\, dz\,(1 + \varepsilon_{zz}) \cong$$

$$\cong \Delta V(1 + \varepsilon_{xx} + \varepsilon_{yy} + \varepsilon_{zz}).$$

(The products of the ε_{ii} values have been neglected.)

From this the relative volume change is

$$\frac{\Delta V' - \Delta V}{\Delta V} = \Theta = \varepsilon_{xx} + \varepsilon_{yy} + \varepsilon_{zz} =$$

$$= \frac{\partial u_x}{\partial x} + \frac{\partial u_y}{\partial y} + \frac{\partial u_z}{\partial z} = \text{div } \mathbf{u}. \tag{B.9}$$

B.2. The stress tensor

The arrangement of the molecules and atoms in an undeformed body correspond to a state of thermal equilibrium. Every part of the body is in mechanical equilibrium, i.e. the force exerted on any volume element is zero. When the body is deformed the equilibrium is disturbed. Internal stresses develop, which endeavour to restore the equilibrium state. No internal stress exists if there is no deformation.

Internal stresses originate from molecular interactions. The effective range of these forces, however, is extremely short, extending in practice to only the neighbouring molecules. Since the theory of elasticity refers to macroscopic bodies and the molecular distances are negligibly small compared with the macroscopic dimensions, the forces operating between the molecules can be regarded as having zero effective range. Consequently, if some volume is chosen within the body, that part of the body which surrounds this volume acts on it only along the surface.

Let us consider the resultant of all the forces exerted on the volume V selected. Let the force acting on the unit volume be $\mathbf{F}$. A force $\mathbf{F}dV$ is then exerted on the volume element dV, and the resultant force is $\int \mathbf{F}dV$, where the integral refers to the whole volume chosen.

The force $\mathbf{F}$ generally consists of two parts: $\mathbf{F} = \mathbf{f} + \mathbf{f}'$. The first part ($\mathbf{f}$) is called the body force (e.g. the weight). This is proportional to the mass dm or since $dm = \rho dV$ (ρ is the density of the body), to the volume dV. The other component of $\mathbf{F}$ is obtained from the surface forces discussed previously.

A surface element is selected inside the volume V. According to Newton's third law, the resultant of the forces at any point of this surface is zero. (On both sides of the surface forces of equal magnitude but opposite direction operate.) Hence the surface forces are non-zero only along the boundary surface of the selected volume element. Thus instead of the integral $\int \mathbf{f}' dV$ it is advisable to apply a new surface integral by introducing a suitable new quantity. Here again we must refer to a theorem of vector or tensor analysis. According to this, if the vector–vector function $\mathbf{f}'(\mathbf{r})$ can be produced as the divergence of tensor $\boldsymbol{\sigma}$ [1], i.e. if

$$\mathbf{f}' = \mathrm{Div}\, \boldsymbol{\sigma} \tag{B.10}$$

then

$$\int_V \mathbf{f}' dV = \int_V \mathrm{Div}\, \boldsymbol{\sigma}\, dV = \int_S \boldsymbol{\sigma}\, d\mathbf{S},$$

where $d\mathbf{S}$ points in the direction of the external normal, and the integral refers to the closed surface of the volume V. (The symbol Div has been introduced to distinguish it from the symbol div which refers to the divergence of vectors.) Hence the resultant force is

$$\int_V \mathbf{F}\, dV = \int_V \mathbf{f}\, dV + \int_V \mathrm{Div}\, \boldsymbol{\sigma}\, dV = \int_V \mathbf{f}\, dV + \int_S \boldsymbol{\sigma}\, d\mathbf{S}. \tag{B.11}$$

The quantity $\boldsymbol{\sigma}$ is termed the stress tensor; its matrix form is

$$[\sigma_{ik}] = \begin{pmatrix} \sigma_{xx} & \sigma_{xy} & \sigma_{xz} \\ \sigma_{yx} & \sigma_{yy} & \sigma_{yz} \\ \sigma_{zx} & \sigma_{zy} & \sigma_{zz} \end{pmatrix}.$$

Let us define the meaning of σ_{ik}. For this purpose let the body force be zero, and let us consider the force acting on a plane of the surface (y, z) whose normal is parallel to the x-axis. Now $d\mathbf{S} = dy\, dz\, [1, 0, 0]$, and thus

$$d\mathbf{F} = \boldsymbol{\sigma} d\mathbf{S} = dy\, dz\, (\sigma_{xx}\, \mathbf{e}_x + \sigma_{yx}\, \mathbf{e}_y + \sigma_{zx}\, \mathbf{e}_z).$$

It can be seen from this relation that the stress component σ_{xx} is the force acting on the unit surface normal to the x-axis in the x-direction, σ_{yx} is the force operating on unit surface normal to the x-axis in the y-direction, and finally σ_{zx} is the force acting on the same surface element in the z-direction. The other components have equivalent meanings (Fig. B.3). Hence the σ_{ik} values at a given point give the forces which act on the planes of unit surface lying across this point with normals pointing in the x-, y- and z-directions. In general $\boldsymbol{\sigma}\, d\mathbf{S}$ is equal to the force acting on a surface element $d\mathbf{S}$ whose normal is $d\mathbf{S}/|d\mathbf{S}|$.

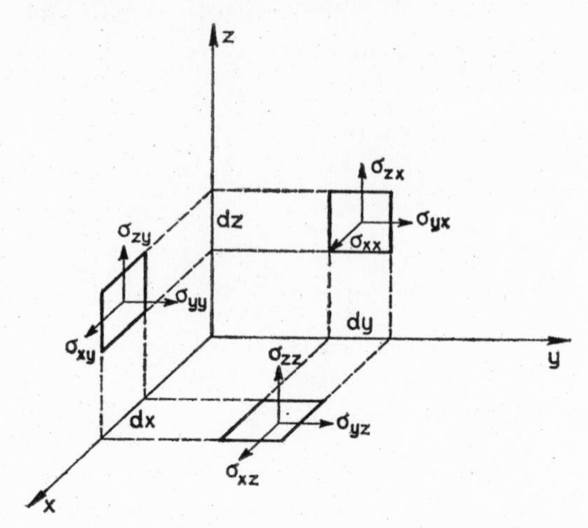

FIG. B.3. The meaning of the components of the stress tensor

B.3. Equilibrium conditions

In mechanical equilibrium the resultant of the forces acting on any volume of a body is zero, as is the torque resulting from all the forces. According to (B.11), the first condition is satisfied if

$$\int_V (\mathbf{f} + \mathrm{Div}\,\boldsymbol{\sigma})\, dV = 0.$$

Since the volume V is arbitrary, it follows that at equilibrium

$$\mathbf{f} + \mathrm{Div}\,\boldsymbol{\sigma} = 0. \tag{B.12}$$

By using the definition of the torque, the second condition takes the following form:

$$\int_V (\mathbf{r} \times \mathbf{f})\, dV + \int_S \mathbf{r} \times \boldsymbol{\sigma}\, d\mathbf{S} = 0. \tag{B.13}$$

The surface integral in relation (B.13) can be transformed in the following way. Denote the row vectors of the stress tensor matrix by $\vec{\sigma}_x, \vec{\sigma}_y, \vec{\sigma}_z$:

$$\vec{\sigma}_x = \sigma_{xx}\,\mathbf{e}_x + \sigma_{xy}\,\mathbf{e}_y + \sigma_{xz}\,\mathbf{e}_z, \quad \text{and so on.}$$

Then

$$\boldsymbol{\sigma} d\mathbf{S} = (\vec{\sigma}_x\, d\mathbf{S})\,\mathbf{e}_x + (\vec{\sigma}_y\, d\mathbf{S})\,\mathbf{e}_y + (\vec{\sigma}_z\, d\mathbf{S})\,\mathbf{e}_z.$$

With these values the x-component of the surface integral is:

$$\int_S (\mathbf{r}\times\boldsymbol{\sigma}\, d\mathbf{S})_x = \int_S (y\vec{\sigma}_z - z\vec{\sigma}_y)\, d\mathbf{S} = \int_V [\operatorname{div}(y\vec{\sigma}_z) - \operatorname{div}(z\vec{\sigma}_y)]\, dV.$$

By applying the identity $\operatorname{div}(u\mathbf{v}) = u \operatorname{div} \mathbf{v} + \mathbf{v} \operatorname{grad} u$, we obtain

$$\int_S (\mathbf{r}\times\boldsymbol{\sigma} d\mathbf{S})_x = \int_V (y \operatorname{div}\vec{\sigma}_z - z \operatorname{div}\vec{\sigma}_y) dV + \int_V (\vec{\sigma}_z \operatorname{grad} y - \vec{\sigma}_y \operatorname{grad} z)\, dV =$$

$$= \int_V (\mathbf{r}\times\operatorname{Div}\boldsymbol{\sigma})_x\, dV + \int_V (\sigma_{zy} - \sigma_{yz})\, dV.$$

Thus, the x-component of (B.13) is

$$\int_V \left\{[\mathbf{r}\times(\mathbf{f} + \operatorname{Div}\boldsymbol{\sigma})]_x + \sigma_{yz} - \sigma_{zy}\right\} dV = 0.$$

The integrand must vanish, and therefore, from a consideration of (B.12) $\sigma_{yz} = \sigma_{zy}$. The y and z components yield similar relations. Hence from the simultaneous realization of equilibrium conditions (B.12) and (B.13), it follows that the stress tensor becomes symmetrical, i.e.

$$\sigma_{ik} = \sigma_{ki}. \tag{B.14}$$

B.4. The relation between the stress and dilatation tensors for isotropic bodies

In most practical cases we have to determine the deformations due to given forces. In order to solve such problems, the relation between the stress and dilatation tensors must be known. In the following discussions we restrict ourselves to isotropic media.

Let us first consider the simple case when a uniform pressure is applied on every side of the body (hydrostatic compression). When the pressure p is directed towards the inside of the body, the force acting on any surface element $d\mathbf{S}$ of the body is

$$-pd\mathbf{S} = -p\mathbf{I}d\mathbf{S} = \boldsymbol{\sigma} d\mathbf{S},$$

where $\mathbf{I}$ is the unit tensor of matrix

$$\begin{pmatrix} 1 & 0 & 0 \\ 0 & 1 & 0 \\ 0 & 0 & 1 \end{pmatrix}.$$

From this equation

$$\boldsymbol{\sigma} = -p\mathbf{I}. \tag{B.15}$$

In the case of infinitesimal deformations, however, the relative volume change Θ of any body is approximately proportional to the change of pressure, i.e.

$$\Theta = -\kappa p, \tag{B.16}$$

where κ is the isotropic compressibility, i. e., the relative volume change per unit pressure change. Hence, in the case of hydrostatic compression,

$$\boldsymbol{\sigma} = -\kappa^{-1}\Theta\mathbf{I}, \tag{B.17}$$

i.e. the stresses are proportional to the relative volume change. It is easy to see that in the present case not only the shear stresses but also the shear strains vanish. As has already been mentioned, ε_{ik} $(i \neq k)$ represents the change of the right angle between the vectors pointing originally to the corresponding coordinate axes. It is quite obvious, however, that the shape of an isotropic body does not change as a result of static compression, but only its volume. Hence, any two vectors which were perpendicular to each other remain perpendicular after the deformation.

In the general case, the stress tensor can be expressed by Hooke's law. According to this, the deformation of an elastic body is proportional to the stress. This law holds in many practical cases. It follows from the relation (B.17) and from Hooke's law, therefore, that the stress tensor generally consists of two parts. One part is proportional to the shear strains, and the other to the volume change. Since the mixed index ε values refer only to the shears, it follows that

$$\sigma_{ik} = 2\mu\,\varepsilon_{ik}, \qquad (i \neq k). \tag{B.18}$$

With diagonal elements, on the other hand, in the general case, in addition to the term $2\mu\varepsilon_{ii}$ originating from Hooke's law, another term is added which is proportional to the volume change [2], i.e.

$$\sigma_{ii} = 2\mu\,\varepsilon_{ii} + \lambda\,\Theta. \tag{B.19}$$

The proportionality factors λ and μ are called Lamé constants. It can be proved that these constants can only be positive.

B.5. Relations between the elastic constants

Let us consider the rod of length l and cross-section q fixed at one end. Let a force Q parallel to the longitudinal axis of the rod act on the other end. There are clearly no forces in the normal direction on the lateral faces of the rod, and consequently here $\boldsymbol{\sigma} d\mathbf{S} = 0$ at every point. If the rod is parallel to the z-axis, $d\mathbf{S}$ is a vector perpendicular to the z-axis at the lateral faces, and thus it follows that every component of the stress tensor is zero except σ_{zz} which is equal to Θ/q. From this it follows according to (B.18) and (B.19) that only the diagonal elements of the dilatation tensor are not zero. Using equation (B.19), their value is

$$\varepsilon_{xx} = \varepsilon_{yy} = -\frac{\lambda}{2\mu\,(3\lambda + z\mu)}\,\frac{Q}{q}, \qquad \varepsilon_{zz} = \frac{\lambda + \mu}{\mu\,(3\lambda + 2\mu)}\,\frac{Q}{q}. \tag{B.20}$$

Considering the meaning of ε_{zz}, the second equation of (B.20) can be written in the following form:

$$\frac{\Delta l}{l} = \frac{1}{E}\,\frac{Q}{q}.$$

This is the well-known form of Hooke's law for extension. E is the elastic stiffness constant or Young's modulus, whose value expressed in terms of λ and μ is

$$E = \frac{\mu(3\lambda + 2\mu)}{\lambda + \mu}. \tag{B.21}$$

The positive ratio of the relative transversal contraction and expansion is called Poisson's number and is denoted by ν. Its value according to (B.20) is

$$\nu = \left| \frac{\varepsilon_{xx}}{\varepsilon_{zz}} \right| = \frac{\lambda}{2(\lambda + \mu)}. \tag{B.22}$$

The deformation due to a force parallel to opposite planes of a homogeneous rectangular prism is said to be a shear, and is usually described (Fig. B.4) by the equation

$$\gamma = \frac{1}{G}\frac{Q}{q}, \tag{B.23}$$

where G denotes the shear or torsional modulus. This type of deformation clearly means that, for example, $\varepsilon_{xy} = \text{constant} \neq 0$, and all the other components of ϵ are zero. It then follows from (B.18) that only σ_{xy} is not zero, and its value is $\sigma_{xy} = 2\mu\varepsilon_{xy}$. However, according to (B.8), $2\varepsilon_{xy}$ is the shear angle, and therefore

$$\gamma = \frac{1}{\mu}\sigma_{xy} = \frac{1}{\mu}\frac{Q}{q}. \tag{B.24}$$

From a comparison of equations (B.23) and (B.24) it can be seen that μ is identical with the torsional modulus. Since λ has no direct physical meaning, ν and μ are usually used instead of λ and μ.

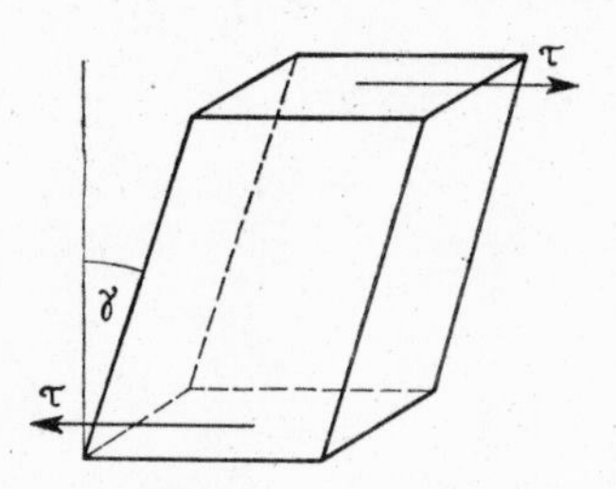

FIG. B.4. The derivation of the shear angle

B.6. Equations of motion and equilibrium conditions of isotropic bodies

A volume element dV inside the body is selected: According to (B.11), the force exerted is $(\mathbf{f} + \operatorname{Div} \boldsymbol{\sigma})\, dV$.

If the density of the body is ρ, then according to Newton's second law

$$\rho dV \frac{\partial^2 \mathbf{u}}{\partial t^2} = (\mathbf{f} + \operatorname{Div} \boldsymbol{\sigma})\, dV,$$

i.e.

$$\rho \frac{\partial^2 \mathbf{u}}{\partial t^2} = \mathbf{f} + \operatorname{Div} \boldsymbol{\sigma}.$$

Using equations (B.4), (B.18), (B.19) and the relation $\Theta = \operatorname{div} \mathbf{u}$, $\operatorname{Div} \boldsymbol{\sigma}$ can be expressed as a function of $\mathbf{u}$:

$$(\operatorname{Div} \boldsymbol{\sigma})_x = \frac{\partial \sigma_{xx}}{\partial x} + \frac{\partial \sigma_{xy}}{\partial y} + \frac{\partial \sigma_{xz}}{\partial z} =$$

$$= 2\mu \frac{\partial^2 u_x}{\partial x^2} + \lambda \frac{\partial}{\partial x} \operatorname{div} \mathbf{u} +$$

$$+ \mu \left(\frac{\partial^2 u_x}{\partial y^2} + \frac{\partial^2 u_y}{\partial x \partial y} + \frac{\partial^2 u_x}{\partial z^2} + \frac{\partial^2 u_z}{\partial x \partial z} \right).$$

Introducing the Laplace operator

$$\Delta = \frac{\partial^2}{\partial x^2} + \frac{\partial^2}{\partial y^2} + \frac{\partial^2}{\partial z^2},$$

the above equation can be written in the following form:

$$(\operatorname{Div} \boldsymbol{\sigma})_x = \mu \Delta u_x + (\lambda + \mu) \frac{\partial}{\partial x} \operatorname{div} \mathbf{u}.$$

The other components of $\operatorname{Div} \boldsymbol{\sigma}$ are similar, and consequently the vector form of the equation of motion of an isotropic body is

$$\rho \frac{\partial^2 \mathbf{u}}{\partial t^2} = \mathbf{f} + \mu \Delta \mathbf{u} + (\lambda + \mu) \operatorname{grad} \operatorname{div} \mathbf{u}. \qquad \text{(B.25)}$$

At equilibrium $\partial^2\mathbf{u}/\partial t^2 = 0$, and thus the condition of equilibrium is

$$\mathbf{f} + \mu\Delta\mathbf{u} + (\lambda + \mu)\ \text{grad div}\ \mathbf{u} = 0. \tag{B.26}$$

B.7. The equilibrium of an elastic body in the case of two-dimensional deformations

The investigation of equilibrium is considerably simpler in the case of deformations, where, for example, the component u_z of the vector of displacement is zero, and the components u_x, u_y depend only upon x and y. In this case it is sufficient to give the stresses in the plane normal to the z-axis as functions of the positions. Hence for two-dimensional deformations,

$$\frac{\partial}{\partial z} = 0, \qquad u_z = 0, \tag{B.27}$$

and so $\sigma_{xz} = \sigma_{yz} = 0$. Disregarding the usually negligible body forces, by using (B.27) one obtains from the equilibrium condition (B.12):

$$\frac{\partial \sigma_{xx}}{\partial x} + \frac{\partial \sigma_{xy}}{\partial y} = 0, \qquad \frac{\partial \sigma_{xy}}{\partial x} + \frac{\partial \sigma_{yy}}{\partial y} = 0. \tag{B.28}$$

The relation between the stresses and the deformation is

$$\sigma_{xx} = 2\mu \frac{\partial u_x}{\partial x} + \lambda \left(\frac{\partial u_x}{\partial x} + \frac{\partial u_y}{\partial y} \right), \tag{B.29}$$

$$\sigma_{yy} = 2\mu \frac{\partial u_y}{\partial y} + \lambda \left(\frac{\partial u_x}{\partial x} + \frac{\partial u_y}{\partial y} \right),$$

$$\sigma_{zz} = \lambda \left(\frac{\partial u_x}{\partial x} + \frac{\partial u_y}{\partial y} \right),$$

$$\sigma_{xy} = \mu \left(\frac{\partial u_x}{\partial y} + \frac{\partial u_y}{\partial x} \right).$$

From the above expressions it can be seen that

$$\sigma_{zz} = \frac{\lambda}{2(\lambda + \mu)}(\sigma_{xx} + \sigma_{yy}) = \nu(\sigma_{xx} + \sigma_{yy}). \tag{B.30}$$

Equation (B.28) is solved by Airy's method [3]. Let us assume that there exists a function χ which satisfies the equation

$$\sigma_{xy} = -\frac{\partial^2 \chi}{\partial x \partial y}.$$

Equations (B.28) then take the form

$$\frac{\partial}{\partial x}\left(\sigma_{xx} - \frac{\partial^2 \chi}{\partial y^2}\right) = 0, \qquad \frac{\partial}{\partial y}\left(\sigma_{yy} - \frac{\partial^2 \chi}{\partial x^2}\right) = 0.$$

From these it follows that the condition for equilibrium is determined by the expressions

$$\sigma_{xx} = \frac{\partial^2 \chi}{\partial y^2}, \qquad \sigma_{yy} = \frac{\partial^2 \chi}{\partial x^2}, \qquad \sigma_{xy} = -\frac{\partial^2 \chi}{\partial x \partial y}, \tag{B.31}$$

where χ is Airy's stress function. Since all stresses originate from the displacement vector **u**, they are clearly not independent of each other. For this reason χ has not only to meet the equilibrium conditions, but must also satisfy certain stress relations. By using the expressions (B.29) and (B.30) the following identity is easily derived:

$$\frac{\partial^2}{\partial y^2}\left[\sigma_{xx} - \nu(\sigma_{xx} + \sigma_{yy})\right] + \frac{\partial^2}{\partial x^2}\left[\sigma_{yy} - \nu(\sigma_{xx} + \sigma_{yy})\right] = $$

$$= 2\frac{\partial^2 \sigma_{xy}}{\partial x \partial y}. \tag{B.32}$$

Substituting the values of (B.31) into (B.32), we find that the function χ satisfies the equation:

$$\Delta^2 \chi = 0,$$

where $\Delta = \partial^2/\partial x^2 + \partial^2/\partial y^2$ is the two-dimensional Laplace operator.

Appendix C

The stress-fields of straight edge and screw dislocations

C.1. Edge dislocations

The calculations are carried out on the basis of the Volterra model. Let us consider the edge dislocations depicted in Fig. C.1, and assume that the dislocation line, i.e. the length of the cylinder is very large compared with b. Since there is no displacement in the z-direction,

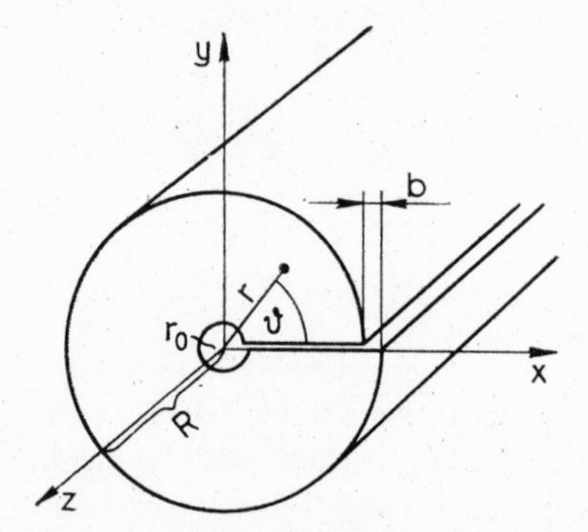

FIG. C.1. The continuum model of the edge dislocation

because of the cylindrical symmetry, we are obviously concerned with the two-dimensional deformation discussed above. The Airy function must be determined from (B.33) for the solution of such problems. The stresses are obtained from relation (B.31). In order to simplify the calculations it is advisable to use cylindrical coordinates. Thus,

$$x = r \cos \vartheta,$$

$$y = r \sin \vartheta,$$

$$z = z,$$

and further,

$$\Delta = \frac{\partial^2}{\partial x^2} + \frac{\partial^2}{\partial y^2} = \frac{\partial^2}{\partial r^2} + \frac{1}{r}\frac{\partial}{\partial r} + \frac{1}{r^2}\frac{\partial^2}{\partial \vartheta^2}.$$

The non-zero stresses in cylindrical coordinates [1] are:

$$\sigma_{rr} = \frac{1}{r}\frac{\partial\chi}{\partial r} + \frac{1}{r^2}\frac{\partial^2\chi}{\partial\vartheta^2}, \qquad \sigma_{\vartheta\vartheta} = \frac{\partial^2\chi}{\partial r^2},$$

$$\sigma_{r\vartheta} = -\frac{\partial}{\partial r}\left(\frac{1}{r}\frac{\partial\chi}{\partial\vartheta}\right), \qquad \sigma_{zz} = \nu\,(\sigma_{rr} + \sigma_{\vartheta\vartheta}). \qquad \text{(C.1)}$$

The equation to be solved is now

$$\left(\frac{\partial^2}{\partial r^2} + \frac{1}{r}\frac{\partial}{\partial r} + \frac{1}{r^2}\frac{\partial^2}{\partial\vartheta^2}\right)^2 \chi = 0. \qquad \text{(C.2)}$$

Solutions must be determined such that the radial stresses (σ_{rr}, $\sigma_{\vartheta r}$) be zero at the boundary surfaces, failing which the dislocation is not stable. By applying the method of separating the variables, one obtains the following solution:

$$\chi = \left(Ar\ln r + Br^3 + \frac{C}{r}\right)\sin\vartheta, \qquad \text{(C.3)}$$

where A, B and C are integration constants. With this equation one obtains the following stresses from (C.1):

$$\sigma_{rr} = \left(A + 2Br^2 - \frac{2C}{r^2}\right)\frac{\sin\vartheta}{r},$$

$$\sigma_{r\vartheta} = -\left(A + 2Br^2 - \frac{2C}{r^2}\right)\frac{\cos\vartheta}{r}, \qquad \text{(C.4)}$$

$$\sigma_{\vartheta\vartheta} = \left(A + 6Br^2 - \frac{2C}{r^2}\right)\frac{\sin\vartheta}{r}.$$

The constants B and C can be determined from the following boundary conditions:

$$\sigma_{rr} = \sigma_{r\vartheta} = 0, \qquad \text{if } r = \begin{cases} r_0 \\ R \end{cases}.$$

After substituting we have

$$B = -\frac{1}{2}\frac{A}{r_0^2 + R^2} \quad \text{and} \quad C = \frac{1}{3}\frac{r_0^2 R^2}{r_0^2 + R^2} A,$$

or since $r_0 \ll R$,

$$B \approx -\frac{A}{2R^2}, \qquad C \approx \frac{A r_0^2}{3}. \tag{C.5}$$

The terms of the stresses relating to these coefficients can be neglected compared with the third term if we are not too close to the surface. Thus the stress-field of an edge dislocation is described in practice by the following expression:

$$\sigma_{rr} = \sigma_{\vartheta\vartheta} = A\frac{\sin\vartheta}{r}, \qquad \sigma_{r\vartheta} = -A\frac{\cos\vartheta}{r}, \tag{C.6}$$

or in rectangular coordinates:

$$\sigma_{xx} = A\frac{y(3x^2 + y^2)}{(x^2 + y^2)^2}, \qquad \sigma_{yy} = -A\frac{y(x^2 - y^2)}{(x^2 + y^2)^2},$$

$$\sigma_{xy} = -A\frac{x(x^2 - y^2)}{(x^2 + y^2)^2}. \tag{C.7}$$

If the stresses are known, the displacements u_x and u_y can be calculated by using equations (B.18) and (B.19) the result is

$$u_x = -A\frac{1 - \nu}{\mu}\arctan\frac{y}{x} - \frac{A}{2\mu}\frac{xy}{x^2 + y^2}, \tag{C.8}$$

$$u_y = A\frac{1 - 2\nu}{2\mu}\ln(x^2 + y^2)^{1/2} + \frac{A}{2\mu}\frac{x^2}{x^2 + y^2}. \tag{C.9}$$

The solution obtained for u_x has a discontinuity along the plane $y = 0$. This expresses the fact that during the generation of the dislocation the cut surfaces formed along the plane $y = 0$ were displaced relative to each other by a finite distance b. Hence the change of u_x while traversing the surface is:

$$\Delta u_x = \lim_{y \to -0} u_x - \lim_{y \to +0} u_x = -\frac{2\pi A(1-\nu)}{\mu} = b,$$

from which the value of the integration constant A is

$$A = -\frac{\mu b}{2\pi(1-\nu)}. \qquad \text{(C.10)}$$

C.2. The stress-field of screw dislocations

The Volterra model is again used in the calculations (Fig. C.2). Accordingly, a screw dislocation is created if the cut surfaces are displaced by a finite distance relative to each other along the axis of the cylinder (the z-axis). The displacement u_z thus obtained does

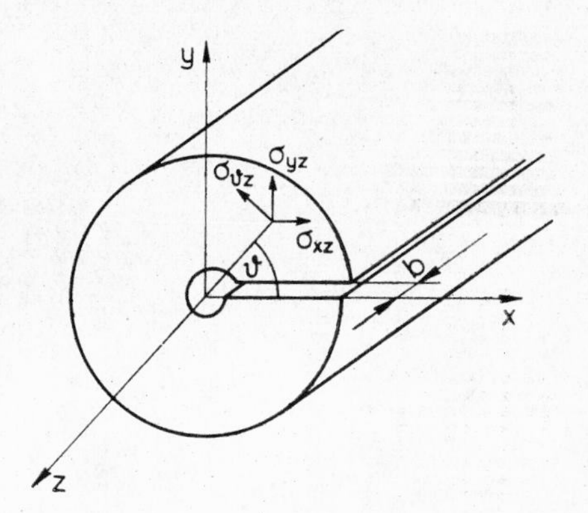

FIG. C.2. The continuum model of the screw dislocation

not depend upon z. The determination of the stress is now considerably simpler, though the relations for the two-dimensional deformations cannot be used. Since only the u_z component of the displacement vector is not zero (and further, this component does not depend upon z), it follows at once that

$$\operatorname{div} \mathbf{u} = 0. \qquad \text{(C.11)}$$

Thus a screw dislocation does not produce a volume dilatation in the material. Using (C.11) the equilibrium condition (B.26) (neglecting the body force) has the following form:

$$\Delta u_z = 0, \tag{C.12}$$

where

$$\Delta = \frac{\partial^2}{\partial x^2} + \frac{\partial^2}{\partial y^2}.$$

If u_z is known, the non-zero stresses are

$$\sigma_{xz} = \mu \frac{\partial u_z}{\partial x}, \qquad \sigma_{yz} = \mu \frac{\partial u_z}{\partial y}. \tag{C.13}$$

Such a solution of equation (C.12) must be determined which changes with b when traversing the cut surfaces, and the relative displacement of the surfaces is again expressed. It is easily seen that the appropriate function is

$$u_z = \frac{b}{2\pi} \operatorname{arc\,tan} \frac{y}{x}, \tag{C.14}$$

which is a right-handed screw in a right-handed system. From (C.13) the stresses are

$$\sigma_{xz} = -\frac{\mu b}{2\pi} \frac{y}{x^2 + y^2}, \qquad \sigma_{yz} = \frac{\mu b}{2\pi} \frac{x}{x^2 + y^2}, \tag{C.15}$$

or in polar coordinates:

$$\sigma_{\vartheta z} = \frac{\mu b}{2\pi r}. \tag{C.16}$$

Consequently, screw dislocations give rise exclusively to shear stresses of cylindrical symmetry. Since σ_{rr}, $\sigma_{r\vartheta}$ and σ_{rz} are everywhere zero, and thus on the boundary surfaces of the cylinder also, it follows that the solution (C.14) also satisfies the boundary conditions.

However, screw dislocations created by displacement exclusively in the z-direction are not stable, because the stresses generated exert a non-zero torque on the cylinder. The value of this torque according to Fig. C.2 is

$$M = \int_{0}^{2\pi}\int_{r_0}^{R} (r\sigma_{\vartheta z})\, r dr d\vartheta = \frac{\mu b^2}{2}(R^2 - r_0^2)\,. \tag{C.17}$$

If no external constraint acts, displacement u_x, u_y is produced so that the torque of the resulting stresses, together with (C.17), amounts to zero. (C.11) must further hold, and therefore from (C.26),

$$\frac{\partial^2 u_x}{\partial y^2} + \frac{\partial^2 u_x}{\partial z^2} = 0\,, \qquad \frac{\partial^2 u_y}{\partial x^2} + \frac{\partial^2 u_y}{\partial z^2} = 0\,. \tag{C.18}$$

The appropriate solutions are

$$u_x = Ayz, \qquad u_y = Bxz. \tag{C.19}$$

Because of the cylindrical symmetry, the stresses cannot depend on z, and therefore,

$$\sigma'_{xy} = \mu\left(\frac{\partial u_x}{\partial y} + \frac{\partial u_y}{\partial x}\right) = \mu(A + B)z = 0,$$

or B must equal $-A$. The stresses resulting from the displacements (C.19) are

$$\sigma'_{xz} = A\mu y, \qquad \sigma'_{yz} = -A\mu x. \tag{C.20}$$

The value of the constant A is so selected that the resultant torque is zero. This is realized if $A = b/\pi(R^2 + r_0^2)$. Thus the total stress is

$$\sigma_{\vartheta z} = \frac{\mu b}{2\pi}\left(\frac{1}{r} - \frac{2r}{R^2 + r_0^2}\right).$$

If r is small compared to R, the effect of the second term is negligible. Consequently, the stress-field of a screw dislocation not too close to the surface of the body is described in practice by the relation (C.16).

References

Chapter 1

1. Clarebrough, L. M., Hargreaves, M. E. and West, G. W.: *Phil. Mag.* **44** (1953) 913.
2. Mackenzie, J. K.: Thesis, Bristol, 1949.
3. Glen, W.: *Phil. Mag.* **1** (1956) 400.
4. Frenkel, J.: *Z. Phys.* **335** (1926) 652
   Wagner, C. and Schottky, W.: *Z. Phys. Chem.* B **11** (1930) 163.
5. Huntington, H. B. and Seitz, F.: *Phys. Rev.* **61** (1942) 315.
   Huntington, H. B.: *Phys. Rev.* **91** (1953) 1092.
   Bauerle, J. E. and Koehler, J. S.: *Phys. Rev.* **107** (1957) 1493.
6. Taylor, G. I.: *Proc. Roy. Soc.* A **145** (1934) 362.
   Orowan, E.: *Z. Phys.* **84** (1934) 634.
   Polányi, M.: *Z. Phys.* **84** (1934) 660.
7. Andrade, E. N. da C.: *Endeavour* LX (1950) 165.
8. Bragg, W. L. and Nye, J. F.: *Proc. Roy. Soc.* A **190** (1947) 474.
   Bragg, W. L. and Lomer, W. M.: *Proc. Roy. Soc.* A **196** (1949) 171.
9. Burgers, J. M.: *Proc. Kon. Ned. Akad. Wet.* **42** (1939) 293, 378.
10. Love, A. E. H.: *A Treatise on the Mathematical Theory of Plasticity*, Cambridge University Press, Cambridge, 1934, Chap. 9.
11. Dash, W. C.: *J. Appl. Phys.* **30** (1959) 59.
12. Vogel, F. L., Pfann, W. C., Corey, H. E. and Thomas, E. E.: *Phys. Rev.* **90** (1953) 484.
13. Nabarro, F.R.N.: *Theory of Crystal Dislocations*, Clarendon Press, Oxford, 1967, p. 44.

Chapter 2

1. Wit, R. de: *The Continuum Theory of Stationary Dislocations*, *Solid State Physics*, vol. 10, Academic Press, New York, 1960.
2. Landau, L. D. and Lifshitz, E. M.: *Theory of Elasticity*, Pergamon Press, London, 1959.
3. Koehler, J. S.: *Phys. Rev.* **60** (1941) 397.
4. Peach, M. and Koehler, J. S.: *Phys. Rev.* **80** (1950) 456.
5. Nabarro, F. R. N.: *Phil. Mag.* **42** (1951) 213.
6. Eshelby, J. D.: *The Continuum Theory of Lattice Defects*, *Solid State Physics*, vol. 3, Academic Press, New York, 1956.

References

7. SEEGER, A.: *Theorie der Gitterfehlstellen, Hdb. der Physik*, VII/1, Springer Verlag, Berlin, 1955.
8. DUNDURS, J.: *J. Appl. Phys.* **39** (1968) 4152.
9. VOLTERRA, V.: *Ann. Ec. Norm. Sup.* **24** (1907) 400.
10. BURGERS, J. M.: *Proc. Kon. Nad. Akad. Wet.* **42** (1939) 378.
11. MANN, E. H.: *Proc. Roy. Soc.* A **199** (1949) 376.
12. ESHELBY, J. D.: *Phil. Mag.* **40** (1949) 903.
13. ESHELBY, J. D. and STROH, A. N.: *Phil. Mag.* **42** (1951) 1401.
14. MOTT, N. F. and NABARRO, F. R. N.: *Report on Strength of Solids*, London, Physical Society, 1940.
15. KROUPA, F.: *Theory of Crystal Defects*, Summer School, Hrazany, 1964. Academia, Prague, 1966.
16. CHEN, H. S., GILMAN, J. J. and HEAD, A. K.: *J. Appl. Phys.* **35** (1964) 2502.
17. HARTLEY, C. S. and HIRTH, J. P.: *Acta Met.* **13** (1965) 79.
18. LI, J. C. M.: *Acta Met.* **13** (1965) 79.
19. LI, J. C. M.: *Acta Met.* **8** (1960) 296.
20. NABARRO, F. R. N.: *Adv. Phys.* **1** (1952) 271.
21. KRÖNER, E.: *Kontinuumstheorie der Versetzungen und Eigenspannungen*, Springer Verlag, Berlin, 1958.
22. CHOU, Y. T. and ESHELBY, J. D.: *J. Mech. Phys. Solids*, **10** (1962) 27.
23. BURGERS, J. M.: *Proc. Kon. Nad. Akad. Wet.* **42** (1939) 293.
24. CHATEL, P. DE and KOVÁCS, I.: *Phys. Stat. Sol.* **10** (1965) 213.
25. HOLDER, J. and GRANATO, A. V.: *Phys. Rev.* **182** (1969) 729.
26. COTTRELL, A. H.: *Dislocations and Plastic Flow in Crystals*, Clarendon Press, Oxford, 1953.
27. ZENER, C.: *Trans. Amer. Inst. Min. Met. Eng.* **147** (1942) 361.
28. CLAREBROUGH, L. M., HARGREAVES, M. E. and WEST, G. W.: *Phil. Mag.* **1** (1956) 528.
29. WIT, G. DE and KOEHLER, J. S.: *Phys. Rev.* **116** (1959) 1113.
30. FOREMAN, A. J. E.: *Phil. Mag.* **15** (1967) 1011.
31. KOVÁCS, I.: *Phys. Stat. Sol.* **3** (1963) 140.
32. ESHELBY, J. D.: *Phil. Trans. Roy. Soc.* **244** (1951) 87.
33. NABARRO, F. R. N.: *Theory of Crystal Dislocations*, Clarendon Press, Oxford, 1967.
34. SIEMS, R.: *Phys. Kondens. Materie* **2** (1964) 1.
35. LOTHE, J.: *Physica Norvegica* **2** (1967) 153.
36. FRENKEL, J. and KONTOROVA, T.: *J. Phys. USSR* **1** (1939) 137.
37. FRANK, F. C.: *Proc. Phys. Soc.* A **62** (1949) 131.
38. ESHELBY, J. D.: *Proc. Phys. Soc.* A **62** (1949) 307.
39. BULLOUGH, R. and BILBY, B. S.: *Proc. Phys. Soc.* B **67** (1954) 615.
40. TEUTONICO, L. J.: *Phys. Rev.* **124** (1961) 1039.
41. HIRTH, J. P. and LOTHE, J.: *Theory of Dislocations*, McGraw-Hill, New York, 1968.
42. BELTZ, R. J., DAVIS, T. L. and MALÉN, K.: *Phys. Stat. Sol.* **26** (1968) 621.
43. WEERTMAN, J.: *Phys. Rev.* **119** (1960) 1871.
44. WEERTMAN, J.: *Phil. Mag.* **7** (1962) 617.

45. ESHELBY, J. D.: *Proc. Roy. Soc.* A **197** (1949) 396.
46. NABARRO, F. R. N.: *Proc. Roy. Soc.* A **209** (1951) 279.
47. LEIBFRIED, G.: *Z. Phys.* **127** (1950) 344.
48. JEFFREIS, H.: *Cartesian Tensors*, Cambridge University Press, 1952.
49. KRÖNER, E.: *Arch. Rat. Mech. Anal.* **4** (1960) 273.
50. DEHLINGER, U. and KRÖNER, E.: *Z. Metallk.* **51** (1960) 457.
51. SEEGER, A.: *Phys. Stat. Sol.* **1** (1961) 669.
52. PFLEIDERER, H., SEEGER, A. and KRÖNER, E.: *Z. Naturf.* **15**a (1960) 758.

Chapter 3

1. HUNTINGTON, H. B. and SEITZ, F.: *Phys. Rev.* **61** (1942) 315.
   HUNTINGTON, H. B.: *Phys. Rev.* **61** (1942) 325.
2. BROOKS, H.: *Impurities and Imperfections*, Amer. Soc. Metals, Cleveland, 1955.
3. AMAR, H.: *J. Appl. Phys.* **33** (1962) 666.
4. BAUERLE, J. E. and KOEHLER, J. S.: *Phys. Rev.* **107** (1957) 1493.
5. (LAZAREV, B. G. and OVCHARENKO, O. N.) ЛАЗАРЕВ Б.Г. и ОВЧАРЕНКО, О. Н.: *ДАН СССР* **100** (1955) 875.
6. DOYAMA, M. and KOEHLER, J. S.: *Phys. Rev.* **119** (1960) 939.
7. FLYNN, C. P.: *Phys. Rev.* **125** (1962) 881.
8. SIMMONS, R. O.: *J. Phys. Soc. Japan* **18** (Suppl. II) (1963) 172.
9. MUKHERJEE, K.: *Phil. Mag.* **12** (1965) 915.
10. LOMER, W. M.: *Defects in Pure Metals, Progress in Metal Physics*, 8, (1959).
11. SEEGER, A.: *Phys. Stat. Sol.* **1** (1961) 669.
12. ESHELBY, J. D.: *The Continuum Theory of Lattice Defects, Solid State Physics*, vol. 3, Academic Press, New York, 1956.
13. DIRAC, P. A. M.: *The Principles of Quantum Mechanics*, Clarendon Press, Oxford, 1947.
14. FLINN, P. A. and MARADUDIN, A. A.: *Annals of Physics* **18** (1962) 81.
15. SIMMONS, R. O. and BALLUFFI, R. W.: *Phys. Rev.* **125** (1962) 862.
16. COTTRELL, A. H.: *Report of a Conference on Strength of Solids*, The Physical Society, London, 1948, p. 30.
17. NABARRO, F. R. N.: see ref. 16, p. 38.
18. COTTRELL, A. H. and BILBY, B. A.: *Proc. Phys. Soc.* A **62** (1949) 49.
19. BILBY, B. A.: *Proc. Phys. Soc.* A **63** (1950) 191.
20. FIORE, N. F. and BAUER, C. L.: *Progress in Materials Science*, **13** (1968) 85.
21. COTTRELL, A. H., HUNTER, S. C. and NABARRO, F. R. N.: *Phil. Mag.* **44** (1953) 1064.
22. WIT, R. DE and HOWARD, R. E.: *Acta Met.* **13** (1965) 655.
23. FISCHER, J. C.: *Acta Met.* **2** (1954) 9.
24. ESHELBY, J. D.: *Ann. der Physik* **7** (1958) 116.

25. ESHELBY, J. D.: *Proc. Roy. Soc.* A **241** (1957) 376.
26. ESHELBY, J. D.: *Progress in Solid Mechanics*, Chap. 3. North Holland Publishing Co., Amsterdam, 1961.
27. KING, H. W.: *J. Mat. Sci.* **1** (1966) 79.
28. BULLOUGH, R. and NEWMANN, R. C.: *Phil. Mag.* **7** (1962) 529.
29. SAXL, I.: *Czech. J. Phys.* B **14** (1964) 381.
30. STEHLE, H. and SEEGER, A.: *Z. Phys.* **146** (1956) 217.

Chapter 4

1. FRENKEL, J. and KONTOROVA, T.: *Phys. Z. Sowjet.* **13** (1938) 1; *J. Phys. USSR* **1** (1939) 137.
2. HOBART, R. and CELLI, V.: *J. Appl. Phys.* **33** (1962) 60.
   HOBART, R.: *J. Appl. Phys.* **36** (1965) 1944, 1948.
   HOBART, R.: *J. Appl. Phys.* **37** (1966) 3573.
3. PEIERLS, R.: *Proc. Phys. Soc. London* **52** (1940) 34.
4. HUNTINGTON, H. B.: *Proc. Phys. Soc. London* B **68** (1955) 1043.
5. NABARRO, F. R. N.: *Theory of Crystal Dislocations*, Clarendon Press, Oxford, 1967.
6. SEEGER, A.: *Theorie der Gitterfehlstellen, Hdb. der Phys.* VII/1, Springer Verlag, Berlin, 1955.
7. NABARRO, F. R. N.: *Proc. Phys. Soc.* **59** (1947) 256.
8. COTTRELL, A. H.: *Dislocations and Plastic Flow in Crystals*, Clarendon Press, Oxford, 1953, p. 98.
9. FOREMAN, A. J., JASWON, M. A. and WOOD, J. K.: *Proc. Phys. Soc. London* A **64** (1951) 156.
10. VITEK, V., PERRIN, R. C. and BOWN, D. K.: *Phil. Mag.* **21** (1970) 1049.
11. HEINRICH, R. SCHELLENBERGER, W. and PEGEL, B.: *Phys. Stat. Sol.* **3** (1970) 493.
12. GEHLEN, P. C. and HAHN, G. H.: *J. Res. Nat. Bur. Stand.* A **73** (1969) 533.
13. SUZUKI, H.: *J. Phys. Soc. Japan* **18** Suppl. I (1963) 182.
14. LABUSCH, R.: *Phys. Stat. Sol.* **10** (1965) 645.
15. STENZEL, G.: *Phys. Stat. Sol.* **34** (1969) 495
16. GUYOT, P. and DORN, J. E.: *Can. J. Phys.* **45** (1967) 983.
17. KUHLMANN-WILSDORF, D.: *Phys. Rev.* **120** (1960) 773.
18. SUZUKI, H.: *J. Res. Nat. Bur. Stand.* A **73** (1969) 548.
19. SCHOTTKY, G.: *Phys. Stat. Sol.* **5** (1964) 697.
20. BARDEEN, J. and HERRING, C.: *Imperfections in Nearly Perfect Crystals*, Wiley, New York, 1952, p. 279.
21. KUHLMANN-WILSDORF, D. and WILSDORF, H. G. F.: *Acta Met.* **10** (1962) 584.
22. VERMA, A. R.: *Crystal Growth and Dislocations*, Butterworth, London, 1953.

23. BURTON, W. B., CABRERA, N. and FRANK, F. C.: *Phil. Trans. Roy. Soc.* A **243** (1951) 299.
24. VERMA, A. R.: *Proc. of the Conference on Silicon Carbide, April 1959*, p. 202, ed. by J. R. O'CONNOR and J. SMILTENS, Pergamon Press, Oxford, 1960.
25. SUITO, E. and UYEDA, N.: *J. Electronmicr.* **8** (1960) 25.
26. GYULAI, Z.: *Magy. Fiz. F.* **2** (1954) 371.
27. KAMIYOSHI, KAN-ICHI, and YAMAKAMI, TSUTOMU: *Sci. Rep. RITU*, A **12** (1960) 258.
28. SINES, G.: *J. Phys. Soc. Japan* **15** (1960) 1119.
29. AMELINCKX, S. *et al.*: *Phil. Mag.* **2** (1957) 355.

Chapter 5

1. WILSDORF, H. G. F. and FOURIE, J. T.: *Acta Met.* **4** (1956) 271.
2. FRANK, F. C. and READ, W. T.: *Phys. Rev.* **79** (1950) 722.
3. LANG, A. R.: *J. Appl. Phys.* **35** (1964) 1956.
4. KOVÁCS, I.: *Phys. Stat. Sol.* **3** (1963) 140.
5. (ORLOV, A. N.) ОРЛОВ, А. Н.: *Физ. металлов и металловидение* **13** (1962) 18.
6. ESHELBY, J. D., FRANK, F. C. and NABARRO, F. R. N.: *Phil. Mag.* **42** (1951) 351.
7. MITCHELL, T. E., HECKER, S. S. and SMIALEK, R. L.: *Phys. Stat. Sol.* **11** (1965) 585.
8. LEIBFRIED, G.: *Z. Phys.* **130** (1951) 214.
9. BILBY, B. and ENTWISLE, A.: *Acta Met.* **4** (1956) 257.
10. (ROZHANSKY, V. N. and INDENBOM, V. L.) РОЖАНСКИ, В. Н. и ИНДЕНБОМ, В. Л.: *ДАН СССР* **136** (1961) 1331.
11. (GRINBERG, V. A.) ГРИНБЕРГ, В. А.: *Физ. тв. тела* **4** (1962) 2593.
12. MITCHELL, T. E.: *Phil. Mag.* **10** (1964) 301.
13. JOHNSTON, W. G. and GILMAN, J. J.: *J. Appl. Phys.* **31** (1960) 632.
14. LOW, J. R., JR. and TURKALO, A. M.: *Acta Met.* **10** (1962) 215.
15. KUHLMANN-WILSDORF, D., MADDIN, R. and KIMURA, H.: *Metallk.* **49** (1958) 584.
16. SAKA, H., DOI, M. and IMURA, T.: *J. Phys. Soc. Japan* **29** (1970) 803.
17. TETELMAN, A. S.: *Acta Met.* **10** (1962) 813.
18. FRANK, F. C.: *Report on Strength of Solids*, London Physical Society, 1948, p. 46.
19. BERGHEZAN, A. and FOURDEUX, A.: *J. Appl. Phys.* **30** (1959) 1913.

Chapter 6

1. HEIDENREICH, R. D. and SHOCKLEY, W.: *Report of a Conference on the Strength of Solids*, Univ. of Bristol, Phys. Soc. London, 1948, p. 57.
2. FRANK, F. C.: *Proc. Phys. Soc.* A **62** (1949) 202.

References

3. SEEGER, A. and SCHÖCK, G.: *Acta Met.* **1** (1953) 519.
4. SEEGER, A.: *Theorie der Gitterfehlstellen, Hdb. der Phys.* VII/1, Springer Verlag, Berlin, 1955.
5. SPENCE, G. B.: *J. Appl. Phys.* **33** (1962) 729.
6. BAKER, C., CHOU, Y. T. and KELLY, A.: *Phil. Mag.* **6**, (1961) 1305.
7. WOLF, H.: *Z. Naturf.* **15**a (1960) 180.
8. HORNSTRA, J.: *J. Phys. Chem. Sol.* **5** (1958) 129.
9. RHODES, R. G.: *Imperfections and Active Centers in Semiconductors*, Pergamon Press, Oxford, 1964.
   ALEXANDER, A. and HAASEN, P.: *Dislocation and Plastic Flow in the Diamond Structure. Solid State Physics*, vol. 22, ed. by F. SEITZ, and D. TURNBULL, Academic Press, New York, 1968.
10. HOLT, D. B.: *J. Phys. Chem. Sol.* **27** (1966) 1053.
11. HOLT, D. B.: *J. Phys. Chem. Sol.* **23** (1962) 1353.
12. KROUPA, F. and VITEK, V.: *Can. J. Phys.* **45** (1967) 945.
13. HIRSCH, P. B.: *Proc. Int. Conf. on the Strength of Metals and Alloys, Tokyo, 1967*. The Japan Inst. of Metals, 1968.
14. VITEK, V., PERRIN, R. C. and BOWEN, D. K.: *Phil. Mag.* **21** (1970) 1049.
15. VITEK, V.: *Phys. Stat. Sol.* **15** (1966) 557.
16. WASILEWSKI, R. J.: *Acta Met.* **13** (1965) 40.
17. SLEESWYK, A. W.: *Phil. Mag.* **8** (1963) 1467.
18. TEUTONICO, L. J.: *Acta Met.* **13** (1965) 605.
19. HARTLEY, C. S.: *Acta Met.* **14** (1966) 1133.
20. HARTLEY, C. S.: *Acta Met.* **14** (1966) 127.
21. ŠESTÁK, B. and ZÁRUBOVÁ, N.: *Phys. Stat. Sol.* **10** (1965) 239.
22. THOMPSON, N.: *Proc. Phys. Soc.* B **66** (1953) 481.
23. HIRTH, J. P.: *J. Appl. Phys.* **32** (1961) 700.
24. HIRTH, J. P.: *J. Appl. Phys.* **33** (1962) 2286.
25. WHELAN, M. J.: *Proc. Roy. Soc.* A **249** (1958) 114.
26. TEUTONICO, L. J.: *Phys. Stat. Sol.* **14** (1966) 457.
27. AMELINCKX, S.: *The Direct Observation of Lattice Defects by Means of Electron Microscopy*, 8th Yugoslav Summer School of Physics, Hercegnovi, 1963.
28. ART, E., DELAVIGNETTE, P., SIEMS, R. and AMELINCKX, S.: *J. Appl. Phys.* **33** (1962) 3078.
29. NABARRO, F. R. N.: *Theory of Crystal Dislocations,* Clarendon Press, Oxford, 1967.
30. HIRSCH, P. B.: *Phil. Mag.* **7** (1962) 67.
31. WEERTMAN, J.: *Phil. Mag.* **8** (1963) 967.
32. HIRTH, J. P. and LOTHE, J.: *Can. J. Phys.* **45** (1967) 809.
33. MARCINKOWSKI, M. J.: *Electron Microscopy and Strength of Crystals*, vol. 5, ed. by G. THOMAS and J. WASHBURN, Interscience, New York, 1963.
34. STOLOFF, N. S. and DAVIES, R. G.: *Progress in Materials Science*, ed. by BRUCE CHALMERS, and W. HUME-ROTHERY, vol. 13, p. 1, Pergamon Press, Oxford, 1966.
35. SCHÖCK, G. and TILLER, W. A.: *Phil. Mag.* **5** (1960) 43.

36. KUHLMANN-WILSDORF, D. and WILSDORF, H. G. F.: *J. Appl. Phys.* **31** (1960) 516.
37. SILCOX, J. and HIRSCH, P. B.: *Phil. Mag.* **4** (1959) 72.
38. BROWN, L. M. and THÖLÉN, A. R.: *Disc. Faraday Soc.* **35** (1964) 35.
39. SIEMS, R., DELAVIGNETTE, P. and AMELINCKX, S.: *Z. Phys.* **165** (1961) 502.
40. HOWIE, A. and SWANN, P. R.: *Phil. Mag.* **6** (1961) 1215.
41. MARCINKOWSKI, M. J. and MILLER, D. S.: *Phil. Mag.* **6** (1961) 871.
42. BROWN, L. M.: *Phil. Mag.* **10** (1964) 441.
43. HÄUSSERMANN, F. and WILKENS, M.: *Phys. Stat. Sol.* **18** (1966) 609.
44. JØSSANG, T., STOWELL, M. J., HIRTH, J. P. and LOTHE, J.: *Acta Met.* **13** (1965) 279.
45. THORNTON, P. R. and HIRSCH, P. B.: *Phil. Mag.* **3** (1958) 738.
46. SEEGER, A., BERNER, R. and WOLF, H.: *Z. Phys.* **155** (1959) 247.
47. HAASEN, P.: *Phil. Mag.* **3** (1958) 384.
48. ROSENFIELD, A. R. and AVERBACH, B. L.: *Acta Met.* **8** (1960) 624.
49. BERNER, R.: *Z. Naturf.* **15**a (1960) 689.
50. BOLLING, G. F., HAYS, L. E. and WIEDERSICH, H. W.: *Acta Met.* **10** (1962) 185.
51. GALLAGHER, P. C. J. and LIU, Y. C.: *Acta Met.* **17** (1969) 127.
52. GALLAGHER, P. C. J.: *Trans. Met. Soc. AIME* **242** (1968) 103.
53. LORETTO, M. L., CLAREBROUGH, L. M. and SEGAL, R. L.: *Phil. Mag.* **10** (1964) 731.
54. AHLERS, M. and HAASEN, P.: *Phys. Stat. Sol.* **10** (1965) 185.
55. RAMSTEINER, F.: *Phys. Stat. Sol.* **15** (1966) K1.
56. VASSAMILLET, K. F. and MASSALSKI, T. B.: *J. Appl. Phys.* **34** (1963) 3389, 3402.

Chapter 7

1. KOVÁCS, I.: *Acta Phys. Hung.* **15** (1962) 11.
2. KAMIYOSHI, K. J. and YAMAKAMI, T.: *J. Phys. Soc. Japan* **15** (1960) 1347.
3. ROBERT, J. M. and BROWN, N.: *Trans. AIME* **218** (1960) 454.
4. STEIN, D. F. and LOW, J. R.: *J. Appl. Phys.* **31** (1960) 362.
5. SUZUKI, T.: *Dislocation Dynamics*, McGraw-Hill, New York, 1968, p. 551.
6. GILMAN, J. J.: *J. Appl. Phys.* **39** (1968) 6086.
7. GILMAN, J. J.: *J. Appl. Phys.* **36** (1965) 3195.
8. BUEREN, H. G. VAN, *Imperfections in Crystals*, North-Holland Publishing Co., Amsterdam, 1960.
9. FELTHAM, P.: *Phil. Mag.* **19** (1969) 635; **21** (1970) 765.
10. BERNER, R. and KRONMÜLLER, H.: *Moderne Probleme der Metallphysik*, Springer Verlag, Berlin, 1965, p. 56.
11. COTTRELL, A. H. and STOKES, R. S.: *Proc. Roy. Soc.* A **233** (1955) 17.
12. EVANS, A. G. and RAWLINGS, R. D.: *Phys. Stat. Sol.* **34** (1969) 9.

References

13. SHERBY, O. D. and BURKE, P.: *Progress in Materials Science*, **13** (1968) 323.
14. SACHS, G.: *Z. VDI.* **72** (1928) 734.
15. BOAS, W. and OGILVIE, J.: *Acta Met.* **2** (1954) 655.
16. MISES, R.: *Z. Angew. Math. u. Mech.* **8** (1928) 161.
17. TAYLOR, G. I. and QUINNEY, Y. H.: *Phil. Trans. Roy. Soc.* A **230** (1932) 323.
18. MACHERAUCH, E.: *Z. Metallk.* **55** (1964) 60.
19. KOCKS, U. F.: *Met. Transactions* **1** (1970) 1121.
20. BISHOP, J. F.W. and HILL, R.: *Phil. Mag.* **42** (1951) 1298.
21. ANDRADE, E. N. DA C.: *Proc. Roy. Soc.* A **90** (1914) 329.
22. SULLY, A. H., CALE, G. N. and WILLOUGHBY, G.: *Symposium on Properties of Metals in Engineering Materials*, 1949, Amer. Soc. Test. Metals.
23. ANDRADE, E. N. DA C.: *Nature* **195** (1962) 991.
24. MCLEAN, D.: *Grain Boundaries in Metals*, Clarendon Press, Oxford, 1952.
25. BLANK, H.: *Z. Metallk.* **49** (1958) 27.
26. WYATT, D. H.: *Proc. Phys. Soc.* B **66** (1953) 459.
27. FELTHAM, P.: *Acta Techn. Hung.* **39** (1962) 243.
28. MOTT, N. F.: *Phil. Mag.* **44** (1953) 742.
29. WEERTMAN, J. and WEERTMAN, J. R.: *Physical Metallurgy*, ed. by R. W. CAHN, Ch. 16, North-Holland Publishing Co., Amsterdam, 1965.
30. FELTHAM, P.: *Deformation and Strength of Materials*, Butterworths, London, 1966.
31. ALDEN, T. H.: *Acta Met.* **17** (1969) 1435.
32. GREENWOOD, J. N., MILLER, D. R. and SNITER, J. W.: *Acta Met.* **2** (1954) 250.
33. HENDRICHSON, A. A. and FINE, M. E.: *Trans. AIME* **221** (1961) 967. KOPPENAAL, T. J. and FINE, M. E.: *Trans. AIME* **224** (1962) 347.
34. JAX, P., KRATOCHVIL, P. and HAASEN, P.: *Acta Met.* **18** (1970) 237.
35. MOTT, N. F. and NABARRO, F. R. N.: *Rep. Conf. Internal Strains in Solids*, Bristol, 1940, p. 86.
36. NAKADA, Y. and KEH, A. S.: *Acta Met.* **16** (1968) 903.
37. MOTT, N. F. and NABARRO, F. R. N.: *Rep. Conf. Strength of Solids*, Bristol, 1948, p. 1.
38. MOTT, N. F.: *Imperfections in Nearly Perfect Crystals*, Wiley, New York, 1952, p. 173.
39. FRIEDEL, J.: *Internal Stresses and Fatigue in Metals*, Elsevier, Amsterdam, 1959, p. 244.
40. SCHOECK, G. and SEEGER, A.: *Acta Met.* **7** (1959) 469.
41. ARSENAULT, R. J.: *Phil. Mag.* **13** (1966) 31.
42. FLEISCHER, R. L.: *Acta Met.* **9** (1961) 996.
43. SAXL, J.: *Czech. J. Phys.* B **14** (1964) 381.
44. FLEISCHER, R. L.: *Strengthening of Metals*, Reinhold, 1964, p. 93.
45. FELTHAM, P.: *Brit. J. Appl. Phys.* **1** (1968) 303.
46. ONO, K.: *J. Appl. Phys.* **39** (1968) 1803.
47. BRINDLEY, B. J. and WORTHINGTON, P. J.: *Metals and Materials and Metallurgical Reviews* **4** (1970) 101.

48. KELLY, A. and NICHOLSON, R. B.: *Precipitation Hardening, Progress in Materials Science*, **10** (1963) 149.
49. *Electron Microscopy and Strength of Crystals*, ed. by G. THOMAS anp J. WASHBURN, Interscience Publisher, New York, 1963.
50. *Oxide Dispersion Strengthening*, ed. by G. S. ANSELL, T. D. COOPER ans F. V. LENEL, Gordon & Breach, New York, 1968.
51. MARTIN, J. W.: *Precipitation Hardening*, Pergamon Press, Oxford, 1968.
52. OROWAN, E.: *Dislocation in Metals*, A.S.M., 1954, p. 131.
53. ASHBY, M. F.: *Acta Met.* **14** (1966) 679.
54. JONES, H.: *Theory of Electrical and Thermal Conductivity of Metals, Encyclopaedia of Physics*, vol. XIX, Springer-Verlag, Berlin, 1956.
55. KOVÁCS, I.: *Acta Met.* **15** (1967) 1731.
KOVÁCS, I. and NAGY, E.: *Phys. Stat. Sol.* **8** (1963) 726.
56. JONGENBURGER, P.: *Appl. Sci. Res.* B **3** (1953) 237.
SIMMONS, R. O. and BALLUFFI, R. W.: *Phys. Rev.* **125** (1962) 862.
57. *Vacancies and Interstitials in Metals*, ed. by A. SEEGER, D. SCHUMACHER, W. SCHILLING and J. DIEHL, North-Holland Publishing Co., Amsterdam, 1970.
58. JONGENBURGER, P.: *Nature* **175** (1955) 545.
59. HUNTER, G. H. and NABARRO, F. R. N.: *Proc. Roy. Soc.* A **220** (1953) 542.
60. SEEGER, A. and BROSS, H.: *Z. Naturf.* **15**a (1960) 663.
61. BASINSKI, Z. S., DUGDALE, J. S. and HOWIE, A.: *Phil. Mag.* **8** (1963) 1989.
62. RIDER, J. G. and FOXON, C. T. B.: *Phil. Mag.* **16** (1967) 1133.
63. SAADA, G.: Thesis, Paris, 1960.
64. NABARRO, F. R. N., BASINSKI, Z. S. and HOLT, B.: *Adv. Phys.* **13** (1964) 193.

Chapter 8

1. *Report on Symposium on Work-hardening, Chicago, 1966*, ed. by J. P. HIRTH, and J. WEERTMAN, Gordon & Breach, New York, 1968.
2. SEEGER, A.: *Kristallplastizität. Hdb. d. Phys.* VII/2, Springer-Verlag, Berlin, 1958.
3. MADER, S. and SEEGER, A.: *Acta Met.* **8** (1960) 513.
4. ESSMANN, U.: *Phys. Stat. Sol.* **3** (1963) 932; **12** (1965) 707 and 723.
5. STEEDS, J. W.: *Proc. Roy. Soc.* A **292** (1966) 343.
6. BAILEY, J. E. and HIRSCH, P. B.: *Phil. Mag.* **5** (1960) 485.
7. OTTE, H. M. and HREN, J. J.: *Experimental Mechanics* (Soc. Exp. Stress Analysis) **6** (1966) 177.
8. KRONMÜLLER, H.: *Z. Phys.* **154** (1959) 574.
KRONMÜLLER, H. and SEEGER, A.: *J. Phys. Chem. Sol.* **18** (1961) 93.
9. KRONMÜLLER, H.: *Can. J. Phys.* **45** (1967) 631.
10. BELL, J. F.: *Phil. Mag.* **10** (1964) 107; **11** (1966) 1135.
11. HIRSCH, P. B.: *Proc. Int. Conf. on the Strength of Metals and Alloys, Tokyo, 1967*, The Japan Inst. of Metals, 1968.
12. ŠESTÁK, B. and SEEGER, A.: *Phys. Stat. Sol.* **43** (1971) 433.

References

13. KOVÁCS, I.: *Acta. Met.* **15** (1967) 1731.
14. HIRSCH, P. B.: *Acta Cryst.* **13** (1960) 1114
15. MOTT, N. F.: *Trans. AIME* **218** (1960) 962.
16. HIRSCH, P. B.: *Phil. Mag.* **7** (1962) 67.
17. HIRSCH, P. B. and WARRINGTON, D. H.: *Phil. Mag.* **6** (1961) 735.
18. COTTRELL, A. H.: *Dislocations and Mechanical Properties of Crystals*, p. 509, ed. P. C. FISCHER, Wiley, New York, 1957.
19. ZSOLDOS, L.: *Phys. Stat. Sol.* **3** (1963) 2127.
20. KUHLMANN-WILSDORF, D.: *Trans. AIME* **224** (1962) 1047.
21. MICHELITSCH, M.: *Z. Metallk.* **50** (1959) 548.
22. KUHLMANN-WILSDORF, D.: *Met. Trans.* **1** (1970) 3173.
23. WOLFENDEN, A.: *Scripta Met.* **4** (1970) 899.
24. SEEGER, A., KRONMÜLLER, H., MADER, S. and TRÄUBLE, H.: *Phil. Mag.* **6** (1961) 639.
25. SEEGER, A., DIEHL, J., MADER, S. and REBSTOCKS, H.: *Phil. Mag.* **2** (1957) 1.
26. HAZZLEDINE, P. M.: *Can. J. Phys.* **45** (1967) 765.
27. SUMINO, K. and YAMAMOTO, M.: *Acta Met.* **10** (1962) 890.
28. HIRSCH, P. B. and MITCHELL, T. E.: *Can. J. Phys.* **45** (1967) 663.
29. WOLF, H.: *Z. Naturf.* **15**a (1960) 180.
30. THORNTON, P. R., MITCHELL, T. E. and HIRSCH, P. B.: *Phil. Mag.* **7** (1962) 1349.

Chapter 9

1. KUHLMANN, D., MASING, D. and RAFFELSIEPER, J.: *Metallk.* **40** (1949) 241.
2. COTTRELL, A. H. and AYTEKIN, V.: *J. Inst. Met.* **77** (1950) 389.
   FELTHAM, P. and MYERS, T.: *Phil. Mag.* **8** (1963) 203.
3. BAILEY, J. E.: *Phil. Mag.* **5** (1960) 833.
4. CLAREBROUGH, L. M., HARGREAVES, M. E., MITCHEL, D. and WEST, G. W.: *Proc. Roy. Soc.* A **215** (1952) 507; **232** (1955) 252; *Phil. Mag.* **1** (1956) 528.
5. BAILEY, J. E. and HIRSCH, P. B.: *Phil. Mag.* **5** (1960) 485.
6. SEEGER, A. and KRONMÜLLER, H.: *Phil. Mag.* **7** (1962) 897.
7. BEUKEL, A. VAN DEN: *Acta Met.* **11** (1963) 97.
8. *Vacancies and Interstitials in Metals*, ed. by A. SEEGER, D. SCHUMACHER, W. SCHILLING and J. DIEHL, North-Holland Publishing Co., Amsterdam, 1970.
9. MANINTVELD, J. A.: *Nature* **169** (1954) 623.
10. MEECHAN, C. J. and SOSIN, A.: *Phys. Rev.* **113** (1959) 422.
11. BAUERLE, J. E. and KOEHLER, J. S.: *Phys. Rev.* **107** (1957) 1493.
    (LAZAREV, B. G. and OVCHARENKO, O. N.) ЛАЗАРЕВ, Б. Г. и ОВЧАРЕНКО, О. Н.: *ДАН СССР* **100** (1955) 875.
    SCHÜLE, W., SEEGER, A., SCHUMACHER, D. and KING, K.: *Phys. Stat. Sol.* **2** (1962) 1199.
12. MESHII, M. and KAUFFMANN, J. W.: *Phil. Mag.* **5** (1960) 687.

Appendix B

1. Borg, S. F.: *Matrix-Tensor Method in Continuum Mechanics*, D. Van Nostrand Co., London, 1963.
2. Sokolnikoff, I. S.: *Mathematical Theory of Elasticity*, McGraw-Hill Co., London, 1946.
3. Sneddon, I. N.: *The Classical Theory of Elasticity*, *Encyclopedia of Physics*, ed. S. Flügge, vol. VI, Springer-Verlag, Berlin, 1958.

Index

OTHER TITLES IN THE SERIES ON NATURAL PHILOSOPHY

Vol. 1. DAVYDOV—Quantum Mechanics
Vol. 2. FOKKER—Time and Space, Weight and Inertia
Vol. 3. KAPLAN—Interstellar Gas Dynamics
Vol. 4. ABRIKOSOV, GOR'KOV and DZYALOSHINSKII—Quantum Field Theoretical Methods in Statistical Physics
Vol. 5. OKUN—Weak Interaction of Elementary Particles
Vol. 6. SHKLOVSKII—Physics of the Solar Corona
Vol. 7. AKHIEZER *et al.*—Collective Oscillations in a Plasma
Vol. 8. KIRZHNITS—Field Theoretical Methods in Many-body Systems
Vol. 9. KLIMONTOVICH—Statistical Theory of Non-equilibrium Processes in a Plasma
Vol. 10. KURTH—Introduction to Stellar Statistics
Vol. 11. CHALMERS—Atmospheric Electricity (2nd Edition)
Vol. 12. RENNER—Current Algebras and their Applications
Vol. 13. FAIN and KHANIN—Quantum Electronics, Volume 1—Basic Theory
Vol. 14. FAIN and KHANIN—Quantum Electronics, Volume 2—Maser Amplifiers and Oscillators
Vol. 15. MARCH—Liquid Metals
Vol. 16. HORI—Spectral Properties of Disordered Chains and Lattices
Vol. 17. SAINT-JAMES, THOMAS and SARMA—Type II Superconductivity
Vol. 18. MARGENAU and KESTNER—Theory of Intermolecular Forces
Vol. 19. JANCEL—Foundations of Classical and Quantum Statistical Mechanics
Vol. 20. TAKAHASHI—An Introduction to Field Quantization
Vol. 21. YVON—Correlations and Entropy in Classical Statistical Mechanics
Vol. 22. PENROSE—Foundations of Statistical Mechanics
Vol. 23. VISCONTI—Quantum Field Theory, Volume 1
Vol. 24. FURTH—Fundamental Principles of Modern Theoretical Physics
Vol. 25. ZHELEZNYAKOV—Radioemission of the Sun and Planets
Vol. 26. GRINDLAY—An Introduction to the Phenomenological Theory of Ferroelectricity
Vol. 27. UNGER—Introduction to Quantum Electronics
Vol. 28. KOGA—Introduction to Kinetic Theory: *Stochastic Processes in Gaseous Systems*
Vol. 29. GALASIEWICZ—Superconductivity and Quantum Fluids

Other Titles in the Series